Evolution and Innovation in Wildlife Conservation

Evolution and Innovation in Wildlife Conservation

Parks and Game Ranches to Transfrontier Conservation Areas

Edited by Helen Suich and Brian Child
with Anna Spenceley

publishing for a sustainable future

London • New York

First published by Earthscan in the UK and USA in 2009

ISBN 978-1-84407-634-5 (hbk)
ISBN 978-0-415-52044-7 (pbk)

Typeset by FiSH Books, Enfield

Cover design by Rob Watts, Rogue Four Design

For a full list of publications please contact:

Earthscan
2 Park Square, Milton Park, Abingdon, Oxfordshire OX14 4RN
Simultaneously published in the USA and Canada by Earthscan
711 Third Avenue, New York, NY 10017
Earthscan is an imprint of the Taylor & Francis Group, an informa buisness

First issued in paperback 2011

Earthscan publishes in association with the International Institute for
Environment and Development

A catalogue record for this book is available from the British Library

Library of Congress Cataloging-in-Publication Data

Evolution and innovation in wildlife conservation : parks and game ranches to
transfrontier conservation areas / edited by Helen Suich and Brian Child with
Anna Spenceley.
     p. cm.
  ISBN 978-1-84407-634-5 (hardcover)
  1. Wildlife conservation. 2. Wildlife conservation—Africa, Southern. I.
Suich, Helen. II. Child, B. (Brian) III. Spenceley, Anna.
  QL82.E85 2008
  333.95'416—dc22

                                                              2008036281

Contents

PART IV – COMMUNITY-BASED NATURAL RESOURCE MANAGEMENT

PART V – INTEGRATING WILDLIFE AND PARKS INTO THE SOCIAL LANDSCAPE

PART VI – CONCLUSION

List of Boxes, Figure and Tables

Boxes

Figures

Tables

Acknowledgements

In editing this book and participating in the annual meetings of the Southern African Sustainable Use Specialist Group (SASUSG), we have been privileged to work with many of the scholar–practioners who have changed the face of conservation in southern Africa. These are busy people with a collective track record of considerable achievement. They are still bursting with ideas and committed to making these ideas work in practice. We hope that this book has captured some of the conceptual knowledge upon which conservation practice in southern Africa is based and shed light on the conservation philosophy that is emerging. This work has drawn heavily on the continued voluntarism that we have come to appreciate within SASUSG, but it could not have been completed without generous funding from the Norwegians which allowed us to meet regularly and compile our ideas. The chapters in this book have been peer-reviewed in the normal fashion of academic literature, albeit almost entirely within the region, and we extend our thanks to all of our reviewers.

We think it is also important that these ideas and hypotheses have been scrutinized within the peer community of southern African conservationists, during many years of testing them in the field. We are glad we could get these important stories written down, and we hope that they are useful to other conservationists.

List of Acronyms and Abbreviations

AA	appropriate authority
AENP	Addo Elephant National Park (South Africa)
AGM	annual general meeting
AP	African Parks Management and Finance Company (Pty) Ltd
APF	Addo Planning Forum
BANP	Bazaruto Archipelago National Park (Mozambique)
BNPB	Bophuthatswana National Parks Board
BPK	Management Planning Committee (*Bestuursplankomitee*)
BRE	Barotse Royal Establishment (Zambia)
CAMPFIRE	Communal Areas Management Programme for Indigenous Resources (Zimbabwe)
CAPE	Cape Action Plan for People and the Environment (South Africa)
CBNRM	community-based natural resource management
CBO	community-based organization
CHA	controlled hunting area
CITES	Convention on International Trade in Endangered Species
CoCT	City of Cape Town (South Africa)
COGEP	Participatory Natural Resource Management Council (*Conselho de Gestão Participativo*)
CPA	Communal Property Association
CPNP	Cape Peninsula National Park (South Africa)
CPPNE	Cape Peninsula Protected Natural Environment (South Africa)
CPR	common pool resource
DEAT	Department of Environment and Tourism (South Africa)
DNAC	National Directorate of Conservation Areas with the Purpose of Tourism (Mozambique) (*Direcção Nacional das Áreas de Conservação*)
DNFFB	National Directorate for Forestry and Wildlife (Mozambique) (*Direcção Nacional de Florestas e Fauna Bravia*)
DNPW	Department of National Parks and Wildlife (Malawi)
DNPWLM	Department of National Parks and Wild Life Management (Zimbabwe)

DUAT	Certificate of Use and Improvement Rights (*Direito de Uso e Approveitamento de Terra*)
DVS	Department of Veterinary Services (Zimbabwe)
DWNP	Department of Wildlife and National Parks (Botswana)
FAO	United Nations Food and Agriculture Organization
FMD	foot-and-mouth disease
gAENP	greater Addo Elephant National Park (South Africa)
GEF	Global Environment Facility
GLTFCA	Great Limpopo Transfrontier Conservation Area
GLTP	Great Limpopo Transfrontier Park
HWC	human–wildlife conflict
ICA	intensive conservation areas
ICEMA	Integrated Community-Based Ecosystem Management (Namibia)
IEMS	Integrated Environmental Management Systems
IFC	International Finance Corporation
IPR	Inhaca and Portuguese Island Reserves (Mozambique)
IRDNC	Integrated Rural Development and Nature Conservation (Namibia)
IUCN	International Union for the Conservation of Nature
JMB	Joint Management Board
JMC	Joint Management Committee
KTFP	Kgalagadi Transfrontier Park
LIFE	Living in a Finite Environment programme (Namibia)
LSU	large stock unit
MADER	Ministry of Agriculture and Rural Development (Mozambique) (*Ministério da Agricultura e Desenvolvimento Rural*)
MET	Ministry of Environment and Tourism (Namibia)
MICOA	Ministry for the Coordination of Environmental Affairs (Mozambique) (*Ministério para a Coordenação da Acção Ambiental*)
MITUR	Ministry of Tourism (Mozambique) (*Ministério do Turismo*)
MoF	Ministry of Finance (South Africa)
MOMS	management-oriented monitoring system
MoU	Memorandum of Understanding
MPA	marine protected area
MUNC	Nature Conservation and the Management and Use of Wildlife (SARCCUS)
MWCT	Ministry of Wildlife, Conservation and Tourism (Namibia)
NACSO	Namibian Association of CBNRM Support Organizations
NDFFC	Nyanga Downs Fly Fishing Club (Zimbabwe)
NGO	non-governmental organization

NKXT	Nqwaa Khobee Xeya Trust (Botswana)
NP	national park
NPB	National Parks Board (South Africa)
NPV	net present value
NPWS	National Parks and Wildlife Service (Zambia)
NRMP	Natural Resources Management Project
NWPTB	North West Parks and Tourism Board (South Africa)
OCT	Okavango Community Trust (Botswana)
PAPA	Principles for Administration of Protected Areas (Mozambique)
PNP	Pilanesberg National Park (South Africa)
PPF	Peace Parks Foundation
PPP	public–private partnership
PWMA	Parks and Wildlife Management Authority (Zimbabwe)
QNP	Quirimbas National Park (Mozambique)
RDC	rural district council
RNP	Richtersveld National Park (South Africa)
ROI	return on investment
SADC	Southern African Development Community
SANDF	South African National Defence Force
SANParks	South African National Parks
SARCCUS	Southern African Regional Commission for the Conservation and Utilisation of the Soil
SASUSG	Southern African Sustainable Use Specialist Group
SDI	spatial development initiative
SMME	small, medium and micro enterprise
STEP	Sub-tropical Thicket Ecosystem Planning (South Africa)
SVC	Savé Valley Conservancy (Zimbabwe)
TBNRM	transboundary natural resource management
TFCA	transfrontier conservation area
TFP	transfrontier park
TMNP	Table Mountain National Park (South Africa)
UNDP	United Nations Development Programme
USAID	United States Agency for International Development
WESSA	Wildlife and Environment Society of South Africa
WfW	Working for Water (South Africa)
WMA	wildlife management area
WWF	World Wildlife Fund/World Wide Fund for Nature
WWF-SARPO	World Wildlife Fund – Southern African Regional Programme Office
WWC	ward wildlife committee
WWMC	ward wildlife management committee
ZAWA	Zambia Wildlife Authority

Part I

Overview

1

Conservation in Transition

Brian Child

This book is a sister volume to *Parks in Transition: Biodiversity, Rural Development and the Bottom Line* (Child, 2004). It chronicles southern Africa's search for a new conservation paradigm that is more politically resilient and relevant to society. The book includes four sections providing case studies that reflect the temporal evolution of these new ideas. Part II describes the evolution of state protected areas using case studies from four countries. The period 1930 to 1970 was one of energetic park building, resulting in globally important parks like Kruger, Hwange, Chobe, South Luangwa, Etosha, Serengeti and Tsavo. Towards the end of this period, conservation was strengthened by recruiting professional conservation managers and researchers. These professionals realized that parks alone could not conserve the region's spectacular wildlife and ecosystems, and off-reserve conservation had to be a part of the conservation matrix. This led to policy experimentation with conservation on privately held land, which is described in Part III. The third phase of experimentation, described in Part IV, was the transfer of the economic model for wildlife utilization developed on private (and, in some cases, state) land to the more complex socio-economic conditions of southern Africa's communal lands. Coming full circle, many of the ideas developed to encourage communities and farmers to manage wildlife sustainably are now being applied back to the management of protected areas, and this is described in Part V, as are other, more recent, innovations in conservations.

These changes have taken the conceptual development of conservation through several phases related to 1) biological conservation, biogeography and the ecosystem approach; 2) economic instrumentalism based on property rights and neo-liberal concepts of resource allocation and market economies; 3) political ecology and political organization related to environmental justice and the challenge of how communities and landholders should be organized to control and manage conservation; and 4) institutional ecology that tries to understand how the way conservation is organized and governed affects power and incentives, and therefore conservation outcomes.

Scholar–practitioners and inductive learning

The purpose of this book is to provide an opportunity for many of the scholar-practitioners who have spent years (or more often decades) developing innovative conservation models to share their experiences. Common conceptual threads weave their way through these case studies, illustrating an hypothesis about the relationship between humans and nature conservation that is still evolving. Put briefly, these are encapsulated by the 'price, proprietorship, subsidiarity hypothesis': in other words, if wildlife is valuable, if this value is captured by the landholder, and if rights are devolved to the lowest level, the probability of successful wildlife and natural resource conservation is greatly increased (Child, 2002). As illustrated throughout this book, this simple statement includes the principles of property rights and liberal economics, the principles of individual discretion and democratization, and concepts relating to the way institutions and organizations are designed.

Rihoy et al (2007) and Hutton et al (2006) comment that the scholarly literature has recently been critical of community conservation, taking the line that simplistic approaches have been applied to complex and multidimensional problems and that, for example, internal differentiation within 'communities' has been glossed over. In responding to these, and earlier, critiques, Barrow and Murphree (2001) agree that donor projects may be inexcusably simplistic, but make the point that this criticism is unfair if it is aimed at practitioners of sustainable use, who are extremely familiar with the complexities of concepts such as community, political economy and the like because they deal with them every day. Indeed they suggest a different problem: that scholars are ignorant of the knowledge being accumulated by practitioners because it resides in oral learning networks and the grey literature of reports and analyses. This knowledge is greatly under-represented in the peer-reviewed literature; indeed, with a few notable exceptions such as Marshall Murphree, universities have played a surprisingly minor role in this learning process. One aim of this book is to begin to close the gap between practice and scholarship.

In 2003, busy practitioners who are members of the Southern African Sustainable Use Specialist Group (SASUSG) and intimately involved with key park, private and community conservation programmes in southern Africa, agreed to write up their experiences. The value of this compilation is that the authors have all been directly involved with the programmes they describe, so their ideas are honed and 'peer-reviewed' in the furnace of making things work. Some of the conceptual thinking related to this compilation was published in *Parks in Transition: Biodiversity, Rural Development and the Bottom Line* (Child, 2004). This volume provides the conservation stories behind these ideas, to illustrate a new way of thinking about conservation that is emerging in the southern African region.

Although there is synchronicity between key international events and the manifestation of community conservation and sustainable use in southern Africa, this book will show that these ideas emerged locally over many years with deep roots in local experiences. They were driven by field scientists working from first principles, with limited exposure to the plethora of international meetings and pronouncements that are such a feature of contemporary conservation. For example, the idea of conserving wildlife through sustainable use emerged in the 1960s, almost two decades before the World Conservation Strategy (IUCN, 1980). Indeed, southern Africa's practical experience influenced, as much as was influenced by, important narrative-setting forums such as the 1982 World Parks Congress held in Bali, the Convention on International Trade in Endangered Species (CITES), and the World Conservation Union's (IUCN's) shift towards people and sustainable use.

Three phases in the political economy of wildlife

This book is about changes in the way wildlife is used and controlled by people; it is about the changing political economy of wildlife, and the policy and management responses to this. It describes devolved, incentive-led conservation and the importance of matching the rising value of wildlife with mechanisms to ensure that these incentives reach and are controlled by landholders. The ideology governing park and wildlife management is not nearly as set in stone as we might imagine. Rather, it is a reflection of the societal priorities of the time that is surprisingly flexible.

Frontier economy

The tone for wildlife conservation in the 20th century was set by the white man's abuse of wildlife in the 1800s. On the frontier of European settlement in southern Africa (with its many parallels to North America) some 20 million wild animals were eliminated between 1780 and 1880, including local and complete extinctions (i.e. bloubok and quagga) (www.kruger2canyons.com). The introduction of technology such as firearms, medicine, fences, railways and markets increased the profitability of hunting in an economy lacking the institutions to control the use of increasingly scarce resources, so that they were seriously over-utilized.[1] Perhaps more importantly, much of the land was converted to agriculture.

Conservation legislation evolved out of the Cape Colony, and entrenched the Roman Dutch system of law (Anderson and Grove, 1988). Fuelled by fears of soil erosion and deforestation, the earliest conservation legislation was promulgated for flora in the Cape Colony in the 1820s (Anderson and Grove,

1988). Following the massive killing of wildlife for food, profit and to make way for agriculture, the Cape Colony protected elephant and buffalo (1858), and the 1886 Cape Act for the Preservation of Game was the first systematic conservation legislation in Africa.

Protected areas, centralized control and banning of commercial use

Concern about the elimination of millions of wild animals on the African frontier led the African colonial powers to draft the 1900 London Convention Concerning the Preservation of Wild Animals, Birds and Fish in Africa. Although never ratified, this sought to standardize game laws across colonial Africa, regulate hunting and establish game reserves (IUCN, 2004). These sentiments were echoed in the more forceful 1933 Convention Relative to the Preservation of Fauna and Flora in their Natural State (the 'London Convention'). This set the legislative tone for African conservation and radically altered the political economy of wildlife into a configuration that reverberates through the conservation community to this day. The London Convention greatly encouraged the setting aside of large areas for game preservation and set in motion a wave of national park establishment. Commercial utilization of wildlife was roundly rejected, presumably because it was associated with pictures of greed and slaughter, but hunting by sportsmen for non-commercial reasons remained acceptable. New laws disenfranchised landholders, both white and black, from using wildlife sensibly.[2] Another recommendation with long-lasting consequences was that the control of wildlife be centralized (Heijnsbergen, 1997).

Consequently, colonial officials became responsible for managing wildlife not only in parks, but also on farmlands and in traditional African areas. Initially, many of these officials were problem-animal control officers, with many game departments growing out of this function. National parks, and wildlife outside of parks, became the responsibility of a single state agency (except in South Africa where wildlife was controlled by the provinces). Well-intended, and appropriate to the frontier economy during and immediately after colonialization, the London Convention implanted governing frameworks that greatly undervalued wildlife: killing animals for sport and (sometimes) for low-value uses like food was acceptable, but using it for financial gain was not.

Across southern and East Africa a world class system of national parks evolved – with Kruger, Hwange, Luangwa, Serengti, Tsavo and Queen Elizabeth national parks, for example, becoming household names associated with Africa's spectacular wildlife. Parks were a very successful innovation. Areas were set aside to protect wildlife (initially from white adventurers), and these areas soon became famous for 'big game' tourism. They also began to attract dedicated and professional managers. Prompted by the emerging science of ecology and the theory of island biogeography (MacArthur and Wilson, 1967), and noticing the importance of off-reserve dispersal areas for migrating herbivores in East African wildlife, and elephants and other large mammals in southern Africa, wildlife authorities concluded that parks were not big enough to conserve nature. Subsequent research into species–area curves suggest that conserving 10 per cent of the Earth's surface falls short of what is necessary to conserve most biodiversity (Hutton and Leader-Williams, 2003). Moreover, from the 1950s onwards a rapid demographic and agricultural transition began placing considerable pressure on land, including land outside national parks. A landscape approach was needed to protect wildlife and wild lands. The problem was that the wildlife outside parks was being rapidly replaced by the cow and the plough; wildlife could neither be owned nor utilized for profit, and could therefore not compete commercially against crops and livestock, and the state could not regulate nor police this problem away. The structure of the rural economy had changed and the ideas in the London Convention were undermining their own goals. It was time for a new political economic model for wildlife conservation. This volume shows how this new approach led, over several decades, to the emergence of the principles of sustainable use (see SASUSG, 1996).

Devolved, incentive-led conservation

Wise use, the soil and accountability of policymakers
Southern Africa has always had a different socio-cultural vision for conservation than Europe or America (Anderson and Grove, 1988). This has been driven by a rural, rather than an urban, economy and political constituency, and has been associated with an obsession with soil erosion and environmental health (Beinart, 1984). It is not surprising, therefore, that the home-grown conservation narrative came to be framed in terms of the wise use of land rather than single species or wilderness conservation. We also observe that this narrative was developed at the nexus between scientists, policymakers and extension agents on the one hand, and land users on the other, in a process of iteratively solving real-life problems that came to be known as adaptive management.

Early experiments in sustainable use and the FAO Special Project
As early as the 1960s, the term 'use it or lose it' gained currency. A series of experiments with game cropping were undertaken in Southern Rhodesia, Northern Rhodesia, Kenya and Uganda (see Parker, 2004 for descriptions of East Africa), spawning a considerable literature about the potential ecological, production and behavioural advantages of wildlife (reviewed in Child, 1988).

In Kenya, the participation of local people (the Walanguli elephants hunters) was proposed (Parker, 2004), while in Northern Rhodesia, the return of benefits to the Nsefu chieftainship was championed (Norman Carr, pers. comm.). Under a United Nations Food and Agriculture Organization (FAO) project, the role of wildlife in the future economy of rural people was promoted scientifically, and Riney and others (for example Talbot et al, 1965; Riney and Hill, 1967; Parker, 2004) developed a philosophy of sustainable use that began to be reflected in the global discourse (for example IUCN, 1980; Brundtland, 1987).

As always, personalities play a significant role in innovation. The renowned mammalogist and Director of Museums in Zimbabwe, Reay Smithers, recognized that wildlife would not survive outside protected areas unless landholders derived economic value from it. Much impetus for a changing conservation philosophy was provided when he invited three Fulbright Scholars (Ray Dasmann, Archie Mossman and Thane Riney, who studied under Starker Leopold at the University of Berkeley) to Zimbabwe (and South Africa) in the late 1950s and early 1960s. The hegemony of the London Convention was first broken by Zimbabwe's Wild Life Conservation Act of 1960, which allowed farmers some freedom to experiment with game cropping and profit-oriented trophy hunting (Child, 1995). This allowed the Fulbright Scholars to experiment with wildlife utilization on ranches in Zimbabwe, where they also introduced new ideas about ecosystem and park management. Dasmann then joined the IUCN and influenced policy through his extensive environmental writing. Riney, who was more field-oriented, undertook a major study of wildlife's potential in Africa through FAO's Africa Special Project (Riney and Hill, 1967), before heading some 80 field-based technical experts in FAO in a programme that played a prominent role promoting conservation through scientific utilization globally.

Southern African Regional Commission for the Conservation and Utilization of the Soil and the need for wildlife to pay its way

Strong linkages between individuals and learning networks have played a prominent role in the evolution of wildlife policy in southern Africa. In the 1960s, the Southern African Regional Commission for the Conservation and Utilization of the Soil (SARCCUS) had grown too broad to fit under the umbrella of regional agricultural ministries. It was reorganized. What is important for this story is that a Standing Committee was set up for Nature Conservation and the Management and Use of Wildlife, where directors and key scientists from regional wildlife agencies began to meet annually for a week to discuss new ideas. These meetings quickly shared information about the emerging science of ecosystem management, and developed new ideas about policy, enabling institutions and wildlife utilization that still reverberate today. Scientists and heads of wildlife agencies, including Zimbabwe (Graham Child),

Namibia (Bernabie de la But) and Mozambique (Ken Tinley), began to articulate and champion the 'use it or lose it' philosophy that underpins the southern African sustainable use movement today. It is no coincidence that legislation encouraging the use of wildlife on private land appeared during this period, including in Namibia, Zimbabwe and Mozambique (see Chapters 5, 6 and 8) and, at the provincial level, in South Africa.

This is not to say that many state agencies were not reluctant to cede control of wildlife to private landholders (see Chapters 5 and 14), and it is not difficult to imagine that innovation could easily have been stifled by the heavy blanket of a controlling state. The wildlife bureaucrats that met at SARCCUS, however, accepted a bold new challenge and opened up policy space for the private sector. This let the genie out of the bottle and private conservation blossomed in Zimbabwe and Namibia. It also took root in South Africa (somewhat unevenly under its provincial structure). Over the next three decades this allowed a robust economic approach to wildlife conservation to be developed on private land. In turn, confidence in this model led to the evolution of rights-based community-based natural resource management (CBNRM) (the subject of Part IV). In the newly independent states of Namibia and Zimbabwe, leading wildlife officials argued that black communities should not be denied similar rights to white farmers, introducing the legislative changes that initiated CBNRM programmes.

Devolution and collective action

The ideas behind the collective management of natural resources and CBNRM have been brewing since the 1940s. They emerged locally and were later validated by global research on common pool resources (for example Ostrom, 1990). As is so often the case in southern Africa, these ideas emerged from concern about the soil. Zimbabwean agriculturalists borrowed the idea of soil conservation districts from institutional solutions to the American dustbowl, but took it a step further, legally empowering landholder communities (rather than officials) to regulate soil conservation measures for themselves through units of collective action called intensive conservation areas (ICAs). Legislated through the Natural Resources Act of 1941, ICAs were voluntary groupings of landholders empowered to deal with environmental issues among themselves that we would now associate with Tragedy of the Commons circumstances (but pre-dating Hardin's 1968 seminal paper). ICAs became a countrywide grassroots structure and the primary institution for soil conservation in Rhodesia (Stocking, 1985), a well-known success that, surprisingly, appears not to have been studied. Zimbabwe's wildlife authorities recognized that regulations invariably exceeded the state's implementation capacity (Murphree, 2000), and instead of assuming the state should control abuses of wildlife, opted to experiment with using peer-based regulation among private landholders with rights to utilize wildlife. The new wildlife legislation (1975) was therefore grafted onto the grass-roots, bottom-up ICA movement (Child, 1995).

This experience with wildlife economics and with collective action informed the next round of innovation – CBNRM. In southern Africa, the term is shorthand for programmes that devolve the rights to utilize and manage natural resources to rural communities. The discussions on conserving wildlife in communal lands that started at SARCCUS (for example Child, 1971) began to be refined in Zimbabwe. These ideas were improved through early experiments in community conservation such as WINDFALL (Child, 1995; Maveneke, 1996) and were brought together with some elegance in the influential Communal Areas Management Programme for Indigenous Resources (CAMPFIRE) documents (Martin, 1984, 1986). At this time private conservation was increasingly recognized as being successful (for example Bond and Cumming, 2004), but communal areas were being severely degraded (Whitlow, 1988) and wildlife was disappearing rapidly. To combat resource degradation and simultaneously improve livelihoods, Martin introduced in considerable practical detail the idea of creating local property rights as a mechanism for internalizing costs and benefits, and thereby optimizing land use choices. Indeed, from the start, CAMPFIRE was a programme designed to develop improved institutions for resource allocation rather than a wildlife conservation programme, emphasizing the importance of local tenurial rights and collective action for improving the pricing and allocation of grazing, forest, water and wildlife resources. That these initiatives were about wise land use rather than wildlife conservation, and are rooted in economic theory, is well illustrated by *The Policy for Wild Life in Zimbabwe* (Minister of Environment and Tourism, 1992). This policy aims to improve the value and sustainability of land use (whether or not this includes wildlife), by setting in place economic institutions that allocate resources to their highest-value uses while mitigating against negative externalities (Minister of Environment and Tourism, 1992). The search for an economic unit equivalent to the private landholder led CAMPFIRE to the issue of units of collective action. A crucial influence was Norman Reynolds, then chief economist in the planning ministry in Zimbabwe. His experience structuring Indian villages as economic units (Reynolds, 2001) introduced the concept of the 'village company' into the CAMPFIRE documents. The concept of 'private community property' still lies at the heart of the CBNRM philosophy. These ideas were informed by, for example, a libertarian view of wildlife institutions by Stroup and Baden (1983) and pre-Ostrom debates about property, pricing, resource allocation and collective action (Martin, 1986).

Although CBNRM developed at much the same time as this scholarly debate, it is more correctly viewed as an example of independent but parallel intellectual evolution, guided by empirical, problem-driven learning. Ideas like

Ostrom's (1990) were used to validate what was being learned, and sometimes to sharpen the language to describe it, rather than being instrumental in the formation and testing of these ideas. Learning was rapid because a diversity of approaches and professional backgrounds were linked by common purpose, meetings and exchange visits. In Zimbabwe, for example, CAMPFIRE was 'designed' by an eclectic collection of conservationists, resource economists, political scientists, development workers, Marxist rural activists and rural politicians. These had close professional and personal ties with, for example, policy makers in Namibia, and project managers and development professionals experimenting with CBNRM in Zambia and Mozambique.

Parks and society

By the 1990s, parks came under pressure and funding declined as the arguments for 'pure' nature conservation seemed of limited relevance in societies that needed to address poverty, economic growth and employment (see Chapter 22). A new relationship between parks and society needed to be renegotiated. Fortunately, the idea that wildlife could pay its way and create employment had been tested on private and communal land. It began to spread to parks, suggesting that many parks could fund themselves, and could also play an important role as an engine of local economic growth with their new-found financial and economic power tending to enhance rather than detract from biodiversity objectives (see Chapter 18). In this way, parks became a beachhead for a much larger economic and environmental landscape, providing the foundation for a wildlife-based tourism economy and enhancing the value of nearby land and the probability of it switching to a nature-tourism based economy. This is described in Part V of this book.

Many of the people who have driven the changes in conservation philosophy that are described in this book (including the authors of most of the chapters) now meet regularly through SASUSG, which is comfortably located within IUCN's Species Survival Commission. Like SARCCUS, its membership originated with government wildlife officials, but broadened rapidly as professional wildlife managers found new homes in non-governmental organizations (NGOs) and as private consultants.

Southern African conservationists were surprised (and angered) by the 1989 ivory ban. They were not aware how many people did not understand, or were opposed to, the philosophy of sustainable utilization that was so familiar to them and seemed so appropriate for aligning conservation policy with the circumstances of developing societies. Trade restrictions were a serious threat to this progress because they reduced the value of wildlife. Within the region, this had the positive effect of bringing many conservation leaders together to articulate their case against the re-imposition of the non-commercial and centralizing ideology of the London Convention. These arguments played out in important forums like CITES, and IUCN's World Conservation Congress. Southern Africans realized that they needed to fight for what they had achieved and to put their knowledge down on paper. Initially the technical principles they were articulating (for example SASUSG, 1996) had some influence in these forums. However, the argument over the ivory trade became so emotional that people stopped listening to technical arguments that showed how successful these new policies were, so SASUSG shifted its message to emphasize environmental justice and poverty reduction and cloaked it in theatre. The play, *Guardians of Eden*, was commissioned by SASUSG. Using actors from wildlife areas to tell their own stories, it toured southern Africa, the European Union, the World Bank and ultimately the World Conservation Congress in Montreal in 1996 (Guhrs et al, 2006). It used actors from rural communities to emphasize environmental justice and poverty alleviation, setting these ideas and emotions in opposition to the equally powerful, if more simplistic, message that killing elephants was bad. This approach found some sympathy and, for example, the ban on trade in elephants was partially over-turned in 1997. Ideas about sustainable use and environmental justice have since gained considerable intellectual traction. The examples provided in this book show that these ideas work in practice – provided, of course, they are implemented properly and tenaciously, and not half-heartedly in the manner of so many development projects. These case studies illustrate different examples of sustainable use on state, private and community-protected areas that are emerging from the same set of conceptual foundations.

The political economy of rural areas has changed rapidly, and the conditions that gave rise to centralized and non-commercial conservation approaches no longer apply. The political economy of park and wildlife conservation needs to adapt to a new globalized and commercialized world, especially to address poverty and political marginalization of parks and of people living with wildlife. This in no way implies any reduction in the importance of sustaining biological diversity and ecological health. Indeed, our intention remains to enhance them. However, we are trying to get to this end-point by changing two things. First, parks and wildlife provide a plethora of environmental goods and services, and we need to select a set of these that better suits our societies. Wilderness, outdoor recreation and public access may be important to richer, urban societies, but in Africa tangible values like jobs and economic growth are important. The emerging value of environmental services such as fresh water and carbon sequestration are also recognized, as is the need to develop mechanisms so that local communities and landholders can capture a fair share of these values.

Second, parks and wildlife have in the past largely served and been controlled by an economic and technical elite. Our hope is to change this by devolving the authority and benefits for hundreds of state, private and community-protected areas to landholders, communities and park managers. Democratizing conservation increases its political legitimacy and sustainability, and reduces the considerable risk of having all the 'conservation eggs' in the hands of a single state conservation agency. Further, it is only at this level that the accelerating complexity and variability associated with conserving parks and wildlife can be managed. By replacing a hegemonic conservation superstructure with a vast number of individual experiments for wildlife conservation, we are in essence invoking Darwin's principles to guide the way that new institutional mechanisms evolve to manage conservation. This seems right.

Notes

1   In a book that influenced the way Zimbabwean wildlife authorities designed new wildlife policy, including CAMPFIRE, Stroup and Baden (1983) explain these processes in economic language in the North American context.
2   The early wildlife legislation was probably less discriminatory than usually reported. Before World War II, the threat to wildlife was identified as coming from sportsmen and commercial hunters, rather than indigenous inhabitants who had coexisted with wildlife since time immemorial. In a number of countries (for example Botswana and Zambia) hunting by local people was specifically allowed and even encouraged, often under the control of chiefs (Astle, 1999; Parker, 2004). Nonetheless, many countries limited hunting to 'civilized weapons' but, often for other reasons, disallowed blacks from owning these and thus effectively alienated them from wildlife (Prendergast and Adams, 2003).

References

Anderson, D. and Grove, R. (eds) (1988) *Conservation in Africa: Peoples, Policies and Practice*, Cambridge University Press, Cambridge

Astle, W. L. (1999) *A History of Wildlife Conservation and Management in the Mid-Luangwa Valley, Zambia*, British Empire and Commonwealth Museum, Bristol

Barrow, E. and Murphree, M. (2001) 'Community conservation. From concept to practice', in Hulme, D. and Murphree, M. (eds) *African Wildlife and Livelihoods. The Promise and Performance of Community Conservation*, James Currey Ltd, Oxford

Beinart, W. (1984) 'Soil erosion, conservationism and ideas about development: A Southern African exploration 1900–1960', *Institute of South African Studies*, vol 2, no 1, pp52–84

Bond, I. and Cumming, D. (2004) 'Wildlife research and development', mimeo

Brundtland, G. H. (1987) *Our Common Future*, World Commission on Environment and Development, Oxford University Press, Oxford

Child, B. (1988) 'The role of wildlife utilization in the sustainable economic development of semi-arid rangelands in Zimbabwe', unpublished DPhil thesis, University of Oxford, Oxford

Child, B. (2002) 'Review of African wildlife and livelihoods. The promise and perform-ance of community conservation', *Nature*, vol 415, no 6872, pp581–582

Child, B. (ed) (2004) *Parks in Transition: Biodiversity, Rural Development and the Bottom Line*, Earthscan, London

Child, G. (1971) *The Future of Wildlife and Rural Land Use in Botswana*, Proceedings of SARCCUS Symposium, Nature Conservation as a Form of Land Use, Gorongoza National Park, Mozambique, September

Child, G. (1995) *Wildlife and People: the Zimbabwean Success. How the Conflict Between Animals and People Became Progress for Both*, WISDOM Foundation, Harare

Guhrs, T., Rihoy, L. and Guhrs, M. (2006) 'Using theatre in participatory environmen-tal policy making', *Participatory Learning and Action*, vol 55, pp87–93

Hardin, G. J. (1968) 'The tragedy of the commons', *Science*, vol 162, pp1243–1248

Heijnsbergen, V. P. (1997) *International Legal Protection of Wild Fauna and Flora*, OIS Press, Amsterdam

Hutton, J. M. and Leader-Williams, N. (2003) 'Sustainable use and incentive-driven conservation: Realigning human and conservation interests', *Oryx*, vol 37, no 2, pp215–226

Hutton, J., Adams, W. and Murombedzi, J. (2006) 'Back to the barriers: Changing narratives in biodiversity conservation', *Forum for Development Studies*, no 2–2005

IUCN (1980) *The World Conservation Strategy*, International Union for Conservation of Nature/United Nations Environment Programme/World Wildlife Fund, Geneva

IUCN (2004) 'An introduction to the African Convention on the Conservation of Nature and Natural Resources', IUCN Environmental Policy and Law Paper No. 56, IUCN, Gland

MacArthur, R. H. and Wilson, E. O. (1967) *The Theory of Island Biogeography*, Princeton University Press, Princeton, NJ

Martin, R. (1984) 'Communal Areas Management Programme for Indigenous Resources', Department of National Parks and Wildlife Management, Government of Zimbabwe, Harare

Martin, R. (1986) 'Communal Areas Management Programme for Indigenous Resources' (April 1986, revised edition) Department of National Parks and Wildlife Management, Government of Zimbabwe, Harare

Maveneke, T. N. (1996) 'Local participation and benefit sharing in wildlife manage-ment: The Zimbabwe CAMPFIRE Programme', World Bank/UNDP Africa Forestry Forum, Nairobi

Minister of Environment and Tourism (1992) *The Policy for Wild Life in Zimbabwe*, Department of National Parks and Wildlife Management, Ministry of Environment and Tourism, Harare

Murphree, M. (2000) 'Constituting the commons: Crafting sustainable commons in the new millennium', paper presented at the 8th Biennial Conference of the International Association for the Study of Common Property, Bloomington, IN, 31 May–4 June

Ostrom, E. (1990) *Governing the Commons: The Evolution of Institutions for Collective Action*, Cambridge University Press, Cambridge

Parker, I. (2004) *What I Tell You Three Times is True: Conservation, Ivory, History and Politics*, Librario, Elgin

Prendergast, D. K. and Adams, W. M. (2003) 'Colonial wildlife conservation and the origins of the Society for the Preservation of the Wild Fauna of the Empire (1903–1914)', *Oryx*, vol 37, no 2, pp251–260

Reynolds, N. (2001) 'Land settlement: Government facilitating a peoples' process. Lessons from Zimbabwe and rights issues in South Africa', mimeo

Rihoy, E., Chirozva, C. and Anstey, S. (2007) '"People are not happy": Speaking up for adaptive natural resource governance in Mahenye', Programme for Land and Agrarian Studies, University of the Western Cape, Cape Town

Riney, T. and Hill, P. (1967) 'Conservation and management of African Wildlife', FAO, Rome

SASUSG (Southern Africa Sustainable Use Specialist Group) (1996) 'Sustainable use issues and principles', IUCN Species Survival Commission, Gland

Stocking, M. (1985) 'Soil conservation policy in colonial Africa', *Agricultural History*, vol 59, no 2, pp148–161

Stroup, R. and Baden, J. (1983) *Natural Resource Economics: Bureaucratic Myths and Environmental Management*, Ballinger Publishing Company, Cambridge, MA

Talbot, L. M., Payne, W. J. A., Ledger, H. P. and Talbot, M. H. (1965) 'The meat production potential of wild animals in Africa', *Commonwealth Agriculture Bureau Technical Communication*, no 16

Whitlow, R. (1988) 'Potential versus actual erosion in Zimbabwe', *Applied Geography*, vol 8, pp87–100

Part II

History of State-Led Conservation

2

The Emergence of Parks and Conservation Narratives in Southern Africa

Brian Child

The idea of what a park is and how it relates to society is constantly changing. This is well illustrated by the chapters in Part II – which is in no way complete and reminds us of how much we do not know. Indeed, the insights provided by Carruthers's chapter (and her other works) emphasizes how much our understanding of parks can be enhanced by an analytical historical perspective.

A desire to protect nature can be traced far back into the southern African past, and to both sides of the racial equation. However, the conceptual rationale for nature conservation has evolved more rapidly than we commonly suppose as a response to changes in economic and political circumstances and in our scientific understanding of nature – the ideas far less an outcome of a purist ideology than the negotiated outcome of shifting configurations of power and ideas between hunters, farmers, townsfolk, scientists and politicians, as well as global narratives and economic forces. Indeed, in Carruthers' (1995) excellent social and political history of Kruger National Park she points out that 'history' is often a politically convenient mythology rather than a true representation of fact.

Conservation narratives

We are interested in the changing relationship between society, local people and protected areas. The narrative that local people were evicted from parks by colonial governments, and are antagonistic towards them, has gained considerable political authenticity (for example Brockington and Igoe, 2006). However, the facts are more complicated than this, and unravelling them can help us improve the way we manage modern conservation policies.

It is certainly true that local people were evicted from parks by colonial wardens in some places, and that antagonism between parks and people exists. However, this relationship is as much a function of the current relationship between people and parks and their managers, than it is of history. For example, outside Tarangire National Park in Tanzania local people practise defensive agriculture, fragmenting the landscape, to protect themselves against a perception that parks will once again expropriate their land (McCabe, pers. comm.). By contrast, outside many parks in southern Africa, landholders and rural communities are setting aside land for wildlife and effectively making parks bigger (Bond et al, 2004; pers. obs.). Another false narrative is that local people are invariably antagonistic towards parks and wildlife – this is certainly not true in parts of southern Africa such as northern Botswana and the Caprivi in Namibia, where young people, in particular, see wildlife as important to their economic future (unpublished survey data). A more accurate assessment may well be that local people, on balance, like wildlife (with its complexity of costs and benefits) but dislike the policies that exclude them from sharing it, and their negative reaction may well be a reaction to the way state wildlife officials treat them rather than a reaction to the wildlife itself. In other words, we need to view the narrative that 'parks are bad' that emerges in the literature (for example Brockington and Igoe, 2006) as measuring the relationship between rural people and the way they are treated by the state and its wildlife officers, rather than as measuring their true relationship with wildlife.

In this regard, the location of authority within the wildlife agency is important. In the days of the early game ranger (often 'dedicated Victorian imperialists' to paraphrase Carruthers's (1995, p36) characterization of Kruger's first warden, James Stevenson-Hamilton), authority was in the hands of field wardens who often had close relations with local communities. Prior to World War II, park wardens often felt that white hunters were far more of a threat to wildlife than African communities. For example, Stevenson-Hamilton felt that the ideal game sanctuary 'should contain as few native inhabitants as possible' but that complete absence of the latter is rather a disadvantage than otherwise (Carruthers, 1995). Although he expelled some 2000–3000 people from the Sabi Game Reserve in 1903, he did not equivocally advocate expelling people because without firearms 'the damage they do in a year will not equal that done by a few Boers in a week' (Carruthers, 1995, p43). Parker (2004) records similar attitudes in Kenya in the 1920s and 1930s – local people had lived with wildlife in perpetuity and the real threat to wildlife was from foreign sportsmen and adventurers.

Nevertheless, the political integrity of national parks has been compromised because Africans were excluded from parks, firearms and hunting, often for reasons more to do with white security or forced labour (in white-owned mines and farms) than for conservation reasons. These practices were by no means ubiquitous: in Zambia (Astle, 1999) and in Botswana (Chapter 4), local people have entrenched rights to hunt with firearms that continue to this day.

Following Independence, African park and wildlife agencies invariably centralized the considerable authority held by field wardens to head offices, which became increasingly politicized[1] (see Gibson, 1999; Mugaranyanga, pers. comm.). Park directors were too distant to cope with the complex of issues surrounding a park, especially in the face of shrinking budgets and political demands. With the waning power of park wardens, some of whom were surprisingly in tune with local communities in the colonial period (see for example Astle, 1999; Parker, 2004), relationships between parks and neighbours suffered. Rectifying the damage to the image of parks caused by political irrelevance and alienation was an important theme in *Parks in Transition* (Child, 2004) and emerges throughout this volume.

South Africa

In common with much of southern and East Africa, the history of conservation in South Africa illustrates how the political economy of parks and wildlife shifted from a frontier economy to state protectionism, and how in response to contemporary conditions the efficacy of the latter is now being questioned.

Prior to colonial settlement, low human populations and old-fashioned hunting technologies generally allowed people to live 'in harmony with nature'. However, even before colonial settlement, the accumulation of wealth and the emergence of social and class divisions were associated with politicized access to wildlife and hunting. Desirable wildlife species and products such as ivory came to be controlled by elites who set aside royal hunting preserves out of bounds to commoners (Carruthers, 1995), an African phenomenon with remarkable similarities to Europe.

By the 1800s, white explorers, hunters, adventurers, missionaries and traders began moving beyond the frontier of white settlement. They exploited wildlife commercially and for sport, often with the assistance of local communities. Although large numbers of animals were killed, this did not generally threaten the survival of species and populations. White settlement was another matter entirely. Carruthers describes the long-established settler attitude, particularly among Boer trekkers, that destroying wildlife was not a criminal offence, indeed it was considered 'immoral and unpatriotic not to exterminate wildlife, because clearing the land in this way encouraged agriculture and expedited the progress of civilization' (1995, p11). White settlers and traders depended heavily on exploiting wildlife for their livelihoods. However, the decimation of once abundant herds led to legislative intervention. The earliest conservation laws in the Zuid-Afrikaansche Republiek (in 1858) were political and utilitarian. They aimed to ensure a sustainable yield of wildlife for economic reasons, while preventing Africans from competing for the resource, despite the similar importance of wildlife to their livelihoods.

Wildlife was exterminated over large areas as farms were formally allocated and surveyed. As towns and an urban and landholding elite emerged, with their philosophy that sport was the only acceptable motivation for hunting, they clashed with landholders who valued their rights to do as they pleased on their own farms – activities that included hunting for their own consumption (Carruthers, 1995).[2]

White settlement was associated with the massive destruction of wildlife. Wildlife protection became a prominent issue in the Western world by 1900. Led by Germany, the European colonial powers convened the London Conference in April 1900. Although the convention arising from this conference was never ratified or implemented, it centralized the control of wildlife in the state, privileged sport over commercial and subsistence uses of wildlife, and encouraged the creation of reserves for preserving game (not entertainment, which Carruthers suggests motivated the Yellowstone model in the US). British administrators began to established game reserves (for example in the Cape Colony (1856) and Zululand (1897)), and three reserves were declared in the Transvaal after the South African War (1899–1902) – Sabi, Pongola and Pretoria – although numerous small game reserves that may not exist today were proclaimed before 1898 by a plethora of pre-South African institutions as a response to the extermination of wildlife by hunting and rinderpest (Carruthers, 1995, p27).

It was decreed that these game reserves would benefit sportsmen and contribute to the economy once wildlife had recovered enough for hunting. Further, game reserves such as those which became Kruger National Park were established on land that was considered to have no 'practical value or utility to man' (Carruthers, 1995, p3), echoing the 'useless land' concept framed by Runte (1979). In her history of Kruger and Dongola, Carruthers postulates that political and cultural support is critical for establishing and managing protected areas, and usually reflects the perceived opportunities and costs of protected areas.

Carruthers (1995) illustrates just how much park ideology emerged as a result of personalities, rather than scientific logic. Thus, Kruger's guiding philosophy evolved very much in line with the personality of the park's pre-eminent warden, the dedicated Victorian imperialist, Stevenson-Hamilton. He insisted that wildlife was a 'heritage' that belonged to the empire as a whole. He also entrenched the concept that no hunting whatsoever should be allowed within the park, but interestingly did not fully endorse the contemporary view that human habitation within a game reserve is inappropriate. Stevenson-Hamilton's tenure (1902–1946) saw a transformation of attitude from wildlife as a utilitarian commodity to a growing sentimental, romantic and aesthetic view of nature, at least among well-to-do whites. By 1914, General Smuts, aligning himself with the trends in the US and elsewhere, began calling for setting aside reserves 'for all time', and their role as natural scientific benchmarks in response to the rapid disappearance of such areas under the advance of civilization (Carruthers, 1995). Between the two world wars, public support for national parks grew rapidly, and parks began to be managed for the benefit of (white) tourists and not purely for increasing the number of animals. These philosophies had long-lasting consequences.

Administratively, the management of parks began to bureaucratize with park wardens increasingly under the control of a centralized state agency. Environmental management policies also evolved rapidly: the culling of predators and vermin to encourage game species (see Chapter 3), was replaced (by 1957) by 'management by intervention' and the active manipulation of water, fire and wildlife populations, and more recently by the concept of systems ecology and resilience theory (du Toit et al, 2003). Tourism facilities aimed at servicing the white middle class expanded rapidly, and people were increasingly confined to their vehicles. The park fenced itself off from the landscape, and perhaps also from its accountability to society, making it vulnerable to political change. Indeed, so irrelevant (if not abhorrent) were parks to the 'new' South Africa that the African National Congress considered converting this land to other uses (see Chapter 22), starkly illustrating the risks of a politically narrow and hegemonic conservation strategy.

South Africa was fortunate that the authority for parks was not monopolized by a single state agency but divided between national, provincial and even local authorities, as well as with the apartheid-derived homeland authorities. This diversity provided the substrate for Darwinian evolution. While much of South Africa's parks estate was managed unimaginatively, bureaucratically and privileged a single ethnic group, the commercially entrepreneurial Natal Parks Board promoted private sector conservation, while Bophuthatswana National Parks Board reconstructed Pilanesberg National Park out of 'broken down goat farms', and harnessed the commercial tourism sector and hunting within parks to drive conservation and to create employment for neighbouring communities (see Chapter 18). Consequently, when Nelson Mandela walked free in 1994, South African National Parks could use these models to improve its accountability to society and its broader economic and conservation roles as we see in Part V.

Botswana

Graham Child's history of state conservation in Botswana focuses largely on the formation of the first parks in the 1960s and the park agency that administered them; prior to 1961, only the Gemsbok Game Reserve (declared in 1932 and managed by South Africa) existed. The emergence of a utilitarian approach to parks and wildlife in the 1960s was linked to the FAO's Africa Special Project (Riney, 1967), which provided technical advisers (including Graham Child), as was the case in Zambia (discussed below). There was also considerable interaction with and sensitivity to local people. Botswana is lightly settled, and the formation of these parks disturbed very few people and sometimes involved hundreds of interviews with local people and groups. Child describes the attention given to people's needs, including legal provisions for subsistence hunting. Sociological considerations also played an important role in setting aside the Central Kalahari Game Reserve, in large part to protect the environment and hunter-gatherer lifestyle of the San from the intrusion of cattle producers. The formation of the Moremi Game Reserve followed passionate civic action by white settlers in Maun who worked in collaboration with local people, and somewhat against the wishes of central administrators. Chobe displaced no people and was immediately used to promote tourism. An interesting historical note is that Botswana, with its plentiful wildlife, was the crucible of southern Africa's safari hunting industry. In the 1970s, half of Botswana's hunting blocks were allocated to 'white hunters' (in which subsistence lifestyles, including hunting, continued) and citizens hunted at a much cheaper price than foreigners. Botswana tried to balance the provision of public hunting with the valorization of the wildlife resource through more commercial policies. In the end, wide-scale citizen hunting that was priced too cheaply undervalued wildlife, proved destructive and was difficult to enforce, problems that Botswana (and Zambia) face to this day, although the rising value of wildlife through commercial tourism and hunting is encouraging a range of corrective measures including community wildlife management.

Although economic arguments were used to support wildlife in Botswana as early as the 1960s, the wildlife resource was devastated by the heavily subsidized livestock sector, which was particularly damaging in Botswana's arid environment. Thousand of migrating wild animals died against veterinary fences, and livestock dominated grazing and water resources. It is taking decades to overcome this policy of predisposition towards livestock, but as the newer generation values jobs over spending weekends at remote cattle posts, the balance may be changing – the new president, it is reported, owns no livestock but is a shareholder in tourism ventures. Child points a finger of blame at the London Conventions for opposing commercial and subsistence uses of wildlife, and rendering wildlife economically less competitive than it should have been.

Zimbabwe

In 1899, a proposal for a park based on the Yellowstone model was rejected by the Legislative Assembly. However, Cecil Rhodes established and funded parks in Matopos and Nyanga in 1902, mixing agricultural, nature conservation and managed landscape objectives. The history of wildlife utilization and conservation in Zimbabwe unfolded much as it did in South Africa and Botswana, with the agricultural community extirpating wildlife, with both individuals and the state being to blame – the veterinary department killed over 659,000 wild animals in the name of tsetse fly eradication, and this was only the beginning.

However, in opposition to this, we see a vibrant conservation movement, albeit on 'useless land' on the periphery of the country. Zimbabwe's first state-initiated game reserve, Hwange, was established in 1928 as a response to public demand, and following the Game and Fish Preservation Act, 1929 it was quickly joined by the Victoria Falls and the Hurungwe Game Reserves in 1931. Nine major game reserves were declared in the 1960 Wildlife Act, which also, for the first time, recognized the need to align conservation with agriculture by permitting game cropping on private land (and supporting this with active research in game cropping). Zimbabwe's game reserves were initially designed to protect wildlife and habitats rather than biodiversity, and it was only in the 1970s that a few small protected areas were established in agriculturally productive regions, often for recreational purposes near dams. The National Parks Act, 1949, described a national park as an area 'for the propagation, protection and preservation therein of wild animal life, vegetation and objects of geological, ethnological, historical or other scientific interest for the benefit, advantage and enjoyment of the inhabitants of the colony'.

The Parks and Wildlife Act of 1975 was a turning point in the philosophical development of park and wildlife policy. Park objectives were clearly rationalized into six categories, an idea that Graham Child (pers. comm.) attributes to the discussions that led to IUCN's nine categories. The Act also took the radical step of devolving authority for wildlife to private landholders, and in 1982 to communal areas, sowing the seeds for private and community conservation in the region. Southern Africa's progressive wildlife policies can be traced to this time, often to this agency, and to networking between scientists and professional bureaucrats at regional forums. This early example of adaptive, practice-driven policy formation is instructive. Policymakers in small, professional agencies worked closely with local landholders and also learned from their peers in other countries; policy was innovative precisely because it worked hand-in-hand with practice, and innovation depended on having quality managers and policymakers in field positions. These ingredients are often missing from the centralized structure of modern conservation, which suggests that they may be unable to respond to the rapid changes facing biodiversity.

Mozambique

In compiling the history of state-led conservation in Mozambique, Bartolomeu Soto argues that early legislation in that country (in 1903) was a response to uncontrolled hunting by Europeans. However, Mozambique's response was utilitarian rather than protectionist in that it encouraged hunting for trade and food as well as sport. In the 1960s and early 1970s, Mozambique began to gazette protected areas. Mozambique also introduced game-ranching regulations early on,[3] though very little (if any) private wildlife activities took place with the nationalization of wildlife, and indeed of the whole economy during Mozambique's socialist period. Wildlife was cropped to feed people, including by government culling schemes (Tello, pers. comm.).

The long-running civil war decimated wildlife and the government's capacity to manage it. When the war ended in 1992, Mozambique faced an immense recovery exercise. Following the regional narrative, Mozambique recognized the importance of wildlife for 'social, ecological and economic benefits of ... the Mozambique people' (Wilson, 2005, p149). However, the absence of an effective parks and wildlife bureaucracy meant that community conservation began to 'bubble up' in various forms in various places (see Chapter 16; Wilson, 2005; see also the discussion of Zambia, below). Mozambique also began to use partnerships with private bodies and NGOs to rehabilitate protected areas. Soto describes these experimental partnerships, noting that Mozambique has to deal with the reality that people live inside many of the protected areas.

Zambia

The sensitivity of wildlife policy in Zambia to local needs serves as a counternarrative to the view that colonial game policies were always insensitive to local people, a narrative emerging largely from political anthropologists working in Tanzania (for example Neumann, 1998; Brockington, 2002; Brockington and Igoe, 2006). According to Astle (1999), on which much of this section is based, Zambia adopted game preservation legislation from the Cape Colony (from 1858 and 1886), creating the Mweru Marsh Game Reserve in 1899 and a reserve to protect giraffe in the Luangwa Valley in 1904. Britain assumed direct administration of what was then Northern Rhodesia in 1924, and administered tribes through 'indirect rule' and native authorities based on tribes. The native authorities were more concerned about people and agriculture than wildlife, and local people's access to wildlife was hardly restricted by law or the pitiful number of game guards. Local people could undertake unlimited hunting in their own areas with a Native Game Licence until 1931, after which they could hunt a quota of animals for a fee worth five weeks of labour in any one year (but for no more than four elephants, four eland, four hippo or two kudu). Natives owned some 30,000 firearms and most people had access to game meat, a feature still common in Zambia today. The British South Africa Company was 'strongly of the opinion that the preservation of game should not be allowed to interfere with the development and prospects of the country' (Astle 1999, p17), but by the 1930s pressure for wildlife conservation was mounting in the British colonies. In 1931, new game reserves were proclaimed including Victoria Falls and Zambezi River, Kafue Gorge, Mweru Marsh and Kafue.

According to Astle (1999) in 1931, Pitman, an experienced game warden from Uganda, provided an influential report that recommended the formation of a game department, and expressed concern that wildlife was being hunted out, blaming muzzle loading guns (and local people) for a 75 per cent decrease in wildlife. The far-sighted head of the Game Department, Vaughan-Jones (in his 1938 proposal) set the tone of conservation in Zambia for the next 50 years. He established reserves where 'fauna apart from potential aesthetic, recreative and scientific value has no direct utility' (Astle 1999, p31). He also established two types of controlled hunting areas, one for 'economic exploitation … to the advantage of the Territory as a whole' and the other with 'exclusive emphasis on native needs' (Astle 1999, p32). Care was taken to ensure that wildlife conservation did not impinge on so-called productive land. The formation of game reserves in Zambia entailed protracted discussions between local people and officials. In the Luangwa Valley, with one exception, nobody was forced to move (Astle, 1999). In Kafue, only five villages were relocated, and 48 families retain traditional fishing rights within the park to this day (Mwima, 2001). Large parks were formed with little effect on local people. The political forces were more complex than is currently assumed. For example, local government (which administered wildlife) was criticized by conservationists for placing more importance on people than on wildlife because of the open-ended licensing system and game laws. The Secretary for Native Affairs 1933 argued that 'the interests and welfare of the native community must be the first consideration' (Astle 1999, p31) and not the preservation of game. Even the Department of Game and Tsetse Control supported some African access to wildlife for the 'rationalization of game meat supplies … for what are, sociologically speaking, still hunting communities' (Gibson, 1999, p17; see also Astle, 1999).

By the 1950s, tensions had emerged between local rights to hunt and overhunting. This led to legislative measures to control licences and quotas, while leaving management with native authorities. Interestingly, the 1952 Ordinance that responded to these tensions introduced the highly innovative concept of private game reserves on private land, or on a chief's land with his consent.

By Zambia's Independence in 1964, some 43 per cent of the country was under some form of protection. Kaunda inherited decentralized and pragmatic wildlife policies. Hunting by local people was widely allowed and locally controlled. South Luangwa National Park hosted an experiment under an FAO-sponsored project to cull elephant, hippo and buffalo and to build roads and an airport to make wildlife into a viable enterprise. However, in the early years of Independence, Kaunda's government introduced two major changes in policy direction: Zambia developed increasingly preservationist polices, and it also centralized the control of wildlife and other natural resources (HURID, 2002). In 1967, Cambridge biologists and Sir Peter Scott lobbied Kaunda towards a more preservationist approach and the Luangwa sustainable use experiment was closed down (Child, G., pers. comm.). The National Parks and Wildlife Act (No. 57 of 1968, and No. 65 of 1970) reflected Kaunda's preservationist and centralizing conservation ideals and firmly established the superiority of the minister over professional civil servants. The state agency monopolized the wildlife sector to the exclusion of other actors. In 1972, Kaunda declared 17 new national parks, although there was never enough money to protect and develop these, a situation that persists into the present (see below). This centralizing legislation was in line with International Conventions for the Protection of African Fauna and Flora (1933), which were built around the principle that game belongs to and is managed by the state. However, this strategy has not served Zambia's wildlife well.

Gibson (1999) describes how wildlife resources were nationalized and became a tool of political patronage in an agency that valued political loyalty over technical capacity. The collapse of copper prices in the mid-1970s, and the simultaneous decline of the wildlife department, led to a bleak period for Zambian wildlife. Elephant populations were decimated, black rhino were extirpated and rampant poaching supplemented the income of military and police employees among others (Gibson, 1999). By 2000, Zambia's national parks had probably lost 80–90 per cent of their original wildlife (unpublished aerial surveys) and several were being encroached by agriculture. Only South Luangwa (60 per cent of carrying capacity) and Mosi-ao-Tunya (heavily overstocked) retained anything like their original wildlife. In desperation, a plethora of experiments began to 'bubble up'. In the mid-1980s several CBNRM programmes were established in and around national parks including ADMADE and Luangwa Integrated Resource Development Project. With considerable funding from Norway, South Luangwa National Park was established as a cost centre managed with some independence from the wildlife agency and elephant began to recover. In the mid-1990s, Zambia began to outsource tourism concessions, which enhanced wildlife protection in Luangwa, Lower Zambezi and Kafue national parks. Similar improvements followed when the management of some national parks was partially or wholly outsourced to NGOs (for example North Luangwa and Kasanka) and a private company (see Chapter 25).

Conclusions

Some general patterns and lessons emerge. Protected area management has gone through three economic stages. With white settlement, wildlife was initially decimated through frontier economic conditions including extensive land conversion and the combination of new harvesting technologies (for example rifles), markets and the absence of rules to control offtake. This was followed by a preservationist phase characterized by periods of intensified park building, especially in the 1960s. Small wildlife agencies were developed to manage these parks and the wildlife outside them, and by the 1970s the level of professionalism in many of these agencies was deliberately increased by hiring professional biologists. Park management improved and was exceptional in some countries. However, some of these professionals were visionary and recognized that the conservation model needed to change again in response to demographic and economic pressures. Working from within government agencies they created the conditions for the emergence of wildlife conservation through sustainable use outside of protected areas, including game ranching and community conservation, intensive crocodile and ostrich farming and the like. Three agencies in particular had highly effective cohorts and played a significant role in regional innovation – the Department of National Parks and Wild Life Management in Zimbabwe, the Bophuthatswana National Parks Board and the Natal Parks Board. In other countries, individual professionals or small groups played significant roles in innovation, with two individuals for example drafting the changes in Namibian legislation that subsequently provided the policy environment for an effective national CBNRM programme. Regional learning networks, including the SARCCUS and the SASUSG, were important mechanisms in accelerating innovation.

Many park agencies lost innovative capacity following Independence, when they increased employment many-fold without any increase in budget, and shifted towards political rather than technical conservation agendas (see Gibson 1999; Grindle and Thomas, 1991), lost key professional staff and came to rely more on a proliferation of donor projects, NGOs and consultants. However, where agencies were able to retain at least a small cadre of experienced professionals in the post-Independence window of flexibility, we often see considerable innovation including the testing of new concepts such as CBNRM in Zimbabwe and Namibia, and current innovations in South Africa.

However, the historical centralization of the control of wildlife in state agencies during the preservationist phase makes it particularly vulnerable to the changes in the competence and agenda of state park and wildlife agencies. Colonial wildlife agencies were small and relied on interpersonal accountability and professionalism rather than bureaucratic checks and balances (Parker, 2004). The rapid expansion of these agencies transforms this informal accountability into serious institutional weaknesses. Moreover wildlife agencies are near-monopolies that control a valuable national resource yet regulate themselves and their competitors. This combination of bureaucratic power and tenuous accountability poses considerable dangers to the wildlife sector, especially in 'fragile states' where weak or politicized agencies paradoxically keep control even closer to their chest, often crowding out innovation and investment in the private sector. Weak agencies are often associated with severely constrained environments for private and community conservation. They tend to look inwards towards their own institutional and personal survival, often at the expense of national development (see Grindle and Thomas, 1991). This is far more costly than is commonly recognized. Sound policy and practice can unlock the many layers of economic activity that depend on parks. For example, allowing communities to retain all the income from wildlife and forestry not only provides them with incentives for conservation, but allows several economic layers to be built on the resources so conserved, creating employment and economic multipliers throughout the economy. Weak agencies often do not understand or care about their considerable negative influence on the national economy, where suspicion of the private and community sectors often leads to poorly conceived regulatory restrictions, exacerbated because fees and licences are managed administratively rather than by market forces.

Southern Africa has, fortunately, partially sidestepped this malady. While regional wildlife agencies have gone through a period of decline despite rapid economic growth in the sector they support (losing budget and capacity), the 'conservation eggs' are no longer in the single basket of state-protected areas. Growth and innovation are occurring rapidly on private and communal land that, in the long run, is likely to benefit park agencies, even if competition leaves these bureaucracies feeling relatively uncomfortable and out of control in the short term. The diffusion of responsibility for wildlife in what we might term private and communal protected areas, has created many new conservation and business models and products, as well as a new way of thinking about wildlife conservation that incorporates economics, governance and political ecology. There is now as much wildlife outside the formal parks estates as inside them, and there are certainly far more commercial activities driving up the range and value of parks and wildlife than there were in 1980 (pers. obs.). State wildlife agencies are now forced to keep up with these changes or lose their relevance. As the series of examples in this section (and also in Part V) shows, many state park and wildlife agencies are adapting to rapidly changing circumstances, some very successfully (for example the North West Parks and Tourism Board and SANParks), and some in difficult environmental and social circumstances. Whatever the case, there is little doubt southern Africa's parks and wildlife would be in a precarious position if they had not challenged the centralist and non-commercial political economic philosophy set out in the London Convention of 1933, and that still infuses (and stultifies) conservation approaches throughout the South. By increasing the value of wildlife and wild land, and by ensuring that there are hundreds, or even thousands, of landholders and communities in the sector, and not just a few dominant wildlife agencies, considerable adaptive capacity has been created within the sector.

In the region, conservation innovation is very much alive. Interestingly, much of the current wave of innovation (see Part V) is occurring in South Africa, presumably because well-paid and enlightened park agencies are crafting sound policies in an environment with strong property and legal rights and a booming wildlife and tourism sector. However, innovation is also happening in other ways, sometimes born more of desperation than design, and often driven by a few dedicated champions and the sharing of ideas within the region through informal learning networks. The bubbling up of CBNRM and park initiatives in Mozambique and Zambia are good examples. The big question is whether these initiatives can be sustained without effective state leadership and capacity, or whether the general dynamism in the region's wildlife sector may provide enough energy to sustain these programmes until they are fully institutionalized. Only time will tell. However, it is clear that the framework for innovation has changed – it is no longer driven mainly by the state, although it seldom happens unless there are a few dedicated people within state agencies to enable NGOs, communities and the private sector to innovate. In other words, new ideas and state capacity are vital ingredients in the growth and maintenance of the sector.

This suggests that southern Africa needs to invest carefully in crafting park institutions and training young professionals. As the examples in this and other parts of the book show, many of the innovations of which the region is proud have been driven by individuals, and the policy legacy of individuals can be remarkably long. For instance, Stevenson-Hamilton played an important role in the emergence of park philosophies in South Africa that continue today, while Riney and his colleagues introduced ideas about sustainable use in the 1960s that are still reverberating through the region. Similarly, private and community conservation, and new park models, have been driven by small cohorts of professionals in many of the countries that we have discussed. Individuals matter, and perhaps the most important indicator of the future of state, private and community-protected areas in southern Africa is whether the sector can attract the brightest young people.

Finally, the way the conservation sector is structured is critical. We have recognized the importance of state leadership in unlocking the potential of the park and wildlife sector, but also noted that many agencies are losing the capacity to create growth. Put briefly, we need to avoid a 'penguin' structure where leadership is weak, where a large bureaucratic middle adds little value, and where the people working on the ground have little authority and experience. We will create far more jobs, innovation and conservation progress with an 'hour glass' structure. We need highly qualified leadership to develop sound policy, to purge bureaucracy and middle management, and to invest very significantly in getting professionals out on the ground (for example park managers, extension works, researchers, community facilitators) because innovation so often arises at the nexus between professional scholar–practitioners and field practice.

Notes

1   Grindle and Thomas (1991, see also Chapter 3) provide a lucid explanation for this phenomenon.
2   Interestingly, Carruthers records that some landholders developed a proprietary interest in wildlife on their farms in what today would be called private game reserves as early as the 1860s and 1870s. This suggests that our current knowledge of the history and emergence of the game ranching sector (see Chapter 10) is limited by our reliance on thin published sources.
3   It is probably no coincidence Ken Tinley, who played a leading role in Mozambican conservation at this time, also attended SARCCUS meetings. His name also crops up in the early formulation of Pilanesberg's policies.

References

Astle, W. L. (1999) *A History of Wildlife Conservation and Management in the Mid-Luangwa Valley, Zambia*, British Empire and Commonwealth Museum, Bristol
Bond, I., Child, B., de la Harpe, D., Jones, B., Barnes, J. and Anderson, H. (2004) 'Private land contribution to conservation in South Africa', in Child, B. (ed) *Parks in Transition: Biodiversity, Rural Development and the Bottom Line*, Earthscan, London
Brockington, D. (2002) *Fortress Conservation: The Preservation of the Mkomazi Game Reserve, Tanzania*, James Currey, Oxford
Brockington, D. and Igoe, J. (2006) 'Eviction for conservation: A global overview', *Conservation and Society*, vol 4, no 3, pp424–470
Carruthers, J. (1995) *The Kruger National Park: A Social and Political History*, University of Natal Press, Pietermaritzburg
Child, B. (2004) *Parks in Transition: Biodiversity, Rural Development and the Bottom Line*, Earthscan, London
du Toit, J. T., Rogers, K. H. and Biggs, H. C. (eds) (2003) *The Kruger Experience: Ecology and Management of Savanna Heterogeneity*, Island Press, Washington DC
Gibson, C. C. (1999) *Politicians and Poachers: The Political Economy of Wildlife Policy in Africa*, Cambridge University Press, Cambridge
Grindle, M. S. and Thomas, J. W. (1991) *Public Choices and Policy Change: The Political Economy of Reform in Developing Countries*, The Johns Hopkins University Press, Baltimore and London
HURID (Institute of Human Rights, Intellectual Property and Development Trust)

(2002) 'Policy and legislation review of the fisheries, forestry, wildlife and water sectors vis-à-vis community based natural resource management', prepared for CONASA, HURID, Lusaka

Mwima, H. K. (2001) 'A brief history of Kafue National Park, Zambia', *Koedoe*, vol 44, no 1, pp57–72

Neumann, R. P. (1998) *Imposing Wilderness: Struggles over Livelihood and Nature Preservation in Africa*, University of California Press, London

Parker, I. (2004) *What I Tell You Three Times Is True: Conservation, Ivory, History and Politics*, Librario, Elgin

Riney, T. (1967) *Conservation and Management of African Wildlife*, FAO, Rome

Runte, A. (1979) *National Parks: The American Experience*, University of Nebraska Press, Lincoln and London

Wilson, K. (2005) 'Of diffusion and context: The bubbling up of community-based resource management in Mozambique in the 1990s', in Brosius, P., Tsing, A. L. and Zerner, C. (eds) *Communities and Conservation. History and Politics of Community-based Natural Resource Management*, Rowman Altamira, Walnut Creek, CA

3

National Parks in South Africa

Jane Carruthers

The US claims to have invented national parks with Congress approval of Yellowstone in Wyoming in 1872 as a 'public park or pleasuring ground for the benefit and enjoyment of the people'. The words 'national park' were, however, first formally applied in legislation in 1879 to an urban public open space in Sydney, in the colony of New South Wales (Griffiths and Robin, 1997). As is well documented, the first national parks in the US celebrated the spectacular landscapes that distinguished North America from Europe, natural scenes that validated nationhood in the absence of cathedrals and other evidence of 'civilization', and that romanticized the settler frontier experience (Ise, 1961; Runte, 1979; Nash, 1982; Everhart, 1983; Beinart and Coates, 1995; Pritchard, 1999). In the first half of the 20th century, the American national park idea 'spread to many parts of the world and changed considerably in the course of its travels' (Harroy, 1972, p9; Nelson et al, 1978). Even though the final product in many parts of the British world bore the name 'national park', the path to protected area philosophy and management did not arise there from the desire to protect unspoilt nature from unbridled capitalist exploitation, as was the case in the US where Niagara Falls had provided an alarming example, but from game preservation – that 'pleasing British characteristic' as colonial game warden C. R. S. Pitman expressed it (Marks, 1984; MacKenzie, 1988). The purpose of this chapter is to provide a brief overview of some of the contextual and other issues – political, personal, scientific and biological – that have had an impact on the genesis and direction of the protected area estate in South Africa. In order to illuminate some of these major transformations, three examples are presented, each from a different era and established in different parts of the country.

Transformations in protected area philosophy

Over more than a century, the world's national parks have evolved to reflect changing ideas around environmental conservation and management, and these and other objectives have altered in specific historical contexts. The once iconic national park – a large, generally unpopulated but carefully managed recreational wilderness – has emerged to become a more flexible form of land use. Once lauded for their preservation of landscape and for the absence of people and almost always established on 'worthless lands' (Runte, 1979), national parks networks are now designed with the aim of achieving well-considered systems that encompass representative biomes in order to provide ecosystem services and provide opportunities for sustainable development (generally tourism), scientific research and biodiversity conservation (IUCN, 1994). Moreover, once regarded as a convenient way of administering national borders, such boundaries are increasingly becoming irrelevant or porous and transboundary or transfrontier conservation areas that straddle them (often called 'peace parks') are planned for many parts of the world. The early history of national parks everywhere in the world demonstrated little concern for the interests of local or indigenous people. Most often, in the US, India and many parts of Africa, communities were forcibly removed from national parks in order to make way for recreational tourism, or tolerated only as a paid labour force (Carruthers, 1993). However, the management of national parks as 'fortresses' that excluded local people, has altered to become a broader 'community conservation' in which all sectors of the population have a stake.

Scientific parameters have changed dramatically too, even to the extent today of incorporating aspects of local knowledge. In the early years of US national parks, for example, all predators were hunted and wildlife protection was not an explicit mandate (Wright, 1992; Sellars, 1997; Pritchard, 1999). When the first national park was established in South Africa in 1926, the concept of biodiversity conservation and the science of ecology were in their formative years. It is not surprising then that there was no mention of scientific study in the legislation that proclaimed the park, because at that time both professional values and scientific politics gave high status to specialized laboratory sciences or to those knowledge areas, such as agronomy, that contributed to economic growth. As a consequence, for many decades, protected areas were considered to be outside the boundaries of any 'science', their purpose quite separate from serious academic study (Carruthers, 2007). Management was left in the hands of military-style administrators, with a general mandate to leave nature alone, although 'vermin' species were often eradicated. In the 1960s in South Africa, the intervention of biologists in national park affairs introduced a more formal policy of 'management by intervention', the result of which was that the environment was actively manipulated by providing water for wildlife, instituting a grass-burning regime and culling animal populations that were considered to be too large. Presently this scientific paradigm is being revisited in the light of new ideas about ecosystem resilience (du Toit et al, 2003).

National park nomenclature and objectives

The national parks were not the only protected areas in South Africa, nor were they the only forum in which nature conservation was discussed. Although South Africa is a unitary state rather than a federation, the national government devolves a number of competencies to provincial governments, and nature conservation is one of these. Each South African province has a nature conservation agency with the authority to establish protected areas and this means that there are provincial national parks as well as national ones. This is also the case in Australia where there are state and federal 'national parks'. So a confusing aspect of the name 'national park' is that some national parks in South Africa are provincial and not national at all. There are, for example, no legally defined national parks in the province of KwaZulu-Natal although there are many protected areas, a number of considerable size and importance, the name of one of them being the Royal Natal National Park in the Drakensberg mountains.

The International Union for the Conservation of Nature (IUCN) recognizes 'wilderness' and 'national parks' as two distinct categories of land use (IUCN, 1994) and the terminology 'national park' has its own history in South Africa. Although both the Kruger National Park and the Dongola Wild Life Sanctuary (see below) had the same legislative status and were established by the South African parliament, one was called a 'national park' and the other a 'wild life sanctuary'. Despite the different names, there was no difference in the legislative establishment or administration of the two protected areas. Given its American and Australian origins, one can appreciate that the terminology 'national park' had connotations of a public facility in which the environment was managed for visitor enjoyment. In the early 20th century there were many wildlife protectionists in the British Empire who considered that the primary purpose of such areas in Africa was not entertainment, but the protection of species of wild animals and that this required a different name. In wildlife journals of the 1920s and 1930s there was energetic debate about the most appropriate discourse. While the idea of legislative entrenchment – an important characteristic of a 'national park' – was attractive because of the permanence of such status, the word 'park' was problematic because it conveyed such strong ideas of public recreation. But also problematic were the titles 'preserve' or 'reserve' because of their medieval origins in preserving game species for the exclusive benefit of the rich or the landowning classes. Well into the late 1930s James Stevenson-Hamilton, the warden of the Kruger National Park and other conservationists advocated naming wildlife reserves in Africa and India (and elsewhere as appropriate) as 'national faunal sanctuaries', 'national wild life sanctuaries' or even just 'animal sanctuaries'. For many people these designations, far more than 'national park', would be preferable to refer to places that were essentially wilderness in which nature and wild animals were more important than people. 'Parks', by contrast, suggested a manipulation of nature, of bowing to the demands of tourists, who agitated for artificial water-points to attract wild animals that might be seen more often and easily, appropriately luxurious accommodation and other recreational and service facilities, ease of access and the like.

In the early years there was no minimum area or even common management strategy that applied to South Africa's national parks and these showed considerable variety. Some of the national parks that were established in the 1930s were identical to zoological gardens or game farms. The Bontebok National Park (1931) and the Mountain Zebra National Park (1937) were extremely small and were founded in order to provide a nursery environment for building up herds of those named endangered species. The Addo Elephant National Park was created in 1931 to enclose the small remaining herd of elephants in the Eastern Cape and to prevent them from marauding on local farms.

Together with the narrative of how and why state land was allocated for certain protectionist purposes is a history of changing scientific management that has impacted on why and where national parks have come into being. When national parks were established in South Africa in the first half of the 20th century, ecology was in its infancy and the only powerful field sciences were those relating to agricultural or veterinary matters. Zoology and botany were studied at universities, not in the field. The task of early wardens and game rangers was a policing one, not a scientific one. By the late 1940s, however, biology had begun to widen into environmental science and it had become important to study wild animals in their natural ecological surroundings. For this reason, attitudes towards 'vermin' species changed and, for example, the anthropomorphic distaste for wild dog, crocodiles and reptiles ended with an appreciation that every creature played a role in what was increasingly regarded as a 'system'. Managing that system through active intervention became the paradigm for South Africa's national parks (Carruthers, 1995a; Hall-Martin and Carruthers, 2003). This entailed culling 'excess' animals, burning the veld and providing water. Thinking about protected areas more as a national system than a miscellany also grew as a philosophy in the 1960s and 1970s and formal plans for establishing national parks so that they better represented the various biomes of South Africa came into play. Marine and Karoo national parks were the result – Wilderness National Park, Karoo National Park, West Coast National Park among them. Prior to this, international initiatives – such as the London Convention of 1900, the American Committee for International Wild Life Protection or the London Conferences of 1933 and 1938 as well as the establishment and directives of the IUCN – were significant factors in national park establishment (Hall-Martin and

Carruthers, 2003). What has so far been lacking is the human and cultural dimensions of national parks in South Africa – ethnology, archaeology and the like – despite a legal mandate to study and care for these matters to the same degree as the attention devoted to wildlife.

Establishing the Kruger National Park

South Africa's first formal national park – the Kruger National Park – was established in 1926 by combining two provincial game reserves in the eastern Transvaal (now Mpumalanga and Limpopo provinces) that had been founded around the turn of the century – the Sabi (1898) and the Singwitsi (1903). They were given legal existence as a national park through an Act of parliament (Carruthers, 1995a, 1995b). The change in status from provincial to national status came about because provincial protection mechanisms – by way of an Administrator's Proclamation – were insecure and could easily be overturned. The conversion of these local game reserves into national parks began after the four colonies of South Africa were united into one country in 1910. Thereafter, with a national interest at stake rather than separate colonial agendas, a number of constituencies interested in protecting wild animals collaborated to support and coordinate the initiatives of certain national government individuals and departments in order to conserve an area in perpetuity through national legislation.

The Kruger National Park was not the first state-owned protected area in South Africa, nor was it the first wildlife preservation scheme in the country. In 1822, Governor Lord Charles Somerset had instituted a colonial game reserve in the Cape, and by the end of the 19th century there were game reserves in the South African Republic (Transvaal), the colonies of Natal and Zululand, with others in the planning phases in the Cape Colony and Bechuanaland (Carruthers, 1985, 1995b; Brooks, 2001). But this kind of Victorian game reserve was not, of course, a national park. The distinguishing feature of a national park relates to state ownership and public access while a game reserve was a medieval type of 'preserve', comprising an area from which the populace was excluded in order that stocks of species of wild animals suitable for recreational sport hunting by the upper classes would increase.

Southern Africa's game reserves of the late 19th and early 20th centuries were established on economically marginal or economically unproductive lands – areas in which serious livestock and human diseases were endemic, where the soil or climate was unsuitable for agriculture or pastoralism, or where minerals had not yet been discovered. Often they were situated in outlying or frontier districts where control of local Africans was difficult and game reserve structures provided formal state authority, the Pongola Game Reserve being an example (Carruthers, 1985, 1995b).

The objectives of these early game reserves were often not clearly articulated but, generally speaking, they were wildlife 'nurseries' in which predators were controlled so that antelope, buffalo and elephant – the sport hunters' quarry – would increase and eventually, when stock was at an appropriate level, would be culled by hunters for a substantial fee (Carruthers, 1995a, 1995b). Management was based on predator control and the prevention of poaching. Matters such as biodiversity, landscape and cultural conservation, recreation, heritage and education were not on the agenda at that time.

The success of converting the Sabi and Singwitsi Game Reserves from their provincial status as closed game reserves to a national park that was open to the public was due to the fact that in the 1920s white South Africans accepted the principle that viewing wildlife was a legitimate and financially viable form of land use, for which state land and state funding should be made available. With the modernization of the country, wilderness areas were becoming rare and there had been a change of environmental attitudes from denigrating such places to valuing the wildlife they contained. Despite objections from some quarters that 'locking up' land for wildlife protection would retard the agriculturally based economy, supporters pointed out that the national parks of the US had proved just how popular national parks might be as recreational destinations that provided the nation's 'pleasuring grounds'. In addition, the US models had also shown that people were prepared to pay to visit national parks and that economic benefits from visitor spending accrued to the surrounding region, not only to a particular park (Carruthers, 1995b).

A recent article stresses that in order to be successful, protectionist initiatives require three components that are in synchronicity with each other – the environmental, political and cultural (Cormier-Salem and Basset, 2007). These came together in the 1920s in South Africa and resulted in the establishment of the Kruger National Park. Environmental values have been mentioned above, but the time was ripe during the 1920s for realigning South African national and cultural symbols. The white population consisted of English and Afrikaans (Dutch), usually referred to as the 'two races', that were divided by history, cultural traditions and language. Their collaboration on the common project of a national park was achieved by employing a discourse of nationalism and this happened at two levels. In the narrower ethnic sense, the Afrikaans/Dutch-speaking community chafed under what they believed were British imperial shackles imposed on them after they lost the Anglo–Boer War of 1899–1902. They longed to have again a republican form of government, independent from Britain, in which they would be free to advance Afrikaner interests above others. Specific Afrikaner symbols were employed in this agenda, including many that suggested a sectarian pride, a romanticized settler past, and recalled or invented the Voortrekker period of the 1830s when many Dutch settlers left the British colonies of the Cape and Natal to establish independent polities in the Orange Free State and Transvaal. At the same time, there was a broader platform of nationalism as many English-speaking South Africans of the 1920s came to appreciate that the time had arrived to set aside the old hostilities between Afrikaners and English-speakers and to put a divided past behind them. Thus there was a general feeling that the country needed to fabricate its own identity and there was a growing political philosophy of 'South Africanism' as new national symbols – such as an anthem and flag – were debated. So great was the sentiment running against the close relationship with Britain at that time that there was a change of government at an election in 1924 when Jan Smuts was ousted as prime minister by J. B. M. Hertzog.

In the circumstances outlined above, it is not surprising that the 'national park' was among the symbols that were enthusiastically accepted as politically viable (Carruthers, 1989, 1995a). Naming the park after Paul Kruger, an Afrikaner hero who had been president of the Transvaal Republic prior to the Anglo–Boer War, not only reflected the national sentiment, but it also ensured the support of many people – particularly the large population of Afrikaans-speaking 'poor whites' – who might otherwise have been antagonistic to wildlife preservation. In later years, after 1948, when the National Party under D. F. Malan came to power and South Africa was firmly on the road to apartheid and to a republican form of government (which was achieved in 1961), situating the Kruger National Park within the Afrikaner fold was further intensified and reinforced. During the 1950s, a good deal of nature protection literature lauded wildlife protection as the moral achievement of Afrikaners in particular, and Paul Kruger was hailed as a major conservationist figure in his own time rather than the political leader whose name had been expediently used in the circumstances of the 1920s (Carruthers, 1994).

The Dongola Wild Life Sanctuary

From its inception, the Kruger National Park was extremely successful. In the late 1920s with the growing popularity of the motor car, visitors arrived in substantial numbers and rest camps, roads and bridges were constructed to meet the tourist demand. But not all South Africa's national parks have shared this record of rapid success. During the 1940s (which, like the 1920s, was a period of political turbulence), the establishment of a national park on the Limpopo River created a political furore that eventually led to its abolition. Dongola, the name applied to a large area west of Musina along the Limpopo River, would have been the first national park in southern Africa to have had landscape, archaeology, botany and ecological processes, and not wild animals, as its core protectionist focus. In addition, it would have been the first formal transfrontier park in Africa, because the neighbouring Rhodesian government and the chartered company of Bechuanaland were willing to cooperate in the venture. Transboundary conservation areas are now one of the major thrusts of the conservation endeavour in the subcontinent (Hall-Martin and Carruthers, 2003) but the first attempt was an abject failure.

The history of the Dongola Wild Life Sanctuary – a national park that existed for a mere two years between 1947 and 1949 – deserves to be better known because it illustrates just how vital it is to obtain political and public support in order for nature conservation ventures to succeed. Dongola had its origins in 1922 when the Dongola Botanical Reserve was established as part of the Botanical Survey of the Union of South Africa. The idea of the reserve was to begin experimental plots, develop new grass varieties and to collect information about the agricultural and pastoral potential of northern South Africa. The scheme was run by Dr I. B. Pole Evans, a plant pathologist in the national Department of Agriculture. Before long, he realized that he had discovered a botanical and archaeological treasure and he agitated for more and more farms to be purchased by the government and added to the reserve. The botany of the area was unusual because it was, in Pole Evans's words, 'in balanced equilibrium' and thus had lessons for the emerging discipline of ecology in which South Africa was then an international leader (Carruthers, 2006). The archaeological gem of the region was Mapungubwe (on the farm Greefswald), a hill that contained the relics of a civilization that flourished at the confluence of the Shashe and Limpopo Rivers between c.900 and c.1290 and that demonstrated state formation, sophisticated metal-working and an extensive Indian Ocean trade (Carruthers, 2006).

By the late 1930s and early 1940s Pole Evans had taken it upon himself to promote the conservation of the Limpopo River valley west of Musina and to transform some 240,000 hectares (ha) into a national park. However, he did not intend it to be a playground for holidaying whites as was the Kruger Park, but rather an area dedicated to wilderness and scientific research. The political underpinning was not well prepared and it seems to have been assumed by the protagonists that their moral and scientific high ground would automatically be acknowledged and accepted. The scheme had strong champions – Pole Evans, Jan Smuts (back in power as prime minister of South Africa after 1939 and during World War II) and Smuts's combative Minister of Lands, Andrew Conroy. But it had determined opponents too. Among them was the National Parks Board that rejected the project out of hand fearing that it might jeopardize the success of the Kruger National Park because of visitor and staff competition. Local farmers were against the scheme because although farming in this remote district was marginal, they were afraid of losing their properties and their hunting rights. A number of government departments were also wary about putting too much land into conservation initiatives and thus withdrawing its economic potential, while the veterinary fraternity fretted about the livestock diseases that might be introduced from across the Limpopo River. Environmentally too, the public was not ready for such a 'scientific reserve' because ecology and African archaeology had not yet been adopted into current thinking.

But more significant than these factors in bringing down the scheme were the opinions of the voters in the area and the need for certain politicians to court them. The main resistance came from Smuts's opposition, the National Party. Most of the local whites were potential National Party supporters and losing their votes might be critical in the forthcoming election. So the National Party fought the project at every turn. The situation was exacerbated when there were rumours that the park might be named the 'Smuts National Park' as a counterpoint to the 'Kruger', for Smuts was anathema to the Afrikaner nationalists of the 1940s. In order to investigate the matter objectively a select committee was appointed and it sat for two years. The 'Battle of Dongola', as it became known, generated some of the longest and most acrimonious debates ever recorded in South Africa's parliament. But when the matter was put to the vote, members of parliament were instructed to vote along party lines and thus despite the fierce objections, on 28 March, 1947 the Dongola Wild Life Sanctuary Act became law. But the following year, when Smuts's government lost the general election, the National Party quickly abolished the national park, which is what it had promised its supporters it would do (Carruthers, 1992). It is somewhat surprising, given the extreme and bitter politicization of nature protection and the enormous publicity it had generated, that Dongola faded so quickly from public memory and it is generally overlooked in accounts of South Africa's national park history (Brown, 1977).

But Dongola has recently been resuscitated and its current fate is of interest because it confirms the fundamental reality that political and cultural support is critical for establishing and managing protected areas. Dongola's downfall – although it could not, of course, be predicted at the time – was that it came in an inappropriate social and scientific environment and from the wrong quarters. But circumstances change. In 1995 – after South Africa achieved a non-racial constitution and a democratically elected government – a small area (7859ha) at the confluence of the Shashe and Limpopo rivers that included Mapungubwe was proclaimed as the Vhembe/Dongola National Park. This was the beginning of a revival that culminated in July 2003 in the acceptance of this area as South Africa's first World Heritage Cultural Landscape (Carruthers, 2006). This prestigious development was greeted with national and international acclaim because the new government's agenda is strongly focused on the African continent and it highlights the African

Renaissance in particular. Mapungubwe thus recalls images of a proud African pre-colonial past that has considerable political currency. The idea of a trans-frontier conservation area has been raised anew and negotiations for its establishment are well under way with the reacquisition of many farms that were once part of the Dongola National Park. This project is one of the most significant national heritage and conservation schemes in southern Africa.

Africans and South Africa's national parks: The Pilanesberg National Park

While many authors have emphasized the ecological innovation of national parks, more recently there has been closer scrutiny of their features as a 'social invention' (Nelson et al, 1978). When South Africa's first national park was established by the parliament of the Union of South Africa in 1926, the enfran-chised 'nation' whose wishes were expressed by the enabling legislation consisted only of adult white males. The franchise base was later extended to white women and, after 1994 with the ending of apartheid to the entire adult population. This has had consequences for the protected area estate.

When formal administration of the Sabi Game Reserve (between the Crocodile and Sabie rivers) began in 1902, the 3000 African residents were summarily evicted by warden Stevenson-Hamilton. But before long it was real-ized that labour was needed and the policy of evictions was reversed. Remaining tenants within the reserves became liable for labour or rents in cash, as was the usual South African custom of the time. Prisoners supplemented the tenant labour workforce; illegal immigrants seeking work on the Witwatersrand gold-fields arrested by reserve officials for trespassing generally received a sentence of a fortnight's imprisonment, during which they worked on road-building and other projects. Other Africans were in direct game reserve employment, as domestic workers, wagon-drivers, handymen, scouts and 'native police' who aided white game rangers in their duties (Carruthers, 1993). Africans thus played a large part in ensuring the success of South Africa's early game reserves. In reconceptualizing African responses to colonial conservation practice, it is as well to recall the caution of John MacKenzie not to cast Africans merely in the role of victims and thus deny them power or agency (MacKenzie, 1994), while Beinart too refers to a 'struggle to free [African] historiography and social studies from narratives of dependence, victimhood and romanticism' (2000, p292).

African–white interaction in matters relating to wildlife preservation has been crucial to the history of South Africa's protected areas and it remains a burning issue. At the heart of it lies access to land. As the population increased in the 20th century and as South Africa's commercial agriculture expanded, the state increasingly restricted African access to land. In this regard, the Native Lands Act of 1913 and the Native Trust and Land Act of 1936 were critical. Measures such as these created a shortage of African land as blacks were confined into smaller and smaller 'reserves'. One of the flashpoints of the 1930s relating to a national park concerned an area between the Luvuvhu and Limpopo Rivers where the Tsonga Makuleke community lived, and which was coveted by the national park authorities of that time because it had an extremely rich riverine vegetation. The tension between the two groups for control of this area lasted for 30 years and the matter was only resolved in 1969 – at the height of apartheid – when the community was forcibly removed from the land they had occupied for at least six generations. The land restitution claim of the Makuleke has since been recognized (de Villiers, 1999) and their example of lodging a land claim against South African National Parks has been followed by other groups, with varying degrees of success, against both the Kalahari Gemsbok National Park and the Augrabies Falls National Park (Hall-Martin and Carruthers, 2003).

Apartheid South Africa spawned its particular variety of national parks. The Pilanesberg National Park, for example, owed its very origins to the 'homelands' policy of the Afrikaner nationalist government of the 1960s. This national park, like Kruger before it, was a means of legitimizing a 'nation', this time of, Bophuthatswana, one of apartheid's 'independent states' and easing it into a place of international respectability. Bophuthatswana was singled out for particular praise for being 'one of South Africa's small black states in the process of establishing a game reserve' – such a 'highly applaudable' project when 'more land for farming is the international cry' (Anon, 1980).

The Pilanesberg National Park was not a conserved natural area but a forced removal, land reclamation and game-stocking project (see Chapter 18). The idea of some kind of wildlife tourist attraction in the unique Pilanesberg volcanic crater of South Africa's North West Province (formerly Bophuthatswana) originated from the extremely conservative Potchefstroom University for Christian National Education in 1969, when self-governing and independent homelands were being mooted as apartheid blueprints for the future. In the early planning stages, opposition to the scheme from the Pilane clan who occupied the land was intense. Perhaps this local resistance had some effect in delaying the project, because it was not until the Sun City hotel, casino and entertainment complex began nearby nearly a decade later that the national park idea became a reality (Brett, 1989; Griffiths and Robin, 1997).

Little of the natural ecology had survived the livestock or cultivation activities of whites and Africans who had farmed in the Pilanesberg crater for more than a century. There were unsuitably sited roads, dams and other structures; invasive exotic plant species had taken hold and soil erosion was considerable. However, in order to enhance the status of Bophuthatswana and to bring in tourist revenue from wildlife viewing, as well as from the gambling and mixed race facilities that were then illegal in 'white' South Africa, there was strong pressure from the South African and Bophuthatswana governments to start a national park, comprising some 50,000ha of the crater and surrounding hills. Scant attention was paid to the concerns of the occupiers, the area merely appearing to outsiders as a dusty, marginal agricultural area with villages filled with the unemployed. Local opposition was unsuccessful and the Pilanesberg community was finally removed to Saulspoort in the late 1980s and early 1990s (Hancock, 1984).

The Pilanesberg National Park thus had a difficult birth, at a time when paramilitary wildlife management and anti-human ecology was powerful in national park dogma. Not only did local people oppose it, but capital invest-ment to begin land reclamation and game introduction projects was not forthcoming from the Bophuthatswana government. In the event, Anton Rupert's South African Nature Foundation, whose close ties with the apartheid government are now well established (Ellis, 1994), came to the rescue with R2 million (US$564,000) that removed the livestock, pulled up the fencing, razed the farm houses and created a national park. Consultants were employed to plan the park, 'Operation Genesis' began and the translocation of wildlife commenced. But opposition from local people continued and early manage-ment was corrupt and incompetent. Moreover, scientists argued about reintroducing wildlife species from other areas of the subcontinent and possi-ble problems resulting from mixing gene pools. When a newly introduced circus elephant killed a Brits farmer, scientists loudly denounced the Pilanesberg park as nothing better than an out-of-control zoo (Brett, 1989).

However, in time, experienced rangers and scientific personnel (principally from Zimbabwe and the then Natal Parks Board) were employed, land recla-mation procedures began to take effect and relocated wildlife populations settled into their restricted habitat. Visitor numbers increased and the coffers of the park swelled. In fact, a new kind of national park had been created in South Africa because, for the first time, the Pilanesberg offered local people an opportunity to benefit from a protected area. The park offered environmental education and financial pay-offs in terms of economic development to neigh-bours and placated some of those people who had been removed from their homes in order to create the homeland national park. Trespassing became less frequent and local communities began to benefit from culling programmes by obtaining meat and other wildlife products. In addition, the local councils received a portion of the fees charged for hunting – an unusual pursuit in a national park, but one that is extremely popular with visitors from Europe. Currently, the Pilanesberg provides a flexible model for a suite of extremely successful protected areas run by provincial authorities in South Africa. But the eviction of the local people has not been assuaged and the Pilane community has instituted a land claim in terms of the restitution legislation – a complex claim that has not yet been finally settled.

Conclusions

In the early 1990s the observation was made of Australia that 'we are seeing ... a basic redefinition of what IS a national park' (Birckhead et al, 1993, p383). Indeed, it is clear from the experience of Australia and elsewhere that the hegemonic Yellowstone model of a wilderness, free from people because it has been artificially 'created by bulldozers and fences, forced migration and resettlement' is inappropriate and that if adhered to, '... the consequences can be terrible' (Birckhead et al, 1993, p363–364).

In South Africa today, national parks and other protected areas have become sites for rural development and capacity-building (see also Chapter 24). While initially national parks had central government landownership as a core requirement, today the notion of who owns the land – whether state, private or community – is less important. South Africa is attempting to create a fresh strategy, although the legacy of apartheid weighs heavily, and some park managers are still uncomfortable with challenges to their beliefs as to what constitutes a national park (Robinson, 1995; Hall-Martin and Carruthers, 2003). National parks, whether initially founded to protect grand scenery from commercial exploitation (as was the case in the western US) or to conserve diverse mega-fauna (as in Africa or India), are currently being reappraised. Integrating people, nature and culture is evidence of the change in environmental ideology from strict preservation to innovative initiatives linked to sustainability.

References

Anon (1980) 'Pilanesberg: A game reserve in a volcano', *African Wildlife*, vol 37, no 6, pp18–21

Beinart, W. (2000) 'African history and environmental history', *African Affairs*, vol 99, pp269–302

Beinart, W. and Coates, P. (1995) *Environment and History: The Taming of Nature in the USA and South Africa*, Routledge, London and New York

Birckhead, J., de Lacy, T. and Smith, L. (eds) (1993) *Aboriginal Involvement in Parks and Protected Areas*, Aboriginal Studies Press, Canberra

Brett, M. R. (1989) *The Pilanesberg: Jewel of Bophuthatswana*, Southern Books, Sandton

Brooks, S. (2001) 'Changing nature: A critical historical geography of the Umfolozi and Hluhluwe Game Reserves, Zululand, 1887–1947', PhD thesis, Queen's University, Kingston, Ontario

Brown, A. C. (ed) (1977) *A History of Scientific Endeavour in South Africa*, Royal Society, Cape Town

Carruthers, J. (1985) 'The Pongola Game Reserve: An eco-political study', *Koedoe, vol* 28, pp1–16

Carruthers, J. (1989) 'Creating a national park, 1910 to 1926', *Journal of Southern African Studies*, vol 15, pp188–216

Carruthers, J. (1992) 'The Dongola Wild Life Sanctuary: "psychological blunder, economic folly and political monstrosity" or "more valuable than rubies and gold"?', *Kleio*, vol 24, pp82–100

Carruthers, J. (1993) '"Police boys" and poachers: Africans, wildlife protection and national parks, the Transvaal 1902 to 1950', *Koedoe*, vol 36, pp11–22

Carruthers, J. (1994) 'Dissecting the myth: Paul Kruger and the Kruger National Park', *Journal of Southern African Studies*, vol 20, pp263–283

Carruthers, J. (1995a) *The Kruger National Park: A Social and Political History*, University of Natal Press, Pietermaritzburg

Carruthers, J. (1995b) *Game Protection in the Transvaal 1846 to 1926*, Archives Year Book for South African History, Pretoria

Carruthers, J. (2006) 'Mapungubwe: An historical and contemporary analysis of a World Heritage Cultural Landscape', *Koedoe*, vol 49, pp1–13

Carruthers, J. (2007) 'Influences on wildlife management and related sciences in South Africa, c.1900–c.1940', *South African Historical Journal*, vol 58

Cormier-Salem, M-C. and Bassett, T. J. (2007) 'Nature as local heritage in Africa: Longstanding concerns, new challenges', *Africa*, vol 77, pp1–17

de Villiers, B. (1999) *Land Claims and National Parks: The Makuleke Experience*, Human Sciences Research Council, Pretoria

du Toit, J., Rogers, K. H. and Biggs, H. C. (eds) (2003) *The Kruger Experience: Ecology and Management of Savanna Heterogeneity*, Island Press, Washington DC

Ellis, S. (1994) 'Of elephants and men: Politics and nature conservation in South Africa', *Journal of Southern African Studies*, vol 20, pp53–69

Everhart, W. C. (1983) *The National Park Service*, 2nd edn, Westview Press, Boulder, Colorado

Griffiths, T. and Robin, L. (eds) (1997) *Ecology and Empire: Environmental History of Settler Societies*, Keele University Press, Edinburgh

Hall-Martin, A. and Carruthers, J. (eds) (2003) *South African National Parks: Celebration*, Horst Klemm, Johannesburg

Hancock, P. (1984) 'Pilanesberg Game Reserve: Four years later', *African Wildlife*, vol 37, no 1, pp47–51

Harroy, J-P. (1972) *World National Parks: Progress and Opportunities*, Hayez, Brussels

Ise, J. (1961) *Our National Park Policy: A Critical History*, Johns Hopkins Press, Baltimore

IUCN (1994) *Guidelines for Protected Area Management Categories*, CNPPA/WCMC, IUCN, Gland, Switzerland and Cambridge, UK

MacKenzie, J. M. (1988) *The Empire of Nature: Hunting, Conservation and British Imperialism*, Manchester University Press, Manchester

MacKenzie, J. M. (1994) 'Edward Said and the historians', *Nineteenth Century Contexts*, vol 18, pp9–25

Marks, S. A. (1984) *The Imperial Lion: Human Dimensions of Wildlife Management in Central Africa*, Westview Press, Boulder, Colorado

Nash, R. (1982) *Wilderness and the American Mind*, 3rd edn, Yale University Press, New Haven

Nelson, J. G., Needham, R. D. and Mann, D. L. (1978) *International Experience with National Parks and Related Reserves*, University of Waterloo, Ontario

Pritchard, J. A. (1999) *Preserving Yellowstone's Natural Conditions: Science and the Perception of Nature*, University of Nebraska Press, Lincoln, Nebraska

Robinson, R. (ed) (1995) *African Heritage 2000: The Future of Protected Areas in Africa*, *Proceedings of the IUCN Commission on National Parks and Protected Areas*, National Parks Board/IUCN, Pretoria

Runte, A. (1979) *National Parks: The American Experience*, University of Nebraska Press, Lincoln, Nebraska

Sellars, R. W. (1997) *Preserving Nature in the National Parks: A History*, Yale University Press, New Haven, Connecticut

Wright, G. (1992) *Wildlife Research and Management in the National Parks*, University of Chicago, Chicago

4

The Growth of Park Conservation in Botswana

Graham Child

Botswana formed a specialized agency for nature conservation in 1961, which was late among southern African countries. This chapter traces the early growth of this agency and the institutions that developed to manage Botswana's spectacular wild resources, inside and outside the impressive spread of conservation areas. It focuses on the period 1960 to 1972, when the bulk of Botswana's present parks and reserves were proclaimed, and hints at the early emergence of a rural economy in which the formalized use of wildlife and wild lands played a significant role, but suffered setbacks in support of a dominant livestock industry.

Most parks and reserves in Botswana came about through a combination of administrative convenience and because the areas supported concentrations of spectacular game, at least seasonally. The majority were created on unoccupied state land that obviated the need for authority from a district council representing the people in a 'tribal area', and the extensive consultation that this involved. The result was that considerable areas of Botswana, amounting to about 17 per cent of the country, were set aside by central government, at minimal social cost, on the advice of bureaucrats.

The early Game Department

The Game Department was created in terms of the Fauna Conservation Act, 1961. Prior to that, wildlife had been under the care of the general administration, and there were no parks or reserves apart from the Gemsbok Game Reserve, which was managed by South Africa in terms of an agreement between the British authorities in Bechuanaland and the South African government. The Fauna Conservation Act was amended extensively in 1979 and was augmented by the Forest Order, 1981.

The Game Department was initially located in Francistown with Mr Pat Bromfield as the first game officer. He had seen service with the Gurkha Regiment during World War II, rising to the rank of major, and brought a typically military flavour to the agency. Initially, his main function was to control troublesome elephant in the Tuli Block along the Limpopo and Shashe Rivers, in the east of the country, on the Zimbabwe and South African frontiers. For this he had a force of scouts, mainly men with elephant hunting experience on secondment from the Zambian Game and Fisheries Department, who helped train early Botswana recruits.

By 1965, the department had three permanent and one temporary wardens. The Chobe Game Reserve had been formed and was staffed by one of the wardens and a detachment of scouts. The author was a wildlife ecologist seconded to this unit as part of the United Nations Development Programme's (UNDP's) multilateral aid programme, which in this case was implemented by the Food and Agriculture Organization (FAO).

Later that year Major Bruce Kinloch undertook an assignment, on behalf of the UK Ministry of Overseas Development, to suggest how the department should be re-engineered and enlarged. Bromfield welcomed Kinloch as they had served together during World War II before Kinloch became chief game warden in (what was then) Tanganyika and Uganda. Among his recommendations, Kinloch suggested replacing Bromfield with himself which, unsurprisingly, was not acceptable to government. Instead, the FAO provided Lawrence Tennant for the post. He was an Edinburgh-trained veterinarian who had grown up in southern Africa and had also been chief game warden in Uganda, and he had Alec Campbell, a naturalized Motswana, as his assistant.

This group formulated much of the early departmental policy, basing it on extensive fieldwork, hundreds of interviews and discussions with local elders, a good understanding of the natural ecology and of the expectations of the rural people who had traditionally lived with and used wildlife. Campbell in particular, had a sound knowledge of the social system and the need to accommodate the prevailing circumstances of communities affected by wildlife policy. For example, many people in the Kalahari were heavily dependent on wildlife, especially small species like springhare, for subsistence in a situation where agriculture was not possible and extensive livestock ranching was marginal (von Richter, 1969a, 1969b, 1976b). This attention to people's needs – by legally allowing people to utilize some species for subsistence purposes – marked an early breakthrough in the inclusion of sociological considerations in the formulation of wildlife policy in southern Africa at least.

The Game Department became the Department of Wildlife and National Parks (DWNP) in 1967, and its head office moved to the new capital, Gaborone. By 1971 there were wardens and support staff stationed at Kasane, Maun, Francistown, Ghanzi, Serowe, Molepolole and Tshabong.

A special environment

The Game Department had jurisdiction over wildlife in a vast country with poor communications but interesting habitats supporting massive wildlife populations. At their peak, the numbers of wildebeest, hartebeest and springbok, at least, were each estimated conservatively at well over 100,000 animals (Newman, pers. comm.; pers. obs.; Child and le Riche, 1969). The south-western Kalahari in Botswana represented some of the least modified true desert anywhere in the world in 1970. Desert or dry semi-desert merging into arid savanna woodland on ancient windblown Kalahari sands covers most of the country. The arid environment is broken by two extensive interlinked areas of wetland, watered by rivers rising in the humid tropics to the north in Angola and Zambia. In the extreme north of the country, the Kwando/Chobe River and Linyanti swamp system is shared with the Caprivi Strip of Namibia, before draining via the Chobe into the Zambezi system. The inland delta of the Okavango River is located in north-western Botswana and drains into Lake Ngami or the Makgadikgadi depressions, to the south and south-east of the delta in good flood years, with limited water escaping to the Kwando/Linyanti system in exceptional years. Both the huge Makgadikgadi Basin and the smaller Lake Ngami lying within it are closed inland drainages.

Although cattle provided the mainstay of the economy, the traditional dependence of poor households on wildlife became even more important as the global terms of trade for livestock declined from the 1970s (Child, 1988; Bailey et al, 1996). The DWNP, with two FAO experts, began to build up Botswana's wildlife industry from about 1967, as it was believed that wildlife would continue to be sacrificed until it could demonstrate its economic competitiveness with livestock as a profitable and environmentally friendly land use.

Child (1970a) conservatively estimated that wildlife was worth US$2.4 million or 5.2 per cent of gross national product (mostly through earnings from hunting) by 1968, and Campbell (1973) put this value at more than double by 1971. While the value of wildlife was growing, the demand for livestock was weakening. Furthermore, the build-up in livestock over the previous one and a half centuries (but particularly in the previous 50 years after the rinderpest pandemic in 1896) had caused widespread habitat modification leading to declining productivity. Loss of perennial grasses and extensive bush encroachment in the best ranching areas become a serious and rapidly spreading threat to the livestock industry and the welfare of rural people (Child, 1970b).

By this time, beef importing countries in Europe had insisted Botswana should be divided up by a series of extended cattle cordon fences before they would allow its beef to be imported, necessary because of the perceived threat of foot-and-mouth disease (FMD), which is endemic in Botswana. FMD was thought to be carried by a variety of game animals before in-country research limited it to buffalo and showed that, like cattle, other species are sensitive to the disease and thus good indicators of an outbreak (Condy et al, 1969; Falconer and Child, 1971; Hedger, 1972). The beef trade was vital to Botswana and fence construction began in the mid-1950s, with new fences still under consideration today. The fences cut across vast tracts of country causing havoc to the large herds of 'migrating' game – mainly wildebeest and red hartebeest – that died in thousands on the fences trying to reach water in the Okavango Delta and Boteti River. Child et al (1970) point out that wildebeest disappeared from the delta when the Kuke fence blocked their moving north to it seasonally.

While the fences were designed to prevent the spread of FMD, buffalo occurred only in the north of the country, so thousands of head of other game animals were needlessly sacrificed further south to save a cattle industry that even in the 1960s appeared neither ecologically nor economically sustainable due to Botswana's low rainfall. Child and Child (1986) showed that it is financially and ecologically hazardous to ranch cattle where annual rainfall is less than 750 millimetres (mm) per annum, as it can not meet the twin objectives of being profitable and sustainable. To this day, control of FMD with cordon fences to prevent the movement of buffalo lacks unequivocal support from objective research, although it has become entrenched international veterinary dogma. Some 19 species of wildlife were also sacrificed for 20 years around the south-eastern fringe of the Okavango Delta in the name of tsetse control (Child et al, 1970) to support the cattle industry.

When the two measures were introduced, Botswana was very poor and ordinary citizens had little influence over affairs of state. The measures were taken in good faith by bureaucrats to save a cattle export industry that was then seen as important, but is now known to be neither ecologically nor economically sustainable (Child and Child, 1986). That the wildlife potential – now known to have been worth millions of dollars – could be sacrificed on such grounds was because game animals had been rendered financially almost valueless by European-driven game laws, which prevented game from being owned privately or utilized and traded commercially. By inhibiting earnings from wildlife, the colonial game laws dissuaded landholders from respecting it or undertaking measures to conserve it. While this was less so in Botswana than in other colonies (possibly because the Game Department was recent and small), it was nevertheless evident and something the new department was determined to minimize when managing wildlife outside parks and reserves.

With hindsight, it is interesting to speculate as to how Botswana might have developed and used the vast and spectacular wildlife resources it had in the 1950s and 1960s to generate greater wealth had there been a more appropriate institutional environment. As it was, the people used wildlife quite freely, but their returns were small due to the distortions in the wildlife market resulting from contemporary laws and their implementation that undervalued wildlife. The high returns now being earned from inferior resources elsewhere in southern Africa, where the vast herds of wildebeest, hartebeest, buffalo, zebra, eland, springbok, gemsbok or ostrich seen in Botswana in the 1960s have not been recorded, suggest that Botswana missed an opportunity to support a substantial industry with the abundant wildlife it had only 60 years ago.

Progress in the 1960s towards using wildlife in support of the economy was retarded by a hangover from the centralized protectionism enshrined in traditional colonial game laws. In retrospect, without a model to emulate, the DWNP's actions, although innovative, were too timid. Lingering protectionist attitudes cost, and continue to cost, Botswana and other parts of the region dearly. Confusing protectionist and utilitarian ideals continue to cloud philosophy guiding institutional development and this leads to indecisive management, often because official attitudes towards wildlife remain ambivalent.

The macro-fauna

By the mid-20th century, Botswana still had some of the richest wildlife resources in Africa in terms of the diversity and numerical richness of the fauna, which combined elements from the mesophytic savannas, the arid south-west of Africa and the extensive wetlands in the north. High numbers of some species were attributable to the naturally fertile and diversified habitats, but those of others apparently reflected a temporary upsurge in populations induced through modification of the desert habitats by livestock, fire and occasionally the provision of water (IUCN, 1987).

Child and le Riche (1969) report what were probably the last two examples of the famous springbok treks (in 1946 and 1950) that took place mainly during the 19th century in the arid interior of southern Africa. They involved hundreds of thousands of animals that moved hundreds of kilometres over a broad front, consuming all edible vegetation in their path. The treks apparently occurred only during one short period of history, having been initiated by a lowering of the successional stages in the vegetation to suit springbok through overgrazing by livestock. Similarly, the vast herds of wildebeest that occurred in the Northern Cape province of South Africa, across Botswana and into Zimbabwe were also a temporary phenomenon of the late 19th to the mid-20th centuries (Child, 1972). Oral tradition suggests that this may have applied, and may still be affecting other species including red hartebeest in the south-west and elephant and buffalo in the north of the country (Campbell and Child, 1971). In the case of elephant, present densities have built up since the mid-1940s, are shared with adjacent parts of Angola, Namibia, Zambia and

Zimbabwe, and seem unsustainable (Child, 1968), except where they have been reduced by poaching in Zambia.

In the arid south-west, in particular, the fauna is characterized either by more or less solitary sedentary forms such as kudu, duiker, steenbok and warthog at low densities, or gregarious mobile forms like wildebeest, red hartebeest, springbok and, to a lesser extent, eland and gemsbok that reach high densities locally for short periods. Movements are triggered by local rainfall and the subsequent growth of forage and water-providing plants such as the tsamma melon (Mills and Retief, 1984; Child and Child, 1986).

Setting aside parks and reserves

The parks and reserves in Botswana are shown in Figure 4.1 and Table 4.1, which indicate their size. They cover approximately 17 per cent of the country and include all major ecological types (von Richter, 1975, 1976a). The mobility of much of the fauna, however, militates against the areas being self-sustaining ecological entities and often requires the integrated management of the nation's wildlife inside and outside the parks and reserves (Child, 1970a, 1972; Campbell, 1973).

Table 4.1 *Partial list of Botswana's conservation areas*

Conservation area	First established	Area (ha)
Chobe National Park	1961	1,059,000
Gemsbok National Park	1932	2,766,500
Mabuasehube Game Reserve	1971	197,200
Nxai Pan National Park	1970	248,000
Makgadikgadi Game Reserve	1970	487,710
Gaborone Game Reserve	1980	300
Moremi Wildlife Reserve	1962	496,800
Central Kalahari (Kgalagadi) Game Reserve	1961	5,180,000
Khutse Game Reserve	1971	255,000
Mannyelang Game Reserve	1985	300
Bathoeng Game Reserve	1992	470
Maun Game Reserve	1975	8500
Mogabana Game Reserve	1992	940

Source: IUCN (1987); Broekhuis (pers. comm.)

Gemsbok Game Reserve

The Gemsbok Game Reserve, declared in 1932, was the first modern reserve in Botswana. It was then only a 40 kilometre (km) wide annex to the Kalahari Gemsbok National Park of South Africa, which had been created a year earlier

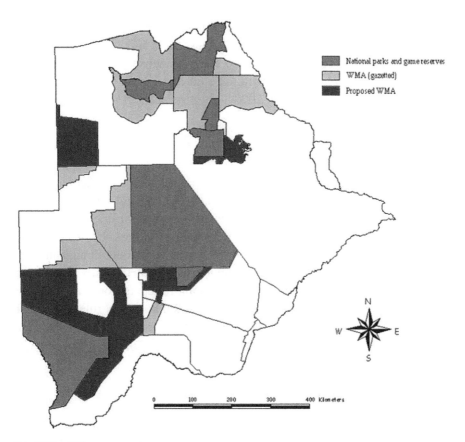

Legend:
- National parks and game reserves
- WMA (gazetted)
- Proposed WMA

Note: WMA: wildlife management area
Source: Broekhuis (pers. comm.)

Figure 4.1 *Botswana's national parks, game reserves and controlled hunting areas*

and from whence the Botswana Reserve was managed under an agreement with the South African Parks Board, now South African National Parks (SANParks). The area included in the complex was undeveloped wilderness, which had been occupied occasionally by itinerant bands of San Bushmen.

The reserve was enlarged and made a national park in 1971, and the Mabuasehube Game Reserve was added in the east. The latter was retained as a game reserve to avoid possible opposition to a package of parks and reserves being proposed to parliament by the DWNP at the time. It was feared that the implicitly greater restrictions of a national park compared with a game reserve might cause objections to the whole package if an area astride the track between the village of Tshane and the district headquarter outpost at Tshabong were given national park status.

The combined area continued to be managed by the South African Parks Board until 1984, when local staffers were posted to it and it was developed for tourism. This arrangement was unusual but remained convenient while the conservation agency was built, virtually from scratch, and allowed the department to concentrate on other priorities. The park, adjacent game reserve and the Kalahari Gemsbok National Park in South Africa were combined into the Kgalagadi Transfrontier Park in 2000 – the first such 'peace' park formally created in southern Africa.

Central Kalahari Game Reserve

The Central Kalahari Game Reserve was declared in 1957 (von Richter, 1976a) with the intention of safeguarding the nomadic hunting and gathering lifestyle of the San Bushmen (Silberbauer, 1965), and to prevent cattle farmers from spreading into the area. In 1971, with modernization of the wildlife legislation, the Khutse Game Reserve was added to the complex, without diminishing the role of the former as a reserve for San Bushmen. In the early 2000s, the Government of Botswana evicted the remaining residents from the park, but in 2006 the High Court decided in favour of the San Bushmen, ruling that their eviction was unlawful and unconstitutional, and confirming they have the right to live on their traditional land inside the reserve. The matter is not yet fully resolved.

Chobe National Park

The Chobe National Park was first created as a game reserve on unoccupied state land in 1961. Following an ecological survey and recommended modification of its boundaries by the FAO (Child, 1968) it became Botswana's first national park in 1968, which was extended in the east in 1974 when the Mikaelelo Game Reserve was added. It had not been possible to protect this extension earlier, as it would have interfered with the stock route by which cattle from Ngamiland were walked some 400km to the Zambezi, for export to Zambia and the then Zaire. These stock routes had been moved progressively further east from the Mababe depression, in the west of the present park, in response to the reinvasion of large areas of northern Botswana by tsetse fly. Each move left serious bush encroachment from the transitory but heavy grazing by the large herds of livestock trekking to market during two or three months each year.

There were a few old settlement sites in, and especially just outside, the reserve. Some time between the mid-1940s and the early 1960s, before the park was created, San Bushmen living near Ngwezumba and elsewhere in or near the present park had chosen to move and settle in the Chobe enclave or at Masuma and Nungu along the Zimbabwean border. Their moves were motivated by a variety of factors – widespread crop failures and stock losses; to be closer to schools, stores and other services; to avoid encroaching tsetse fly; and because of tribal politics. In addition to modification of the habitats by agriculture and the stock routes that had traversed the reserve, there had been scattered pastoralism and logging, and much of the east was burned every year. By the time the reserve was established, the only settlement in the park was at Serondella, a disused sawmill from the 1930s.

In the mid-1960s, the reserve had 38 large mammal species, including white rhino that had been reintroduced from Natal in South Africa. This is the highest number of jackal-sized or larger mammals known in any park in the world, but since then the diversity has probably declined, especially along the Chobe River. Both species of rhino were probably eliminated by hunting, in the case of relict black rhino in the south of the park, since 1968. In contrast, species such as elephant, buffalo, impala and perhaps puku and kudu have increased and spread along the Chobe River front near Kasane. Observations 40 years apart suggest that the ecosystems in this part of the park are being grossly simplified by overuse by elephant (Child, 1988) and perhaps other bulk roughage feeders such as buffalo or the numerous impala, superimposed on past burning, logging, and overgrazing by cattle.

The park was staffed almost from its inception. The first warden, Patrick Hepburn, was an experienced stockman and led a team of scouts seconded from the strongly paramilitary Zambian Game and Fisheries Department, who helped him train the first Batswana recruits. Park headquarters were established inside the park at Kasane and an all-weather road with game viewing off-shoots was constructed along the Chobe River between Kasane and Ngoma in 1963–1964. Other roads in and through the park remained sand tracks, while half a dozen boreholes to supply game and tourists with water were sunk in 1967.

The park was opened to the public in 1964 and immediately became a tourist destination of some note, attracting over 2500 visitors in its first year. By 1971 it supported two large hotels, several safari camps, a camping ground and park accommodation facilities. Since then, tourism has grown significantly, with positive spin-offs in terms of local infrastructure development, including national highways, a bridge across the Chobe River into Namibia, and the Kasane airport. More importantly, the park has become the Botswana core of a buoyant regional tourist industry covering the north of the country, the Caprivi Strip in Namibia, and adjacent parts of Zambia and Zimbabwe, centred on Victoria Falls.

Moremi Wildlife Reserve

The Moremi Wildlife Reserve was the first reserve to be created on 'tribal land' in Botswana. June Kay (1970) describes the saga that led to the Batawana tribe (then the Ngamiland Tribal Authority, now the North West District Council) declaring this jewel a wildlife reserve in 1961 (Robbel and Child, 1976). Public pressure in Maun built in favour of a wildlife sanctuary in the Okavango Swamps during the late 1950s, spearheaded by prominent black and white citizens and expatriates living in the town, then a remote outpost. Those in favour of a sanctuary wished to preserve the wilderness qualities of the delta, but also wanted to develop its commercial potential for tourism to grow the local economy and diversify it away from domination by cattle ranching.

The progressive action catalysed by these public-spirited individuals and led by Chief Moremi was opposed by the colonial administration, which perceived wildlife as a hindrance to taming the wilderness for agriculture and saw no merit in preserving untamed tsetse-infested land in such a remote situation for future tourism. However, supporters of the reserve eventually won, and the reserve was declared by Legal Notice 14 of 1965. When the game-rich area for the reserve was selected, between two major distributaries in the north-east of the Okavango delta, it had dense tsetse and was unoccupied except occasionally by a small group of itinerate San Bushmen, said to number between 40 and 60 individuals. Being infested with tsetse fly, it was totally unsuitable for cattle, so that the tourist benefits from the reserve, which grew quite rapidly, were a welcome add-on to the local economy. In 1976, the reserve was enlarged to incorporate Chief's Island and a link with the Chobe National Park. Today, Maun has a thriving tourist industry and the cattle industry has withered to a shadow of its former magnitude, largely due to past overgrazing and poverty-induced disease.

The Fauna Conservation Society of Ngamiland was set up to administer the area (Campbell, 1973), but had difficulty raising sufficient revenue to manage it adequately, even with voluntary help from members. By 1965, the reserve was staffed with the society's employees, mostly former residents of San Bushman extraction. These game guards were joined by central government scouts in about 1967 and, briefly in 1969, by an FAO ecologist. Jack Ramsden, a Maun resident and one of the leading personalities in the creation and early management of the reserve, later became a warden in DWNP, epitomizing the close relationship that generally existed in Ngamiland between the different tiers of government. The reserve is now said to being managed under a co-operative agreement with the DWNP, with its much greater resources.

Nxai Pan National Park and Makgadikgadi Game Reserve

The Nxai Pan National Park and Makgadikgadi Game Reserve formed an impressive complex in the north-west of the Makgadikgadi depression. In the mid-1960s, Nxai Pan supported impressive seasonal herds of wildebeest, zebra, gemsbok and springbok. Most of these animals moved to the western

Makgadikgadi and watered along the Boteti River during in the dry season, after the small temporary waterholes on Nxai pan dried up and seasonal heavy dew ceased. Eland were also common and there were smaller numbers of red hartebeest, giraffe, lion, cheetah, spotted hyaena, bat-eared fox and black-backed jackal, among others. Both areas were proclaimed in 1970 as two separate entities, divided by the main Francistown to Maun highway and the Kanyu flats, to allow for a stock route between Ngamiland and eastern Botswana. They have since been joined by a southward extension of the park.

Child (1968) attributed the heavy bush encroachment west of Nxai Pan, inside and outside the park, largely to overgrazing by cattle – resulting from a temporary cattle post and stock route to the Zambezi. Open parkland was converted to dense thicket covering some 1300 square kilometres (km^2), in which a number of wildlife species dwindled towards zero in response to the dense bush, through which a route had to be bulldozed to allow the cattle to pass.

Other parks and reserves

A number of lesser parks and reserves (see Table 4.1) have been created with government support to conserve special features or provide special services, including readily accessible outdoor recreation to towns and areas of high population density. (Although local endeavours may not be funded by government, they still have to be approved by local authorities and legalized by government.) Some examples include the Nata Sanctuary, created to protect water birds (particularly flamingo and pelican); the Khama Rhino Sanctuary, created to breed rhino for restocking; and Mokolodi Nature Reserve, created for conservation education and through which thousands of school children have passed and been taught about conservation.

Management of wildlife outside parks and reserves

With such a small agency responsible for wildlife, the government could do little, until recently, to manage the resource outside parks and reserves, other than develop institutions aimed at persuading the general public to respect the resource and use it wisely. This included the promotion and rationalization of high fee-paying recreational hunting by tourists and subsistence hunting by the local people. Land outside parks and reserves was divided into 40 hunting blocks that supported a three-tiered hunting programme. Local people resident in a block could hunt there if they held a licence that was free for many animals, but made a small charge for valuable trophy species sought by visiting sportsmen, who paid much higher fees. Residents of Botswana were given preference over non-residents to hunt in the 22 blocks not leased as concession areas to safari operators, and they paid cheaper licence fees than visiting hunters for the limited duration hunts. Quotas were set for each block and central government shared the revenue from blocks in tribal land with the appropriate local district council (Child, 1970a). This has evolved into community-based wildlife management, at least in the north of the country, although a major share of the revenue from hunting species such as elephant still goes to government (see Chapter 15).

It is interesting to recall that safari hunting in southern Africa began in earnest in Botswana. Following a scandal involving President Kenyatta's wife in the sale of ivory, Kenya, the cradle and Mecca of outfitted hunting, suspended all hunting indefinitely in about 1962. Many of the established and highly reputed 'white hunters' left Kenya, taking their skills and their foreign clients in a highly competitive market to hunting grounds elsewhere in Africa. Some of the best-known hunters ended up in Botswana where they were joined by Batswana, Namibian and Zimbabwean capital and hunters, and the last two soon repatriated the expertise gained from the Kenyans. From a tentative beginning in 1962–1963, safari hunting was into its stride in Botswana by 1965. Much of the hunting sold as individual hunts that could be booked in advance, in the 22 blocks not reserved for safari operators, was taken up by South Africans.

As in Zimbabwe, high fee-paying hunting for foreigners – which is less capital intensive and requires more modest wildlife resources than photographic tourism – has sometimes paved the way for non-hunting or photographic safaris in suitable areas. Some of these activities depend on the proximity of a park or reserve, but a few, at least, are on land where wildlife is being managed by a community for the revenue it generates from hunting.

Conclusions

Botswana was among the first countries in southern Africa to formally recognize the value of using wildlife to promote its conservation, by channelling benefits from it to people living with wildlife on their land.

With the importance of wildlife to rural subsistence, and as a resource of last resort, it seemed wise to ensure that it survived for those purposes at least. From the mid-1960s, official policy recognized that to do so, the returns from wildlife had to be improved, while still providing cheap subsistence hunting for the local population. These objectives have largely been met, though argument persists over whether or not a higher proportion of wildlife-generated income should be returned to residents in the areas where it is generated.

Early policy required large areas of land to be protected from agricultural expansion to create wildlife reservoirs and to provide spillover for hunting. It also required that these areas should be a representative sample of habitats, ecosystems and geological formations that would facilitate the preservation of all existing wildlife species, so far as this was possible. Policy also noted that the development of these areas needed to take account not only of wildlife conservation, but also the preservation of sites of historic and cultural value, as well as preserving wilderness and encouraging tourism (Child, 1970a; Campbell, 1973).

The parks and reserves created did not displace many people but under-pinned wildlife conservation in the country and served people by contributing to the local and national economy. The creation of parks and reserves was not limited to central government, but was also undertaken by the local people in the cases of the Moremi and Khutse Reserves, although conservation-conscious expatriates assisted and provided information regarding the land use options. It is interesting to note that both these reserves, now devoted to tourism, are on land that had been tribal hunting areas.

With a mobile large mammalian fauna, it was clear from the outset that wildlife management inside and outside of protected areas needed to be harmonized. As a consequence, with the possible exception of the Gemsbok Game Reserve, which was created in response to the wishes of a friendly neighbouring state, the parks and reserves in Botswana were created with a strongly utilitarian motive. They were created to conserve wildlife and its habi-tats (what today we would describe as biological diversity), which is essential if Botswana's rural areas are to remain able to adapt to a changing world.

Setting aside land for nature conservation was also seen as important for diversifying the economy in the face of the widespread simplification of habits and loss of productivity being caused by the dominant livestock industry. This concern has been fully vindicated by the decline in the global terms of trade for beef and the rapid growth in nature-based tourism, although it was not fore-seen that the switch from cattle to game would be driven by economic rather than ecological forces.

These policies were given legal effect through the National Parks Act in 1967. It was also proposed that all management should be based on scientific research – accepted as meaning scientific monitoring and the adaptive management arising from it. By the late 1960s, the central authorities were managing ecological reserves and wildlife for people but were already showing a willingness to share these responsibilities with local people. In the case of the Moremi Wildlife Reserve, central government provided the policy guidelines and legislation and cooperated in the management of the reserve, but encour-aged the district council to take the lead in decision making and to retain the revenue the area generated to manage the reserve. The partnership had minor difficulties but generally worked well.

Similar partnerships were less relevant on state land where there were no district councils and little, if any, settlement. With few people living near the reserves there was, in effect, no one with whom to cooperate. This was because the Batswana traditionally lived in large villages and towns, most of which, with the exception of Maun situated on the south-eastern fringe of the Okavango Delta, were located in the east of the country.

Severe bush encroachment over large areas was attributable to a form of shifting pastoralism, where land was abandoned for new pastures when grazing around an existing cattle post became depleted. Deteriorating habitats were characterized by a loss of perennial grass, moderate to dense bush encroachment and changes in the dependent macro-fauna. Roan, sable and tsessebe were usually the first sensitive grazing species to decline, and have disappeared from large parts of their former range. They were followed by other grazing species, some of which were apparently temporarily favoured by the range conditions that accompanied stages in the ecological decline. Such changes are apparent in many national parks and ecological reserves throughout southern Africa (and beyond), but their causes and the effects they induce are seldom as recent and starkly obvious as they were in the arid sandy environments in Botswana.

Botswana provides special opportunities for understanding the short- and long-term effects of land use, particularly extensive pastoralism in dry sandy deserts. This is because, in the 1960s, some parts of the country remained some of the least modified of such ecosystems anywhere in the world, coupled with the fact that changes in either direction tend to be rapid in sandy ecosystems. The changing ecology presents park managers with many immediate challenges if they are to preserve the inherent richness and special adaptations of the rich indigenous fauna in these harsh environments. It is surely worth the effort as such habitats are deteriorating almost universally, often beyond the potential for rehabilitation, because they are clearly unsuited to prevailing livestock practices. Ascertaining how to manage them correctly for indigenous wildlife offers important leads as to how they might be managed sustainably, at a global scale, to maximize long-term human welfare.

Botswana was one of the first countries to experiment with better approaches for making ecological reserves more socio-economically compatible and acceptable in the areas in which they were embedded. With South Africa, it created the first transfrontier park in southern Africa, the Kgalagadi Transfrontier Park in the south-west of the country. The agricultural potential is also low in most of the north-east of the country around Kasane and adjacent parts of Namibia, Angola, Zambia and Zimbabwe. Here, the opportunity exists for Botswana to use its innovative leadership to catalyse the proposed Kavango–Zambezi Transfrontier Conservation Area in the region. Botswana has already joined leaders in the region in demonstrating the advantages of a service-driven wildlife and tourism economy over commodity-driven agriculture. Greater international cooperation could greatly improve the lot of poor people living in the region as a whole, and through them their national economies. It could also be tailored to facilitate better wildlife management in general, and in particular of the vast shared elephant population of some 250,000 animals, to elevate the region to become a sustainable tourism jewel in the African crown.

References

Bailey, D. W., Goss, J. E., Laca, E. A., Rittenhouse, L. R., Coughenour, M. A., Swift, D. M. and Sims P. L. (1996) 'Mechanisms that result in large herbivore grazing distribution patterns', *Journal of Range Management*, vol 49, no 5, pp386–400

Campbell, A. C. (1973) 'The national park and reserve system in Botswana', *Biological Conservation*, vol 5, pp7–14

Campbell, A. C. and Child, G. (1971) 'The impact of man on the environment in Botswana', *Botswana Notes and Records*, vol 3, pp91–110

Child, B. (1988) 'The role of wildlife utilization in the sustainable economic development of semi-arid rangelands in Zimbabwe', DPhil thesis, University of Oxford, Oxford

Child, B. and Child, G. (1986) 'Wildlife, economic systems and sustainable human welfare in semi-arid rangelands in southern Africa', report to the FAO/Finland Workshop on Watershed Management in Semi-Arid and Arid Zones of SADCC Countries, April 1986, Maseru, Swaziland

Child, G. (1968) *Report to the Government of Botswana on an Ecological Survey of North Eastern Botswana*, FAO, Rome

Child, G. (1970a) 'Wildlife utilization and management in Botswana', *Biological Conservation*, vol 3, no 1, pp18–22

Child, G. (1970b) 'Ecological constraints on rural development in Botswana', *Botswana Notes and Records*, vol 3, pp157–164

Child, G. (1972) 'Observations on a wildebeest die-off in Botswana', *Arnoldia (Rhod)*, vol 5, no 31, pp1–13

Child, G. and le Riche, J. D. (1969) 'Recent springbok treks (mass movements) in south-western Botswana', *Mammalia*, vol 33, no 3, pp499–504

Child, G., Smith, P. and von Richter, W. (1970) 'Tsetse control hunting as a measure of large mammal trends in the Okavango Delta, Botswana', *Mammalia*, vol 34, no 1, pp34–75

Condy, J. B., Herniman, K. A. J. and Hedger, R. S. (1969) 'Foot-and-mouth disease in wildlife in Rhodesia and other African territories', *J.Comp. Path*, vol 79, pp27–31

Falconer, J. and Child, G. (1971) 'A survey of foot-and-mouth disease in wildlife in Botswana', report to the Government of Botswana, Gaborone

Hedger, R. S. (1972) 'Foot and mouth disease and the African buffalo (*Syncerus caffer*)', *J. Comp. Path.*, vol 82, pp19–28

IUCN (1987) IUCN *Directory of Afrotropical Protected Areas*, IUCN, Gland

Kay, J. (1970) *The Thirteenth Moon*, Hutchinson, London

Mills, M. G. L. and Retief, P. F. (1984) 'The response of ungulates to rainfall along the riverbeds of the southern Kalahari', *Koedoe Supplement*, pp129–151

Robbel, H. W. O. and Child, G. (1976) *Notes on the Ecology of the Moremi Wildlife*

Reserve, Department of Wildlife, National Parks and Tourism, Gaborone

Silberbauer, G. B. (1965) *Bushman Survey Report*, Government Printer, Gaborone

von Richter, W. (1969a) *A Survey of the Wild Animal Hide and Skin Industry*, report prepared for the FAO, Rome

von Richter, W. (1969b) 'Wildlife and rural economy in south-western Botswana', *Botswana Notes and Records*, vol 2, pp85–94

von Richter, W. (1975) 'Major vegetation types in relation to existing national parks and game reserves in Botswana', Proceedings of the 6th SARCCUS Meeting for Nature Conservation, Wildlife Utilization and Management, Kasungu National Park, Malawi

von Richter, W. (1976a) *The National Park and Game Reserve System of Botswana*, FAO, Rome

von Richter, W. (1976b) 'The utilization and management of wild animals as a form of land use in marginal areas of Africa', *Resources and Development*, vol 10, pp93–102

5

The Emergence of Modern Nature Conservation in Zimbabwe

Graham Child

Zimbabwe's formal commitment to nature conservation commenced with the Game Law Amendment Act, 1891, a year after white settlement began in what was then Rhodesia (Thomlinson, 1980).[1] Previously the indigenous people had had mostly weak conventions that helped to conserve wild animals. They had also set aside religious sanctuaries and the Ndebele, at least, had created a large royal hunting preserve along the Shangani River (Child, 1995). The first formal reserves set aside after the establishment of white government were the Rhodes Matopos and Rhodes Nyanga national parks, created in 1902 on land purchased privately by Cecil John Rhodes and entrusted to the nation in terms of his will.

The Rhodes Estate

In one of his many philanthropic legacies, Rhodes donated scenically spectacular and historically and prehistorically significant land with a high intrinsic biological diversity in the eastern border mountains (Nyanga) and the Matopo hills, for the citizens of Zimbabwe to enjoy. Nyanga was provided with a hotel, and Matopos two. A rail link – required by Rhodes – enabled the citizens of Bulawayo 'to enjoy the glory of the Matopos from Saturday to Monday' (Tredgold, 1956, p6). Besides outdoor recreation, Rhodes decreed that his land should be used for agricultural research and training and should be planted with all types of trees, including exotic species. This extended to a zoo at Matopos. The zoo housed sable antelope, which apparently provided the founders of the park's present population, having escaped when the zoo's fence was washed away by the Maleme River.

The mixing of agricultural, nature conservation and managed landscape objectives suggest that Rhodes was concerned to provide areas that served people, both productively and recreationally. He does not seem to have been influenced by the Yellowstone model, which was known in Zimbabwe by then. In 1899, a member of the Legislative Assembly, Dr Sauer, proposed that an area of land should be reserved in the country in the manner of the Yellowstone National Park, but his suggestion was rejected by the Assembly as conflicting with the rights of the British South Africa Company (Cumming, 1981).

Rhodes provided an endowment for each park to be managed in terms of his will. In the case of Matopos, this was to yield at least GB£4000 per year. The agricultural commitment was satisfied initially by the Matopos agricultural and Nyanga horticultural research stations and by the Matopo Agricultural College, which proved unviable and was later converted into a primary school. The land bequeathed by Rhodes at Matopos and Nyanga has been added to by successive governments, to form two of the prime national park complexes for recreation and the conservation of biological diversity in Zimbabwe.

Initially the parks were managed by the Rhodes Trustees, but the responsibility passed to the administration in 1918, by order of the Legislative Council.[2] The areas were called national parks and administered by local committees for almost half a century before their management was passed to the Department of National Parks and Wild Life Management (DNPWLM). Following the transfer, the special management committees were retained for both parks, and Nyanga, in particular, is managed in accordance with a budget that is largely discrete from that for the rest of the wildlife and parks authority. The two parks, being at least partially on trust land, became Special National Parks by Proclamation No. 48 of 1953.

In the meantime, the land around the Matopos Dam had apparently received some sort of protection in 1926 (IUCN, 1987). The old Mtarazi National Park, contiguous with Nyanga was added to the Special National Park for administrative convenience. In 1975, the Matopos complex was divided into the national park and a recreation park, incorporating the Matopos Dam and an area with camps for Boy Scouts and Girl Guides, on the land where Baden-Powell conceived the worldwide scouting movement (Baden-Powell, 1897). The Tshabalala Sanctuary, on the boundary of the city of Bulawayo on land that had been owned by Rhodes, was added to the complex in the 1970s, when leases to private individuals on seven farms owned by the Rhodes Trustees were terminated and the land was split between the parks estate and the Matopos Agricultural Research Station on the advice of a parliamentary committee.

Taming the wilderness

White settlement brought population growth and a determination to tame the wilderness. In rural areas this led to the rapid intensification of crop and livestock production and the opening up of numerous mines. In this context, most rural people viewed wildlife and ecological reserves as unaffordable luxuries that were incompatible with mainstream progress. As a result, reserves demanded by a few persistent conservationists were created only in vacant or sparsely settled areas, on land judged as unsuitable for peasant or commercial agriculture. Elsewhere, it was unofficial government policy to eliminate wildlife where it was perceived to threaten agriculture.

The most wasteful anti-wildlife measures were government actions to control tsetse fly (particularly to protect white-owned commercial ranches) and the shooting out of buffalo in an unsuccessful attempt to eliminate foot-and-mouth disease. Both activities were undertaken without scientific evidence and showed a disregard for the considerable damage they caused to the environment, particularly wildlife.

Over 659,000 game animals were shot in relatively confined areas in the Zambezi and Savé valleys as part of tsetse control measures between 1919 and 1957–1958 (Child and Riney, 1987) without eliminating any reasonably common species. Black rhino were almost exterminated from the hunting areas and sensitive grazers and eland were shot in declining numbers as the hunting programme progressed, but the most favoured tsetse food animals (warthog, bushpig, kudu and bushbuck) (Weitz, 1956) held their own or increased during 38 years of being hunted. This applied to several other species including elephant and buffalo. More significantly, tsetse control led to habitat deterioration, some of it in parks and reserves. Early seasonal veld burning was undertaken to facilitate the hunting and protect insecticides (like DDT and Dieldrin) from later burns, suppressing the perennial grasses on which stability in savannas depends. This encouraged bush encroachment, leading to accelerated loss of topsoil. Other tsetse control measures included the destruction of the woody riparian vegetation along watercourses and the destruction of the near endemic ironwood forest patches and seasonal pans in the Gonarezhou. The pans, in particular, were essential habitat for many non-target species, including an endemic killifish, whose global range is limited to a few of the pans.

Most tsetse control in Zimbabwe represented misdirected public sector investment. It was undertaken by competent, honest hard-working civil servants, unwittingly causing serious environmental degradation and foreclosing options to use wildlife, mostly in communal areas. Fragmentation of government bureaucratic responsibilities that led to tsetse control authorities being unaccountable for the state in which they left land, or for ensuring that its subsequent use was sustainable, cannot be blamed entirely for the fiasco. Public objections to the programme were ignored by the responsible authority, which did good research on tsetse but failed to examine the broader ecological, economic and social merits of its programme. Some of the criticism was driven by sentiment, but much arose from sound logic with national interest in mind.

Tsetse control was a prime example of government agencies having mandates that affect land quality and long-term productivity, but remaining unaccountable for ensuring that land quality and productivity were maintained. Unfortunately nature conservation agencies and their fellow travellers who see their mandate as the production of large numbers of herbivores in response to public demand, irrespective of the effect on habitats, are not immune from the same malady.

The greater parks and wildlife estate

Early conservation legislation in Zimbabwe was concerned primarily with protecting game and regulating hunting. The Hwange Game Reserve was created in 1928 in anticipation of a governor's proclamation under the Game and Fish Preservation Act 1929, which enabled him to 'define reserves within which it shall not be lawful without special permission ... to hunt'. The proclamation of the Victoria Falls Game Reserve (Taylor, 1990), which incorporated a previously proclaimed national monument extending over the falls themselves, and the Hurungwe Game Reserve followed in 1931. The first national parks, other than Matopos and Nyanga, were declared in the 1950s and 1960s in terms of the National Parks Act, 1949. This Act described a national park as an area 'for the propagation, protection, and preservation therein of wild animal life, vegetation and objects of geological, ethnological, historical or other scientific interest for the benefit, advantage and enjoyment of the inhabitants of the colony'. The early days of Hwange are traced by Ted Davison, the founder warden, who describes how the original park was expanded through the addition of contiguous commercial farms (Davison, 1967).

The setting aside and the key attributes of these and subsequent ecological reserves in Zimbabwe is indicated in Table 5.1, which shows the legal status of the areas in terms of the Parks and Wild Life Act, 1975, which rationalized the classification of protected areas. The Act followed the 1969 Wild Life Commission (Petrides and Pienaar, 1969), which found that even enlightened members of the Zimbabwean public and stakeholders in the wildlife sector had only a hazy idea of the contemporary international perception of a national park. This is not surprising, as many prime areas were still classified as game reserves and a number of the reserves then classed as national parks were used for more extensive recreation around large dams close to urban centres.

The confusion arose partly because jurisdiction over ecological reserves devoted to wildlife and recreation was still split between the federal and territorial governments. Game reserves and controlled hunting areas were created

Table 5.1 *Zimbabwe's national parks, botanical reserves and gardens, sanctuaries, safari and recreation areas*

Protected area	Area (ha) [Area of water, ha]	Main features
National parks		
Chimanimani	17,100	Mountainous area
Chizarira	191,000	High plateau and low veld with game
Gonarezhou	496,400	Semi-arid low veld with game
Nyanga	32,132	Mountainous area
Kazuma Pan	31,290	Open plains with game
Mana Pools	220,000	Zambezi valley with game
Matopos	44,200	Kopjes, biological diversity and history
Matusadona	137,000	Kariba escarpment, lake and game
Victoria Falls	1900	Waterfall and spray forest
Hwange	1,462,000	Semi-arid savanna and game
Zambezi	56,400	Upper Zambezi River and game
Total	**2,689,422**	
Botanical reserves and gardens*		
Pioneer	38	Fuchsia trees, *Scotia brachypetala*
Tolo River	44	*Acacia galpinii* trees
South Camp	26	Nyala berry trees, *Xanthocercis zambesiana*
Chirinda	949	Mountain forest
Banti	175	Yellow woods, *Podocarpus milanjianus*
Stapleford	212	Cycads, *Encephalartos manikensis*
Sebakwe I	60	Acacia Karoo
Sebakwe II	163	Great Dyke flora
Sebakwe III	53	*Brachystegia glaucescens*
Bvumba	201	Botanical garden, mountain forest
Chisekere	94	Hot springs and mangrove fern
Miware	34	Raphia palm, *Raphia farinifera*
Tingwa	290	Raphia palm, *Raphia farinifera*
Mazoe	8	*Brachystegia glaucescenes*
Bunga	1558	Mountain forest
Shashi	4300	Mulala palm, *Hyphaene banguellensis*
Chingwarara	256	Cycads, *Encephalartos concinnus*
Kubarakwawena	2	Bamboo, *Oxytenanthera abyssinicai*
Marishira	2	Bamboo, *Oxytenanthera abyssinicai*
Nyanzwe	8	Low-altitude evergreen forest
Pungwe Bridge	14	Low-altitude evergreen forest
Rumise	39	Low-altitude evergreen forest
Tegwe	4	Bamboo, *Oxytenanthera abyssinica*
Haroni	20	Low-altitude evergreen forest
Rusitu	150	Low-altitude evergreen forest
Ewanrigg	286	Botanical garden
Harare	67	Botanical garden
Total	**9053**	

Table 5.1 *Continued*

Protected area	Area (ha) [Area of water, ha]	Main features
Sanctuaries		
Chimanimani	1200	Conservation of relict eland herd, protection of town water supply and Bridalveil falls
Manjinji	300	Bird sanctuary on oxbow lake of Wanezi River; riverine vegetation
Mbaze	40	Bird sanctuary on oxbow lake on Shangani River
Mushabdike	12,900	Staff training college
Tshabalala	1200(?)	Game park on the outskirts of city of Bulawayo
Total	**15,640**	
Safari areas		
Chete	108,100	Single safari hunting concession
Hartley A	71,597	Uncertain – previously leased to hunting association
Hurungwe	288,000	Individual hunts and hunting association lease
Chewore	339,000	Several hunting concessions
Sapi	118,000	Individual hunts
Tuli	40,400	Previously individual hunts
Doma	76,400	Previously leased to hunters association
Chipinge A	26,100	Previously leased to hunters association
Mufurudzi	76,000	Non-hunting tourism concession
Chirisa	171,000	Wildlife research, hunting concession
Malapati	16,200	Used as part of Gonarezhou National Park
Dande	52,300	Safari hunting in community-based natural resource management programme concession
Charara	170,000	Individual hunts and game viewing
Matetsi	292,000	Seven safari hunting concessions
Total	**1,845,097**	
Recreation parks		
Kyle	16,900 [9105]	Impoundment in attractive kopjes, game park and research, angling, commercial fishing
Chinhoyi	148 [0]	Dolomite caves
Binga	2 [0]	Hot spring (small geyser)
Lake Chivero	6100 [2630]	Impoundment near Harare, game park, angling, commercial fishing, boating, bird viewing and research
Ngezi	5800 [580]	Great Dyke flora, some game, quiet recreation
Sebakwe	2700 [1518+]	Yachting, flora, Great Dyke
Bangala	2800 [1133]	Undeveloped
Manjirenji	3501 [2023]	Undeveloped
Umzingwani	1233 [456]	Undeveloped
Kariba	283,000 [c.280,000]	Boating, angling, commercial fishing and research, crocodiles, safari camps and hotels
Total	**322,184** **[297,445]**	

Note: * About 2496ha are in the parks and wildlife estate, of which 2112ha are in separate areas outside other reserves

Source: Adapted from Child (1977)

by the Southern Rhodesian Territorial Government, mainly in the early 1960s, and were administered by the Southern Rhodesian Game Department. The national parks were recognized earlier and were administered by the Federation of the Rhodesias and Nyasaland Government, through its National Parks Department (which had jurisdiction only in Southern Rhodesia). Other reserves were managed by the Southern Rhodesian Forestry Commission, the Ministry of Internal Affairs (responsible for administering the African Area covering 42 per cent of the country), as well as private game reserves that fell under the control of the private sector.

Sociologists are sometimes critical when land is kept from peasant settlement for nature conservation, so it is interesting to point out that relatively few people were displaced by the formation of the parks and wildlife estate in Zimbabwe. It is also useful to recall that the human population has grown to three times the size it was when most areas were set aside in the early 1960s. At that time, the areas were mostly rejected as unsuitable for settlement and were largely unoccupied, although a number including the Mana Pools, Gonarezhou, Matopos and Nyanga national parks had had a history of peasant settlement and the Matetsi area and parts of Hwange had been white-owned commercial farms.

There was considerable rivalry and a number of turf battles between the government agencies responsible for ecological reserves, particularly among staff at the field level. This largely ceased with the break-up of the federation, when the national parks and wildlife conservation departments were amalgamated in 1963, to form the DNPWLM (Gibson, 1963). What were legally 'private' game reserves in the African Area were absorbed into the parks and wildlife estate in 1975 (Child, 1975), although the Internal Affairs Game Section, responsible for wildlife outside parks and reserves in communal areas was not incorporated into the department until after Independence in 1980.

The classification of ecological reserves (known jointly as the Parks and Wild Life Estate (see Figure 5.1) into national parks, botanical reserves and botanical gardens, sanctuaries, safari areas and recreation parks was undertaken in terms of the 1975 Act, and has changed little since then. There is a gradation in the emphasis placed on maintaining areas in a natural state from national parks, safari areas (and to a lesser extent because of their small size, botanical reserves) to sanctuaries, recreational parks and botanical gardens, and an opposite gradient in the extent to which the areas can be utilized for outdoor recreation. However, all are important to both the national programme for conserving biological diversity and for tourism. Each class of area is managed in terms of an appropriate section of the Parks and Wild Life Act, according to a specific policy that describes the management objectives for the area and how the Act should be applied in it (Child, 1975).

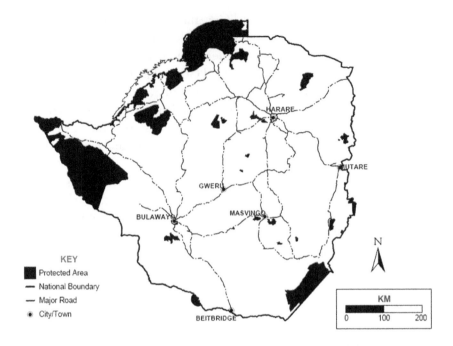

Source: WWF-SARPO (2008)

Figure 5.1 *Parks and wildlife estate of Zimbabwe*

An outline plan accompanies most area policy documents, explaining the management objectives for the park or reserve and describing how it is to be divided into zones of permissible use. In the main, these are: special conservation areas (mostly small areas to protect especially sensitive biological features, or for scientific research with no public access, at least seasonally); wilderness areas (areas with minimal development in which the level of visitation is strictly limited, but in which visitors are allowed considerable freedom of action); wild areas (the main zones of public use with good access in which visitor activities are more prescribed); and development areas (where there may be considerable development of staff facilities and visitor amenities).

The estate, with minor exceptions, is administered by what is now known as the Parks and Wildlife Management Authority (formerly the DNPWLM). The aim of the tourism and visitor use policy is to provide an integrated combination of areas in which a reducing intensity and diversity of recreational opportunities is accompanied by an increasing quality of the wilderness experience, from recreation parks through to sanctuaries, botanical gardens and safari areas. In all cases, the aim has been to optimize the permitted throughput of visitors using an area, while preserving the natural qualities prescribed for that area. This has involved regulating visitor usage to minimize their impact on the environment and each other, and constructing visitor amenities of natural materials like stone and thatch, and using neutral colours to minimize visual impact. Within these limits, visitors are offered a variety of activities and allowed freedom and contact with nature.

Sport hunting has been permitted in most safari areas since their inception as non-hunting areas, game reserves or controlled hunting areas. This introduces an anomaly. The IUCN classification of so-called 'protected areas' (IUCN, 1994) places safari areas in a lowly Category VII, implying a much lower conservation status than national parks in Category II, because of hunting (see for example IUCN, 1987). This is in spite of the fact that, in Zimbabwe, the two classes of land are managed similarly to conserve natural values, and that accommodating non-hunting tourists requires greater infrastructure and makes much heavier demands on the environment than catering for hunting tourists. In addition, the Sapi and Chewore Safari Areas are part of a natural World Heritage Site. Nevertheless, the difference in the IUCN status has prompted several commentators (for example Huntley, 1987) to omit safari areas from Zimbabwe's list of internationally recognized prime conservation areas.

As in many nations, Zimbabwe's ecological reserves were created to preserve land with a strong human appeal, perceived to be of particular natural excellence, with tourism and outdoor recreation important justifications for setting aside 12.7 per cent of the country by 1981 (Lightfoot, 1981). Thus the estate includes samples of much of the nation's spectacular scenery, geological formations, wildlife assemblages and even its natural curiosities. Although, the estate is located mostly on land with a low agricultural potential, rejected for any kind of farming when the areas were proclaimed, Child and Heath (1992) found that the parks, wildlife and forestry estates together sampled all but 3 of the 25 major ecotypes in the country, recognized by Wild and his co-workers (a group of eminent botanists who demarcated the ecotypes) (Wild, 1965; Wild and Fernandes, 1967; Wild and Grandvauz Barbosa, 1968) (see Table 5.2). In fact, they argued that the greatest threat to the country's biological diversity came from an overabundance of elephant in at least 53 of the 88 reserves in the estate, and in a number of forestry areas. It is a measure of failed conservation that the elephant population has been allowed to increase to a conservative estimate of 100,000 head when sound evidence (Craig, 1989; Martin, 1989) indicates Zimbabwe can support no more than 50,000 head sustainably.

The relationship between state-sponsored wildlife conservation and the overall wildlife sector

Zimbabwe built up an impressive commercial wildlife sector from small beginnings in the late 1950s, but this has been severely harmed since 1999.

Table 5.2 *Representation of vegetation types in ecological reserves, Zimbabwe*

Vegetation class	Vegetation type	Parks and wildlife estate	Forest reserves	Total	No. of areas with dense elephant populations
Forest	Med. Altitude closed	3	2	5	0
	Moist montane	2	2	4	0
	Dry montane	0	0	0	0
Thicket	*Commiphora–Combretum*	6	0	6	6
Woodland	*Baikiaoa*	4	10	14	4
	B. spiciformis on Kalahari sand	0	5	5	0
Savanna Woodland	*B. spiciformis–J. globiflora*	8	4	12	1
	B. Boehmii–J. globiflora	13	4	17	7
	J. globiflora	6	0	6	5
	C. mopane	17	8	25	15
Tree Savanna	*Baikiaea–Burkea–C. mopane*	1	2	3	1
	Parinari	0	0	0	0
	P. angolensis–Pericopsia–Acacia	1	0	1	1
	Commiphora–Combretum	1	1	2	1
	Terminalia serioea	5	2	7	3
	Acacia spp.*–Albizia–Bolusanthus*	7	0	7	1
	Adansonia–Sterculia–Kirkie	1	0	1	1
	C. mopane	3	0	3	2
	A. nigrescense–Combretum	0	0	0	0
	Acacia–L. nelsii	1	2	3	1
Shrub	*C. mopane*	2	0	2	2
Grassland	*Loudetia*	2	1	3	2
	Andropogon	2	0	2	0
	Hyparrhenia	0	1	1	0
	Montane	3	5	8	0
No. of types in area		88	49	137	53

Source: After Child and Heath (1992)

Despite this, for over 40 years the sector has provided many lessons of the ingredients for sound conservation and use of wild resources. Dr Reay Smithers, then Director of the National Museums, initiated questioning of the potential (but institutionally suppressed) value of large game animals and the many biological research opportunities that they offered. He catalysed a visit to Zimbabwe by three eminent US Fulbright scholars in 1958 to help exploit these opportunities. At that time, knowledge of Africa's big game animals had scarcely proceeded beyond the taxonomy and biogeography of the species (see for example Roberts, 1951) and hunters' anecdotes of their ecology and behaviour (for example Shortridge, 1934).

In the three years they were in Zimbabwe, the three Fulbrights planted many innovative ideas and stimulated local wildlife personnel to re-engineer the wildlife industry and the institutional framework guiding it. The small Department of Wild Life Conservation (subsequently the DNPWLM) with its two terrestrial biologists and progressive director were convinced that wildlife was ecologically superior to livestock and offered an opportunity to diversify ranching and reverse the widespread ongoing environmental deterioration resulting from a cattle monoculture. They were also convinced that, unless landholders wanted wildlife to be present on their land, there was nothing government could do to prevent its disappearance outside of parks and reserves. As many ranchers were cash-strapped, it followed that unless the ecologically more desirable game animals could at least compete financially with livestock, they would not be retained in significant numbers. Subsequent research demonstrated that it was the economic properties of wildlife, more than its ecological attributes, that gave it a comparative advantage over cattle. This is because wildlife enterprises tend to add economic tiers (for example tourism and hunting) to a normal animal production system, and depend on human initiative and effort and not local environmental energy. This can result in wildlife outcompeting livestock by a factor of four or more, both financially and economically (Child, 1988, 1990a, 1990b, 1990c, 1995).

The Wild Life Conservation Act, 1960, introduced to consolidate and update the existing game laws, allowed game ranching under permit. It was the test bed for the subsequent Parks and Wild Life Act, 1975, which accorded landholders considerable freedom to use wildlife (other than 'Royal Game') on their properties and to trade freely in the resulting products. Cautious adaptive institutional reform under the 1960 Act, and more rapid progress after 1975, which grew from the confidence gained from successes during the reform phase, vindicated the original belief in the merits of wildlife. These, however, depended on getting proprietorship and prices right (Child and Chitsike, 2000; Child, 2000a, 2000b). The institutional arrangement that emerged was strengthened by linking wildlife to the natural resources movement, with its proven record of promoting soil conservation and sound agricultural practices, and made landholder communities the first level of regulatory control over their members.

The importance of conserving biological diversity, the use of what has come to be called adaptive management, and research in the DNPWLM were but some of the more momentous and innovative outcomes championed by the three Fulbrights that were implemented by far-sighted local officials.

When the existing colonial-type protectionist game laws embodying centralized management of the resource were challenged and defeated in the courts, new legislation became essential (Child, 1995). A man was prosecuted for hunting Royal Game on his farm, but argued that it was eating natural forage which was his property, which he had the right to protect. The judge agreed and thus recognized that wild animals carry an opportunity cost and this cost of conserving them could not be imposed on landholders by society. The decision meant the authorities could no longer prevent landholders from hunting browsing and grazing animals on their land, so the Parks and Wild Life Act, 1975 delegated control over wildlife to private landholders, making them the custodians of the resource. The Act encouraged them to use most species sustainably and to maximize their profit from doing so. With that, the countrywide decline in wildlife outside ecological reserves was halted and reversed, as the growth of the wildlife industry over the next couple of decades in both communal and commercial farming areas demonstrated. The impressive increase in land devoted to wildlife and in the number of registered wildlife producers after 1975 were powerful surrogate measures of the rapid increase in wildlife numbers that became obvious in many parts of the country (for example Child, 1988, 1995; Cumming, 1989; Skinner, 1989).

Strengthening institutions to conserve and use wildlife outside the estate was highly cost effective and lent considerable credibility and political leverage to the wildlife agency. Many wild animals used to restock depleted habitats came from the estate, and the DNPWLM conducted much research that was useful to the wildlife sector. Trading ineffective power and dependence on regulation and law enforcement for effective influence based on landholder-friendly institutions and encouraging wildlife producers led to the growth of a strong wildlife sector in which the DNPWLM was a key player, both as a major producer and the regulatory authority.

It was not long before the private sector began challenging the supremacy and quality of the tourism and hunting provided by the estate and generating relatively more income than the DNPWLM from lesser wildlife. This occurred even though many of the private enterprises were in competition with subsidized government ventures, which undersold the same products (Child and Heath, 1990). Private entrepreneurs were, however, able to exploit the large markets developed by the state through its underpricing of hunting and other services in the estate. Providing high quality cheap tourism was one of several measures taken by government to stimulate a quality nationwide industry that could compete with the best in the region – during a period of international sanctions – to attract valuable foreign currency, create jobs and persuade citizens to holiday in the country.[3] It is interesting that, in the early 1970s, the recurrent costs of the departmental division responsible for tourism were met from revenue generated by low tariffs and gate charges.

The role and influence of the services offered in the estate by the DNPWLM diminished after Independence in 1980, at the same time that the private sector burgeoned in response to the rapidly growing economy and the lifting of international sanctions. From being the third largest 'hotel group' in the country, the DNPWLM declined into a relatively minor player, as it was required to restructure to facilitate black advancement. Furthermore, Treasury allocations for its tourism services did not keep pace with either the costs of providing the services or the demand for them. During this period, performance and the revenue generated by individual agencies had little influence on Treasury allocations, which were subject to across-the-board cuts for large swathes of government, as post-Independence inflation gathered momentum. This undermined staff morale as the department's corporate culture became more *laissez-faire* under a socialist government and the quality of services declined. The department had always lacked the information needed to operate efficiently in a free market situation, but a further decline of its entrepreneurial skills after the mid-1980s severely weakened its ability to compete on reasonable terms with the private sector. As a result, it was greatly handicapped when it was thrown in the deep end and expected to live from its own income from 1999, on advice from the World Bank.

Conclusions

Nature conservation featured early in the colonial history of Zimbabwe, when formal provisions were introduced to supplement, and then replace, the weak measures required by a sparse indigenous population. Until the middle of the 20th century, no single government agency was charged with the implementation of national nature conservation policy. This changed and the responsibilities were centralized with the creation of a Federal National Parks Department in about 1949 and a Southern Rhodesian Game Department in 1952. The parks and wildlife estate was greatly expanded during the early years of these agencies, culminating in the areas added with the promulgation of the Parks and Wild Life Act in 1975. After that, most of the conservation areas, such as the Lake Kariba Recreation Park, were created in order to formalize previous administrative arrangements.

Amalgamation of the two departments in 1963 coincided with considerable political controversy over the way in which government was managing wildlife. This acrimony culminated in parliament setting up a Wild Life Commission to examine and report on the way it was being managed, and to suggest improvements. The commission's report in 1970 did much to clear the air, as it led to a more scientific approach to wildlife management inside and outside the estate, which set in motion a period of rapid growth and development in the wildlife sector. It commenced with the expansion of tourist facilities in state-protected areas and extended to the wildlife industry outside the estate. Building on the confidence gained from 15 years of institutional evolution under the Wild Life Conservation Act 1961, the DNPWLM devolved proprietorship over most wildlife outside the estate to landholders in terms of the 1975 Act. The department was able to shed much day-to-day management of wildlife over which, in any case, it had had little direct influence, without abrogating its countrywide responsibilities for the resource. With the professional leadership advocated by the Wild Life Commission and the reduced public controversy, this allowed the DNPWLM to concentrate on managing the estate, which it saw as its key function and which it did reasonably well (MacKinnon and MacKinnon, 1986). Management of ecosystems was prioritized and based on adaptive management and included curbing dominant herbivore populations to maintain (or restore) the physiognomic character of the habitats in a state as similar as possible to that in 1900, before the increasing human population began modifying them rapidly.

The DNPWLM catalysed and benefited from the emergence and growth of an increasingly strong and influential private wildlife sector. Devolving proprietorship over wildlife to landholders allowed and encouraged them to maximize their profits from using the resource sustainably. Additional measures taken by government included incremental measures to remove artificial constraints, such as fixed hunting licence fees, which had inhibited wildlife from realizing its true market value. These were replaced by free market mechanisms such as tenders and auctions for hunting and other products offered by the state, wherever practicable. It also laid the foundations for a vigorous and successful community-based wildlife management programme in communal areas (see Chapter 13) (Child, 2000b).

With these successes, the value of wildlife and the land supporting it appreciated immediately. Benefits were felt by the resources, the responsible agency, the wildlife sector, the national economy and human welfare. The wildlife industry quickly became a significant and rapidly growing element in the national economy, creating employment and generating substantial foreign earnings. DNPWLM staff became front-runners in international conservation forums which, together with the favourable impressions of a growing number of foreign visitors and knowledgeable commentators, reflected credit on the country.

Political interference in the leadership and running of the DNPWLM was kept in check between 1971 and 1980 and for some years after Independence, but began to be felt in large measure from the mid-1980s to the early 1990s. As it gained momentum from around 1990, it accelerated the loss of skills which, incidentally, sped up the decline in services the department could offer and undermined its corporate resolve to succeed, in what became a self-perpetuating destructive vortex attributable to poor leadership. As the leadership lost confidence in itself, it sought to restrict the freedom allowed to the private sector, and many of the actions that had stimulated growth in the wildlife sector (to the advantage of the DNPWLM) were reversed. For example, unnecessary permitting and petty restrictions were reintroduced. Predictably, as bureaucratic road-blocks were erected, transaction costs increased and growth in the sector slowed. It was further reversed following the redistribution of land that commenced in 1999, with much wanton destruction of wildlife and its habitats. Having been a global example of effective nature conservation, especially outside its well-run parks and reserves, Zimbabwe slid to become an embarrassing non-entity in conservation circles in southern Africa and beyond.

The history of the fluctuating success of nature conservation in Zimbabwe is a clear demonstration of the power of institutions and their implementation on conservation outcomes. While conservation depends on natural laws to produce valuable goods and services, it usually succeeds only if these occur in a socio-politically and economically benevolent setting. Conserving and using wildlife inside and outside parks and reserves represents two sides of the same coin. The Zimbabwean experience suggests that it is probably the single most optimistic land use for enhancing human well-being on a broad front in the arid and semi-arid lands that dominate southern Africa, much of the rest of the continent and far beyond. Unlocking this potential requires that governments get the institutions right and apply them effectively and efficiently in land-holder-friendly but disciplined ways that are transparent.

Notes

1   This was an amendment by virtue of the fact that Zimbabwean common law is based on the Cape Colony statutes of 1890.
2   Rhodesia – now Zimbabwe – never had a colonial government. The British South Africa Company governed the country under a Royal Charter from Queen Victoria until 1923, when it relinquished authority to the British Dominion office.
3   Tourism grew to record levels under sanctions (when the Rhodesian government 'stamped' foreign passports on a slip of paper), until it was curbed as the civil war intensified.

References

Baden-Powell, R. S. S. (1897) *The Matabele Campaign 1896*, Methuen, London
Child, B. (1988) 'The role of wildlife utilization in the sustainable economic development of semi-arid rangelands in Zimbabwe', DPhil thesis, University of Oxford, Oxford
Child, B. (1990a) 'Assessment of wildlife utilization as a land use option in the semi-arid rangelands of southern Africa', in Kiss, A. (ed) *Living with Wildlife: Wildlife*

Resource Management with Local Participation in Africa, World Bank, Washington DC

Child, B. (1990b) 'Wildlife utilization and management systems', in Kiss, A. (ed) *Living with Wildlife: Wildlife Resource Management with Local Participation in Africa*, World Bank, Washington DC

Child, B. (1990c) 'Economic analysis of Buffalo Range Ranch', in Kiss, A. (ed) *Living with Wildlife: Wildlife Resource Management with Local Participation in Africa*, World Bank, Washington DC

Child, B. (2000a) 'Making wildlife pay: Converting wildlife's comparative advantage into real incentives in African savannas, case studies from Zimbabwe and Zambia', in Prins, H. H. T., Grootenhuis, J. G. and Dolan, T. T. (eds) *Wildlife Conservation by Sustainable Use*, Kluwer Academic Publishers, Dordrecht

Child, B. (2000b) 'Application of the southern African experience to wildlife utilization and conservation in Kenya and Tanzania', in Prins, H. H. T., Grootenhuis, J. G. and Dolan, T. T. (eds) *Wildlife Conservation by Sustainable Use*, Kluwer Academic Publishers, Dordrecht

Child, G. F. T. (1975) *Report of the Director of National Parks and Wild Life Management for 1979*, Government Printer, Salisbury

Child, G. F. T. (1977) 'Problems and progress in nature conservation in Rhodesia', *Koedoe Supplement*, pp116–137

Child, G. (1995) *Wildlife and People: The Zimbabwean Success*, WISDOM Foundation, Harare

Child, G. and Chitsike, L. (2000) '"Ownership" of wildlife', in Prins, H. H. T., Grootenhuis, J. G. and Dolan, T. T. (eds) *Wildlife Conservation by Sustainable Use*, Kluwer Academic Publishers, Dordrecht

Child, G. and Heath R. A. (1990) 'Underselling national parks in Zimbabwe: The implications for rural sustainability', *Society and Natural Resources*, vol 3, pp215–227

Child, G. and Heath, R. A. (1992) 'Are Zimbabwe's major vegetation types adequately protected?', *Geographic Journal of Zimbabwe*, vol 23, pp20–37

Child, G. and Riney, T. (1987) 'Tsetse control hunting in Zimbabwe, 1919–1958', *Zambezia*, vol 14, no 1, pp11–71

Craig, C. G. (1989) 'A simple model of tree/elephant equilibrium', in Martin, R. B., Craig, C. G. and Booth, V. R. (eds) *Elephant Management in Zimbabwe*, Department of National Parks and Wild Life Management, Harare

Cumming, D. H. M. (1981) *National Parks and Zimbabwe: Ideals, Concepts and Ideals for the 1980s*, Department of National Parks and Wild Life, Harare

Cumming, D. H. M. (1989) 'Commercial and safari hunting in Zimbabwe', in Hudson, R. D., Drew, K. R. and Baskin, L. M. (eds) *Wildlife Production Systems: Economic Utilization of Wild Ungulates*, Cambridge University Press, Cambridge

Davison, T. (1967) *Wankie: The Story of a Great Game Reserve*, Books of Africa, Cape Town

Gibson, J. N. (1963) *Report of the Director of National Parks and Wild Life Management*, Government Printer, Salisbury

Huntley, B. (1987) 'Ecosystem conservation in southern Africa', in Wergner, J. A. and van Bruggen, C. R. (eds) *Biogeography and Ecology of Southern Africa*, Junk, The Hague

IUCN (1987) *IUCN Directory of Afrotropical Protected Areas*, IUCN, Gland

IUCN (1994) *Guidelines for Protected Area Management Categories*, IUCN, Gland

Lightfoot, C. J. (1981) 'The national parks of Zimbabwe: Their uses and justification', unpublished report for the Department of National Parks and Wild Life Management, Harare

MacKinnon, J and MacKinnon, K. (1986) *Review of the Protected Area System in the Afrotropical Realm*, IUCN/UNEP, Gland and Nairobi

Martin, R. B. (1989) 'Relationship between elephant and canopy cover', in Martin, R. B., Craig, C. G. and Booth, V. R. (eds) *Elephant Management in Zimbabwe*, Department of National Parks and Wild Life Management, Harare

Petrides, G. A. and Pienaar, U. de V. (1969) *Report of the Wildlife Commission*, Government Printer, Salisbury

Roberts, A. (1951) *The Mammals of South Africa*, Central New Agency, Cape Town

Shortridge, G. C. (1934) *The Mammals of South West Africa*, Heinemann, London

Skinner, J. D. (1989) 'Game ranching in southern Africa', in Hudson, R. D., Drew, K. R. and Baskin, L. M. (eds) *Wildlife Production Systems: Economic Utilization of Wild Ungulates*, Cambridge University Press, Cambridge

Taylor, R. D. (1990) 'Zimbabwe', in Allen, C. W. (ed) *International Handbook of National Parks and Nature Reserves*, Greenwood Press, Connecticut

Thomlinson, D. N. S. (1980) 'Nature conservation in Rhodesia', *Biological Conservation*, vol 18, pp159–177

Tredgold, R. (1956) *The Matopos*, Federal Department of Printing, Salisbury

Weitz, B. (1956) 'The feeding habits of *Glossina*', *Bulletin of the World Health Organisation*, vol 28, pp711–729

Wild, H. (1965) 'Vegetation map of Rhodesia', in Collins, M. O. (ed) *Rhodesia: Its Natural Resources and Economic Development*, Collins, Salisbury

Wild, H. and Fernandes, A. (1967) *Flora Zambeziaca: Supplement to the Vegetation Map of the Flora Zambeziaca Area*, Collins, Salisbury

Wild, H. and Grandvauz Barbosa, L. A. (1968) *Vegetation Map of the Flora Zambeziaca Area*, Collins, Salisbury

WWF-SARPO (2008) 'Map of protected areas, Zimbabwe', prepared by WWF-SARPO, Harare

6

Protected Areas in Mozambique

Bartolomeu Soto

Mozambique is a vast country of about 800,000km², with a range of ecosystems and significant cultural diversity. The management of protected areas has been influenced by the political situation of the country, the culture of the people involved, as well as, more recently, world policies for protected areas. In Mozambique, protected areas in the form of forest reserves are distributed throughout the country, many of which have hardly been managed since their proclamation. In recent decades, some management of forest resources, through community-based natural resource management programmes has begun, focusing on the exploitation of resources. The protected areas that were created with wider objectives, including the protection of wildlife, habitats and ecosystems are seen as the source of multiple benefits, including social and economic benefits (derived mainly from tourism activities), and are the focus of this chapter. In Mozambique, there are 6 game reserves, 6 national parks and 12 controlled hunting areas or *coutadas* (see Table 6.1).

Pre-colonial and colonial eras

There is very little information available regarding wildlife and protected areas in Mozambique in the pre-colonial era. Although there were no written natural resource management policies or legislation, the use of natural resources was governed by traditional controls through chiefs and village headmen who regulated access to, and use of, resources in their territories. Ritual prohibition of hunting, or harvesting resources in some areas (such as sacred forests), magic and taboos relating to certain wildlife species regulated the number of people that used these resources (Berry and Petty, 1992).

From the 10th century, long before the Portuguese arrived, there was a commercial relationship between African tribes and the Swahili Arabs, and wildlife products such as ivory and rhino horn were traded. The onset of colonization in Africa by Europeans in the 1500s marked a period of intense, unsustainable exploitation of natural resources. The Portuguese were established in Sofala as traders by 1505 and were present across Mozambique as colonizers until Independence in 1975 (História de Moçambique, 1974).

Three periods can be distinguished on the basis of the major goods traded: the gold trade phase between the 10th and 17th centuries; the ivory trade phase during the 17th and 18th centuries; and the slave trade phase during the 18th and 19th centuries. Intensive elephant hunting took place by local people along much of the coastal regions for ivory to exchange for Indian-made materials and for products such as glass beads from Venice, particularly during the middle phase (História de Moçambique, 1974; Newitt, 1995). Rhino horn and hippopotamus teeth were also traded.

The first wildlife legislation in Mozambique was issued in 1903 in response to uncontrolled hunting by Europeans, and focused mainly on regulating hunting in the Maputo area. The decree was revised in 1909 to cover the whole of the country. Legislation throughout the colonial period was inclined towards wildlife utilization – for trading, food and for sport – rather than strict protection. Of the 92,200km^2 set aside as wildlife areas, 61 per cent were *Coutadas Oficiais* (controlled hunting areas, in which individuals could hunt if they had a government-issued licence) and 39 per cent as wildlife reserves, the management rights of which were granted by government to a specific entity, and in which some sport hunting was also allowed, as was hunting for the supply of meat for farm workers (Dias et al, 1963).

The Portuguese administration remained uncommitted to wildlife preservation until after the 1933 London Convention on the protection of the African wild fauna and flora.[1] However, the educated classes of Portuguese society did not exhibit any conservation culture, and the colonial government did not ratify the convention or declare their support for conservation until late 1948 (de Sousa, 1938; Pro Natura, 1973). It was not until 1955 that another decree outlined the first principles aimed at the protection of soil, fauna and flora: that the state should conserve wildlife, which should be state property and utilized under government licence; that endangered and threatened wildlife species should receive special treatment; and that the state should create protected areas to achieve these conservation goals (Rosinha, 1973).

In 1964, the colonial government issued a new decree that defined five types of wildlife protection zone: national parks, integral natural reserves, partial reserves, special reserves and zones of special vigilance. In national parks and integral natural reserves, the law did not allow resource utilization, however in partial reserves, special reserves and zones of special vigilance, hunting was allowed if authorization from provincial authorities had been granted (Governo Geral em Lourenço Marques, 1964). It was primarily during the 1960s and early 1970s that national parks, game reserves and *coutadas* were gazetted (see Table 6.1). The establishment of forest reserves followed a similar pattern – where the decision to create some reserves appeared to centre on their potential for the economic exploitation of resources within them (Black and Schafer, undated).

Table 6.1 *National parks, reserves and* coutadas

Name	Province	Area (km²) [% of surface area]	Year of proclamation (amendment)
National Parks			
Gorongosa National Park	Sofala	5370	1966 (1967)
Bazaruto National Park	Inhambane	1600	1971
Zinave National Park	Inhambane	6000	1973
Banhine National Park	Gaza	7000	1973
Limpopo National Park	Gaza	10,000	2001
Quirimbas National Park	Cabo Delgado	7506	2002
Total area		**37,476 [4.68%]**	
National Reserves			
Maputo Special Reserve	Maputo	700	1960 (1969)
Pomene Reserve	Inhambane	200	1964
Marromeu Buffalo Reserve	Sofala	1500	1960
Gilé Reserve	Zambézia	2100	1960
Chimanimani Reserve	Manica	1740	2003
Niassa Reserve	Niassa	42,200	1964
Total area		**48,440 [6.06%]**	
Coutadas			
Coutada 4	Manica	12,300	1969
Coutada 5	Sofala	6868	1972
Coutada 6	Manica	4563	1960
Coutada 7	Manica	5408	1969
Coutada 8	Sofala	310	1969
Coutada 9	Manica	4333	1969
Coutada 10	Sofala	2008	1961
Coutada 11	Sofala	1928	1969
Coutada 12	Sofala	2963	1969
Coutada 13	Manica	5683	1960
Coutada 14	Sofala	1353	1969
Coutada 15	Manica	2000	1969
Total Area		**49,717 [6.21%]**	

Source: Adapted from DNAC (2006)

The Portuguese government believed that agriculture and livestock farming were the key to economic development, and with widespread wildlife populations, farmers were authorized to shoot wildlife considered to be pests and disease carriers. Between 1949 and 1969, the government undertook its own culling schemes, such as the operation to eliminate about 180,000 wild herbivores around the Savé River to control tsetse in the south of the country (DNFFB, 2001). Meat hunting was also significant – between 1940 and 1960 the Marromeu area served as a meat factory, with 13 professional hunters, each with an annual quota of 200 buffalos and 35 elephants, producing about a 1000 tonnes of meat to supply the workers of agricultural plantations in Zambézia (DNFFB, 1991).

At the same time, government control of wildlife utilization was very weak; wildlife was being poached throughout the country, with commercial species such as rhino and elephant the main target. In Tete Province in 1950, 100 rhino horns were found in the warehouse of a trader, to be sold in Asia (Dias et al, 1971). Black and Schafer (undated) also note that between the 1930s and 1960s officials frequently complained about having inadequate funds to carry out their duties, of high staff turnover and a lack of staff training.

Inspired by the development of game ranching in South Africa and Zimbabwe, the government introduced regulations with the idea of promoting ranching as a means of meat production. Despite the fact that it stopped issuing licences to meat hunters during the 1960s, very little development of game ranching occurred, primarily due to the fact that the investments required meant that ranch-produced meat was considerably more expensive relative to that hunted within *coutadas* (which required low investments, had high potential and provided quick returns).

Post-colonial era

The Mozambican Constitution, adopted at Independence in 1975, nationalized all natural resources and brought them under the direct control of the state. The government revised part of the inherited colonial wildlife legislation, '*Legislação Sobre as Actividades da Caça*' (legislation for hunting activities) in 1977. The revisions enabled the use of natural resources for the socio-economic improvement of the population, as wildlife was one of the richest renewable natural resources of the country (though it had been subject to indiscriminate utilization). The revisions provided local communities with rights to legally participate in wildlife utilization, predominantly through (meat) hunting. To access the rights, communities had to be organized into cooperatives (institutions that were working well at the time for agrarian activities), which were then charged low fees for hunting licences. However, while there were a few cases in Marromeu District where communities were issued rights to hunt for meat, the reality was that cooperatives did not work for wildlife utilization, primarily due to constraints relating to their formation.

In addition to this legislation, the government wished to increase access to game meat, and between 1976 and 1981 it conducted a major wildlife culling operation in the Marromeu area. The 'Buffalo Operation' was conducted on the assumption that the Marromeu buffalo population exceeded carrying capacity and could sustain an annual extraction of 3000 animals, with the main objective of producing meat. Waterbuck were also culled but in smaller numbers (Bielfuss et al, 2001; Sansao Bonito, pers. comm.). The government, through the state wildlife utilization company (*Empresa Moçambicana de Fauna*), conducted similar operations in several parts of Mozambique, including culling nyala and impala in the far south of the country (the meat was sold to hotels in Maputo) and for hippopotamus and crocodiles along the Zambezi River. Hunting operations were also organized by district-level wildlife authorities to supply meat, mainly to soldiers and local communities.

Wildlife was also viewed as an asset for developing tourism and the economy, but the government did not formulate a clear policy and framework of how this was to be achieved before the 1980–1992 civil war virtually eliminated the government's capacity to manage wildlife. During the war, the government, the opposition and the general public used wildlife resources indiscriminately – the prevailing perception was that it was a 'free good' to be used for food supplies and trophies (ivory and rhino horns), which were smuggled and illegally traded to pay some of the costs of the war (McGregor, 1998; Schafer and Bell, 2002, cited in Virtanen, 2005). The result was that wildlife numbers were dramatically reduced – some species such as rhino, cheetah and giraffe became locally extinct. In Marromeu, while 30,000 animals were culled between 1978 and 1985, wildlife populations remained strong. However, by the end of the civil war in 1992, buffalo, waterbuck, reedbuck, hippo and zebra had been reduced by 95 per cent, with the last two of these close to local extirpation (Beilfuss et al, 2001). In addition to the decimation of wildlife populations, much of the infrastructure within the protected area network was damaged or destroyed, particularly where they had been used as bases or where armed forces were stationed nearby for lengthy periods, for example in Gorongosa (which was used as a rebel base for some time, and both sides laid land mines within the park at different times) and Moribane (Hutton et al, 2001; Black and Watson, 2006).

In contrast, vegetation in many areas flourished as a result of the war – where farms and villages were deserted and commercial timber extraction was reduced (or halted altogether), vegetation could re-establish itself, except close to areas of high human population density, where charcoal production was high to meet household demand for fuel (Hutton et al, 2001).

Post-civil war era

In 1992, the year the Peace Accord was signed and the development of 'The general guidelines toward policy for wildlife conservation' took place

(DNFFB, 1992). The guidelines were developed by headquarters staff with no consultation of stakeholders – most conservation areas were inaccessible at the time making it exceptionally difficult to engage local people. The guidelines were based on scanty information about the wildlife situation in the country; hence they simply formed a framework for a policy that was developed in subsequent years. The guidelines focused on wildlife resource surveys, emergency rehabilitation programmes for the main protected areas (for example Gorongosa National Park), wildlife management training, logistical support to wildlife services, establishment of pilot community-based natural resource management (CBNRM) projects, and the development of protected area management plans.

Following the formulation of these guidelines, the government initiated a process of developing a comprehensive forest and wildlife policy (DNFFB, 1996). This revision of wildlife policy was seen as a priority because of the need to rehabilitate conservation areas and restock depleted protected areas. Wildlife in this process was seen as an asset that would stimulate the development of rural areas by involving rural communities in the rehabilitation and management of natural resources (DNFFB, 1996).

The drafting process emphasized the importance of non-governmental stakeholder participation in natural resource management, as government recognized its prior inability to successfully manage the protected area network and the difficulty it had in finding sufficient funding for the maintenance and development of protected areas. Donors involved in the policy development played a role in emphasizing participation in natural resource management, and the collaboration between sovereign states in transboundary natural resource management also gained prominence.

The new policy entered into force in 1996 and remains valid, and its goal is to conserve, utilize and develop forests and wildlife resources to gain social, ecological and economic benefits for present and future generations of Mozambicans (DNFFB, 1996). The policy encourages the formal and informal private sectors to participate and invest in the conservation and management of forest and wildlife resources, in all categories of land, including state-protected areas. It encourages potential concessionaires to enter into direct negotiations with recognized communities (people living in the area with a representative committee, registered at the nearest government administration) and requires that tenders for concession allocations incorporate community agreements that guarantee existing customary rights. The policy encourages NGOs to assist in capacity-building at local levels, helping rural communities to organize themselves and supporting the development of management skills, as well as facilitating funding from international donors.

The policy empowers communities by affirming existing customary rights and ensuring exclusive access to natural resources in customary areas and permitting sustainable commercialization of resources for community benefit. It also provides for the involvement of rural communities in the management of state-protected areas through such mechanisms as biosphere reserves and concession agreements with government, as well as promoting and enabling the development of community resource management regimes involving recognized communities and reflecting the role of women in resource use and management.

The policy was also the basis of government activities promoting partnerships in the management of protected areas during the 1990s. Subsequently (in 2006) the government approved the 'Principles for Administration of Protected Areas' (PAPA), a framework guiding protected area management, consolidating the 1996 Forest and Wildlife Policy and providing for better collaboration among protected area management stakeholders. PAPA affirms that Mozambicans should be encouraged to participate and invest in conservation, and encourages the development of protected areas through public–private partnerships (PPPs). PAPA is intended as the first step in developing a national conservation policy, which is intended to resolve the current fragmentation and occasional contradiction in current policies relating to different natural resources.

Mozambique's protected area system

National parks are defined by the forest and wildlife law as zones of total protection for the propagation, protection, conservation and management of vegetation and wildlife, and for the protection of local landscape and geological formations of particular scientific value, cultural and aesthetic areas of interest, and for public recreation representative of national heritage (see Figure 6.1). National reserves are defined as zones of total protection for the protection of rare, endemic and/or endangered species of flora and fauna, and of fragile ecosystems (for example wetlands, dunes, mangroves and coral reefs) and the flora and fauna present in these ecosystems. The third category is zones of use and historical/cultural values, whose purpose is to protect forests of religious interest and other sites of historic importance and cultural use, according to the norms and customary practices of local communities. These zones can be declared by provincial governors, although even if they are not formally declared, communities still have rights over the area (so the declaration is merely a formality).

The national Ministry of Tourism is responsible for national parks and reserves, while the local communities are directly responsible for the administration of the zones of use and historical/cultural value.

While the legal definition of national parks and reserves suggests that there should not be people living within their boundaries, all protected areas of these categories have people resident within them in Mozambique. The average

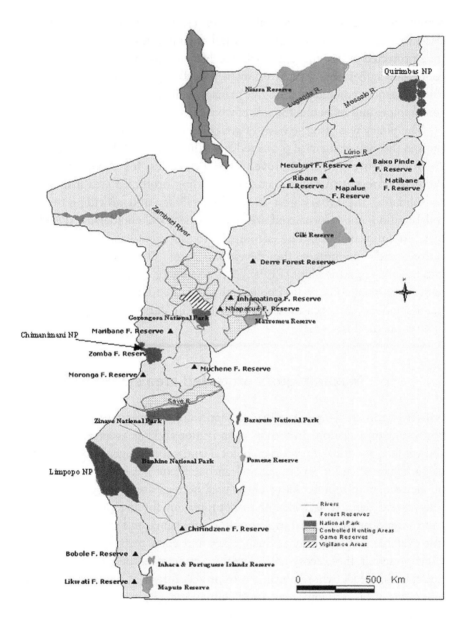

Source: DNAC (2006)

Figure 6.1 *National parks, game reserves and* coutadas

number of people living within the boundaries is 3000 people, though
Quirimbas National Park has considerably more, with particularly high human
activity inside the park and high levels of human–wildlife conflict. Residents
practise various activities, most of them incompatible with the purposes of the

protected areas, including slash and burn agriculture, livestock rearing and wildlife poaching. Nevertheless, the forest and wildlife law remains silent on the issue of protected area residents, and the land law says that within protected zones there is no right of use of land unless government provides a special licence or concession for a specified amount of time.

Coutadas have the status of partial protection zones, which appears somehow ambiguous, as government may still allocate land for activities other than conservation. The government has already authorized a cotton plantation within *Coutada* 8, *Coutada* 14 is being pressurized by a sugar cane plantation (which already occupies about 20 per cent of the area), a petroleum gas survey has been authorized within *Coutada* 5, as has logging within *Coutada* 6. All *coutadas* suffer from human encroachment and a resulting increase in the use of resources, as communities can settle within them without restrictions. The impacts of such activities can result in the destruction of the potential of the area for sport hunting, and *coutadas* therefore require an urgent revision of their legal status and an improvement in law enforcement to ensure the maintenance of their biological viability. However, as the demand for land increases in the country, and considering the large land area occupied by the *coutadas*, the zoning of *coutadas* to allow multiple use of resources (including settlements) is perhaps the most appropriate management option.

In contrast, some areas of Cabo Delgado and Niassa still have (currently unprotected) areas with considerable populations of wildlife that could be declared as *coutadas* or game farms, which would bring the existing sport hunting by foreigners that occurs in some of these areas within the law (currently sport hunting by foreigners can only be legally undertaken on game farms or in *coutadas*). In Cabo Delgado, an area of about 90,000ha adjacent to Niassa Reserve has been identified by government as an area to be declared as a *coutada*.

Institutional arrangements

The responsibility for protected areas and wildlife originally fell under the Veterinary Department until 1975, when it was moved to the National Directorate for Forestry and Wildlife (*Direcção Nacional de Florestas e Fauna Bravia*, DNFFB) in the Ministry of Agriculture and Rural Development (*Ministério da Agricultura e Desenvolvimento Rural*, MADER). In 2001, the government transferred the responsibility for 'conservation areas with the purpose of tourism' from MADER to the National Directorate of Conservation Areas with Purpose of Tourism (*Direcção Nacional das Áreas de Conservação*, DNAC) in the Ministry of Tourism (*Ministério do Turismo*, MITUR).[2] MADER resisted the transfer and retained the responsibility for authorizing game farms, licensing safari hunting on game farms and licensing safari hunting conducted by Mozambicans outside controlled hunting areas, which has resulted in confusion within government and outside of it. In 2006, central government named MITUR as the agency responsible for dealing with human–wildlife conflict, though as noted, it had previously assigned responsibility for wildlife outside of protected areas to MADER. After several months of discussion, the government decided finally to return the main responsibility of managing human–wildlife conflict to MADER.

Theoretically, the 2002 forest and wildlife regulations also gave MADER the responsibility for zoning forest and wildlife areas, which left ambiguity about whether MITUR was able to propose the proclamation of protected areas. Since the responsibility for protected areas was transferred to MITUR, two proclamations of national parks have taken place – the Limpopo and Quirimbas national parks – as well as the Chimanimani National Reserve and 18 zones of use and historical/cultural values surrounding the Chimanimani Reserve. All of these processes were led by MITUR, and the practice seems to be well accepted and will probably determine the procedures in the future.

MITUR developed a tourism policy and implementation strategy in 2003, which states the main environmental objectives of the agency as being to ensure that tourism and the environment are mutually supportive; to promote a proactive approach by all tourism stakeholders to develop, market and manage the sector in a responsible and integrated manner; to prioritize the preservation of the quality and sustainability of biodiversity; to contribute to the rehabilitation, conservation and protection of ecosystems and natural heritage; and to promote the development of natural resources, especially those that possess ecological or historical values in recreational, aesthetic and/or socio-cultural ways. Furthermore, the policy establishes protected areas as one of the priority areas for intervention.

Challenges remain, however, as protected area management is a complex field that requires a diversity of expertise, academic qualifications and experience, and the capacity of human resources within Mozambique remains a particular problem. While progress has been made in recent years, the levels of qualifications and experience of personnel remains inadequate at all levels within DNAC headquarters and in protected areas. The promotion of the sustainable use of forest and wildlife resources is impossible under such capacity constraints, and consequently illegal use of resources is still a serious problem, regardless of the intentions of policy and legislation.

Institutional innovations to improve management

The allocation of state financial resources to the management of protected areas has been declining over recent years. Since the transfer of protected areas from MADER to MITUR and the creation of DNAC in 2005, protected areas have received about US$1.2 million from government each year to manage the protected area network. Donors contribute a larger share, directing funds to the management of specific protected areas and the transfrontier conservation areas programme. The contribution of donors in 2005 and 2006 was approximately US$4 million annually (DNAC, 2006).

To ease the constraints of limited financial resources and implementation capacity, the government is encouraging partnership arrangements among different stakeholders in managing forest and wildlife resources. Various models are being tested in a number of protected areas. Gorongosa National Park is currently co-managed by the government and the Carr Foundation; Banhine and Zinave national parks, the Maputo Special Reserve and the Chimanimani National Reserve are currently being supported by the Transfrontier Conservation Areas and Tourism Development Project; and the Quirimbas and Bazaruto National Parks are managed in partnership with the World Wide Fund for Nature, with funding from the Agence Française. The partnerships in Niassa Game Reserve and Gorongosa National Park are approved by the Council of Ministers and oblige the partners to invest large sums of money, while in other partnerships it is the government (through funding provided by donor agencies) that invests the largest portion of the funds involved.

Partnerships for protected area management

In 1994, the Group Madal[3] requested a safari hunting concession to the east of the Niassa Reserve, one of the protected areas least affected by the civil war. Having noticed that Niassa Reserve was underfunded, the group subsequently put together a proposal to include the reserve within the concession area. The intention of the proposal was to use funds made available by one of the prominent members of the group to mobilize additional donor funds to improve conservation management within the reserve. In 1996, a two-year management contract was granted to Group Madal to manage and develop the Niassa Game Reserve. During the contract period, a management plan was developed that recommended the formation of a 'development society', in which the government, private company and local communities would be shareholders.

The government approved the idea of forming the development society, though the Group Madal pulled out, creating a new company (Niassa Investments) that became the private sector partner. In 1998, the Development Society of Niassa was established and was granted a concession for ten years, renewable five times. 51 per cent of the shares of the society belong to the government (which made available the natural resources) and 49 per cent to Niassa Investments (which contributed money and equipment). The contract states that over time these two members are required to reduce their shares to accommodate other shareholders, so that eventually the government will hold only 30 per cent, Niassa Investments will hold 36 per cent and the remaining 19 per cent will be released other Mozambican investors, with local communities able to access to up to 15 per cent (through legal institutions representing them). The Development Society of Niassa is expected to help the communities to establish these institutions and will facilitate their participation, but so far it has failed to do so due to the complexity of defining the communities and encouraging collaboration, and difficulties with logistics because of the scattered location of the communities. Furthermore, the society is not satisfied with its legal status (which is interpreted as commercial) and is examining alternatives, as they are not producing profit and are to some extent dependent on donor funding.

The main activities of the society will be to promote community development and the socio-economic activities needed to improve the quality of life of local communities, to be achieved partly through the sustainable commercial use of resources in the reserve. The society will also promote the rehabilitation, building, development and use of necessary tourism and conservation infrastructure. In order to achieve these goals, the society is free to sublet areas for the establishment of lodges, hunting and photographic safaris, and so on.

The partnership has faced a number of difficulties since 1998, and as a consequence few activities have been undertaken so far. Most activities to date have been concentrated in the eastern part of the reserve and on three of the five hunting blocks that are in the buffer zone (Soto, 2007). Although the partnership has improved the management of the reserve, the reserve remains only partly funded and managed. To date, the investment in management has yielded very small returns, hampering the development of the reserve.

As it is located in one of the most remote parts of Mozambique, and due to extremely poor infrastructure, tourism potential at present is very low. The society did not expect to have to fund tourism development for the period of the agreement, hoping that funds for such developments would be forthcoming from donors. Another complication has been the delay in releasing shares to the Mozambican private sector and the local communities.

The government recently approved a management plan for the reserve for 2007 to 2012, which emphasizes three areas. The first is to improve community participation in natural resource management and reduce human–wildlife conflict, zoning the areas where people live, promoting modern agricultural techniques, providing social infrastructure (for example schools, hospitals, roads, commerce facilities), job creation in tourism ventures and providing access to subsistence hunting. The second is the establishment of safari hunting within the reserve in addition to the hunting that already occurs in the buffer zone. This will be the first time safari hunting will be undertaken in the core area of the reserve and will be conducted to increase the income generated by the reserve for use in management activities. Institutional reform is also highlighted – as already noted, the current structure of the society is a commercial organization, which has caused difficulties in mobilizing donor funding for the management of the reserve.

In another experience of park management partnership, in 2004 MITUR signed a 12-month Memorandum of Understanding (MoU) to co-manage the Gorongosa National Park with the Gregory C. Carr Foundation. The MoU was subsequently extended indefinitely (until a long-term agreement could be signed). On 1 January 2008 MITUR and the Carr Foundation signed a long-term agreement for 20 years. According to the agreement, the Carr Foundation will work with the government in the joint administration of the Gorongosa National Park to ensure that the ecosystem is preserved and a sustainable tourism industry is established. The Carr Foundation will contribute US$1.2 million per year and government will provide US$158,000 for at least the first three years, with park revenues to be used for park development.

In order to improve partnerships in other reserves and parks in the future, mechanisms to guide the formation of such partnerships need to be developed, ensuring a clear understanding of the roles and responsibilities of each of the stakeholders involved. Training programmes for stakeholders entering such partnerships are also recommended.

Community-based natural resource management

The first CBNRM programme launched by the government was in 1995 and was the Tchuma Tchato project, with funding from the Ford Foundation, in a multiple use area in the Magoé District of Tete Province (see also Chapter 16 for more information about CBNRM in Mozambique).[4] The project was the first that allowed (through government decree) part of the money generated by sport hunting to be allocated directly to local communities. Prior to 1995, legislation did not allow any money generated by any wildlife-related activity to be allocated to local communities. The decree also provided the opportunity for the communities of Magoé to be allocated a quota for hunting for their own consumption.

Based on some of the lessons learned from the Tchuma Tchato project, current regulations of the forest and wildlife law state that, of the revenue collected by government for allowing access to forest and wildlife products (in protected areas and commercial forest), 20 per cent must be given to local communities. The implementation of these regulations remains deficient, as some of the institutional and procedural issues have not been completed throughout the country (for example the creation of community legal entities and the opening of community bank accounts), and the disbursement of funds is slow. Despite problems, there are currently more than 60 CBNRM initiatives in Mozambique (Magane, 2001; Nhantumbo, 2004).

Transfrontier conservation areas

In 1997, the Government of Mozambique initiated a 15-year transfrontier conservation area (TFCA) programme. This programme represented the first big post-war programme for wildlife conservation, and its implementation has strengthened wildlife conservation institutions, increased the level of trust of donors in the government, and has had the added benefit of leveraging funding from additional donors in areas outside of TFCAs.

The first two phases of the TFCA programme have aimed to establish collaborative and cooperative conservation practices between Mozambique and her neighbours. Three TFCAs have been established with neighbouring countries: the Lubombo TFCA, the Great Limpopo Transfrontier Park and the Chimanimani TFCA. The Lubombo TFCA includes the Maputo Elephant Reserve and the Futi Corridor in Mozambique and Tembe and Ndumo Reserves in South Africa, and was established in 1999. The Great Limpopo Transfrontier Park includes the Limpopo National Park in Mozambique, the Kruger National Park in South Africa and the Gonarezhou National Park in Zimbabwe (see also Chapter 23). According to the treaty signed in 2002, these three countries will seek to include adjacent areas to these parks to establish the Great Limpopo TFCA. The Chimanimani TFCA was established in 2001 and includes part of the Sussundenga District in Mozambique (the Chimanimani Reserve) and the Chimanimani National Park and adjacent areas in Zimbabwe. In addition to these, the country is planning to establish two more TFCAs – one along the Zambezi valley of Mozambique, Zimbabwe and Zambia, and the Niassa TFCA, which, though still in the early stages of definition, will probably include the Niassa Reserve in Mozambique and neighbouring areas of Tanzania.

Because TFCAs incorporate formally protected areas and land that is not formally protected, different institutions will be required to participate in the integrated management of natural resources according to the nature of land uses. Management practices and laws will also be subject to harmonization between the countries involved.

Conclusions

In recent years, partnerships in the management of protected areas have been adding value to Mozambique's protected area network and are occurring even without detailed establishment and management procedures. The challenge remains for government to strengthen its regulatory role and establish an appropriate framework to enable the participation of all stakeholders, each with clear roles and responsibilities. Although there is a good legal framework to empower communities, the weakness of their institutions and the dominance of the private sector have not yet enabled communities to receive the management rights and benefits that they deserve. Implementing partnership agreements is complex; training of staff and the establishment of positive incentives appear to be the major challenges to the success of protected areas management in Mozambique. However, the country is on the path to improving the management of its protected areas and has become a leader in experimentation and innovation.

Notes

1   The convention was attended by the colonial administrations of Africa to discuss the continued decline of flora and fauna resources on the continent (de Sousa, 1938).
2   Mozambique established the MITUR in 2000 in order to unlock tourism business potential through ensuring strong linkages with wildlife and conservation areas. These, in addition to the coastal areas, are considered to be the main resources that tourism can utilize for its development.
3   Group Madal is a private Mozambican company devoted to coconut plantation and cattle farming. One of the owners decided to inject his own money into Niassa Reserve in behalf of the company.
4   According to Mozambican legislation, a multiple use area is a non-gazetted area, i.e. it is not an officially proclaimed protected area.

References

Beilfuss, R., Moore, D., Bento, C. and Dutton, P. (2001) 'Patterns of vegetation change in the Zambezi Delta, Mozambique', Working Paper No. 3, Program for the sustainable management of Cahora Bassa dam and the lower Zambezi valley, unpublished paper, www.savingcranes.org/conservation/our_projects/program.cfm?id=23 (accessed 19 March, 2007)
Berry, V. and Petty, C. (1992) *The Nyasaland Survey Papers. 1938–1943. Agriculture, Food and Health*, Academy Book Limited, London
Black, R. and Schafer, J. (undated) 'Colonial forestry and national resource management policy in Mozambique: Full report of research activities and results', www.sadccitizen.net/regiondocs/6000205.pdf (accessed 23 July 2007)
Black, R. and Watson, E. (2006) 'Local community, legitimacy and cultural authenticity in post-conflict natural resource management: Ethiopia and Mozambique', *Environment and Planning D: Society and Space*, vol 24, pp263–282
de Sousa, A. G. (1938) 'Parques nacionais: colónia de Moçambique', *Boletim Economico e Estatístico de Mozambique*, Número 16
Dias, A. H. G. S., Dias, J. A. T. and Pereira, M. G. C. (1963) 'O Problema do Aproveitamento Racional dos nossos recursos faunísticos naturais com forma de se obviar a uma pecuaria convencional deficitária (reflexões sobre uma visita de estudo a África do Sul e Rodésia)', *Separata do Boletim da Sociedade de Estudos de Moçambique*, Lourenço Marques, Moçambique
Dias, J. A. T., Dias, A. H. G. S. and Rosinha, A. J. (1971) 'Alguns Aspectos da Caça

Clandestina em Moçambique e Sugestões para se por Cobro a Destruição do Nosso Patrimonio Cinegético', *Revista de Ciências Veterinárias*, vol 4, série A, pp101–125

DNAC (2006) 'Áreas de Conservação para fins Turísticos', Ministério do Turismo, República de Moçambique, Maputo

DNFFB (1991) 'Subsídios para o Relatório Nacional do Sector de Fauna Bravia', unpublished report for the DNFFB, Maputo

DNFFB (1992) 'Linhas Gerais para uma Política de Fauna Bravia', unpublished, report for the DNFFB, Maputo

DNFFB (1996) 'Forest and wildlife policy and strategy', unpublished report for the DNFFB, Maputo

Governo Geral em Lourenço Marques (1964) 'Diploma Legislativo número 2496', Governo Geral, Lourenço Marques, Moçambique

História de Moçambique (1974) *Primeiras Sociedades Sedentárias e Impacto dos Mercadores*, República de Moçambique, Maputo

Hutton, J., Couto, M. and Oglethorpe, J. (2001) *Biodiversity and War: A Case Study of Mozambique*, Biodiversity Support Programme, Washington DC

Magane, S. (2001) 'Directório de iniciativas de envolvimento comunitário na gestão dos recursos naturais do país', unpublished report for the DNFFB, Maputo

McGregor, J. (1998) 'Violence and social change in a border economy: War in the Maputo Hinterland 1984–1992', *Journal of Southern African Studies*, vol 23, no 2, pp187–203

Newitt, M. (1995) *A History of Mozambique*, University of Witwatersrand, Johannesburg

Nhantumbo, I. (2004) 'Maneio Comunitário em Moçambique: Evolução e Desafios para o Futuro', paper presented at the Third National CBNRM Conference, 21–23 July 2004, Maputo

Pro Natura (1973) *Zonas de Protecção da Fauna Selvagem, Discussão, Recomendações Conslusões*, Sá da Bandeira, Angola

Rosinha, A. J. (1973) 'Bases da protecção da natureza no Ultramar Português', in *Fauna Selvagem e Protecção da Natureza*, Legislação, Agência Geral do Ultramar, Lisboa

Schafer, J. and Bell, R. (2002) 'The state and community based natural resource management: The case of the Moribane Forest Reserve, Mozambique', *Journal of Southern African Studies*, vol 28, no 2, pp401–420

Soto, B. (2007) 'Ministério do Turismo Memorando. Processo de Pedido de Extensão dos Termos de Autorização para a Gestão da Reserva do Niassa Durante 25 Anos e sua Revisão do Plano de Maneio', República de Moçambique, Maputo

Virtanen, P. (2005) 'Community-based natural resource management in Mozambique: A critical review of the concept's applicability at local level', *Sustainable Development*, vol 13, pp1–12

Part III

Conservation on Private Land

7

Private Conservation in Southern Africa: Practice and Emerging Principles

Brian Child

The chapters in this section describe how private conservation emerged in southern Africa. It expands on the literature with considerable personal knowledge because the significant conversion of private land to wildlife is remarkably under-documented. Two chapters provide a general description of private land conservation in South Africa and Namibia, a third describes the economic and institutional conditions for its emergence in Zimbabwe in the 1980s, and the fourth describes collective landholder action in southern Zimbabwe.

Private landholders have played an important role in the growth of wildlife conservation in southern Africa, creating new economic and business models that have in turn affected both state and community wildlife management. Despite its importance to biodiversity conservation and the regional economies of southern Africa, wildlife production, or 'game ranching', is vastly under-researched.

In introducing the reader to private conservation in the region, this chapter synthesizes several of the key lessons from the four case studies. The key message is that wildlife has an economic comparative advantage in certain environments. However, astute policy that devolves rights to landholders and that encourages value-adding uses of wildlife, including trade, is necessary to translate this theoretical advantage into wildlife conservation and economic growth. Indeed, game ranching is a recent phenomenon that only emerged from insightful policy. Moreover, as recently as the 1980s and 1990s, questions were still being raised about whether wildlife could pay for itself or was a legitimate form of land use, while the effectiveness of extensive wildlife production in biodiversity conservation is still questioned, despite evidence that it often works (Bond et al, 2004).

Economics and institutions

This section introduces two theoretical questions. First, does wildlife have a comparative ecological and economic advantage, and where is this most likely to occur? Second, what is needed institutionally to unlock this potential? We define the 'economic' value of parks and wildlife properties in its technical sense so that it includes the full range of values that accrue at the level of society, and includes use values, intangible and existence values, including biodiversity and ecosystem services. However, payments are rarely made for intangible values that, like carbon sequestration, can be by-products of wildlife enterprises. Even when the sum of benefits to society of wildlife conservation and use is very positive, land will only be allocated to wildlife by individual landholders if it can 'pay' for itself (including monetary and non-monetary values); in other words if it is financially viable.

This question is best answered by the private sector, where we can best measure enterprise profit and where we can see if macro-economic trends such as the shift in land use towards wildlife and tourism confirm our conclusions. Action in the private economy reflects the costs and benefits of wildlife much more accurately than state bureaucracies or open-access property regimes in the communal sector. The private sector also keeps reasonably good financial records. By contrast, it is difficult to assess the economics of wildlife in protected areas, which combine for-profit (tourism) and not-for-profit (conservation, education) activities in the same set of accounts. Having said this, private investment in wildlife properties is done for a number of reasons, many of which are not financial.

The motivations of private wildlife producers are not well understood. Some landholders rely on wildlife for their financial survival, while others place a higher value on non-financial benefits (for example aesthetic and cultural values), particularly in South Africa. The chapters in this section suggest that wildlife has a comparative advantage in many agriculturally marginal habitats. However, this is not a sufficient condition for its emergence as a land use option, as demonstrated in other countries in the region – including those with similar or better wildlife resources, where private extensive wildlife production has not become an important land use option. The key in southern Africa has been to give landholders strong use rights and to encourage the commercialization of wildlife and the development of new wildlife products. The fugitive nature of wildlife and the need to control misuse and economic externalities has often been used by state conservation agencies to justify strong top-down regulation. These case studies show that this is not necessarily the best solution: South Africa and Namibia have controlled externalities using game fencing, while Zimbabwe avoided the financial and ecological disadvantages of fences with a rather elegant common property solution.

South Africa

Bothma, Suich and Spenceley's review of extensive wildlife production in South Africa presents much of the recent literature on the subject. Until the 1980s, the heavy hand of the South African state and its agricultural subsidies aimed at the conservative farm vote, constrained the emergence of wildlife enterprises. However, where once wildlife was devastated by agricultural expansion there are now some 10,000 wildlife properties together with 4000 mixed wildlife/livestock enterprises.

In a country where wildlife outside parks is controlled by provincial rather than national legislation, the historical record about when and why rights to wildlife utilization were devolved to farmers is thin. We do know that use rights are acquired through a 'certificate of enclosure'; in other words, rights are established by game fencing. One suggestion is that the enlightened Natal Parks Board created some momentum for game ranching by restocking farms in the 1970s with excess wildlife from parks and by providing quality extension services to encourage it. We also know that game ranching was certainly occurring in the Transvaal in the early 1960s (see also Riney and Kettlitz, 1964), and there are even suggestions that a few gentrified landholders conserved wildlife in the late 1800s (Carruthers, 1995).

Bothma, Suich and Spenceley suggest that private wildlife properties cover 16.8 per cent of the country, compared to 6.1 per cent in provincial and national protected areas. Despite this, its impact on biodiversity is poorly documented but thought to be largely positive and improving as managers gain experience. However, we need to know more about the potential problems associated with extensive wildlife production, such as habitat fragmentation, genetic isolation and overstocking.

Information regarding the wildlife economy is extremely piecemeal. From the information available, the rapid expansion of this sector implies that extensive wildlife production is profitable. Several studies also suggest that it provides more jobs at higher wages than the land uses it displaces (mostly livestock production). The extensive wildlife sector is extremely varied, with landholders combining a range of wildlife-related activities including tourism, hunting, live-game and venison sales, as well as property development and other enterprises. In many ways, wildlife production is based on tourism experiences rather than commodity production. So far, meat production has not emerged as a growth industry, which scientists in the 1960s hoped it would (see Walker 1979; Child, 1988), and while a lot of meat is shared informally, commercial meat cropping is restricted to parts of South Africa and Namibia, at least at the moment.

Namibia

Namibia's sector has followed a similar path to South Africa, with the utilization of wildlife on private land evolving from traditional biltong hunting and venison sales by livestock ranchers to higher-value uses such as recreational hunting and later photographic tourism. Regulatory changes in 1968 and legislative changes in 1975 allowed landholders to use wildlife (but with more restrictions than parallel changes in Zimbabwe's legislation), and use rights in Namibia are contingent on fencing, in common with South Africa. The importance of relationships between wildlife directors cannot be underestimated; they met regularly at SARCCUS to discuss and instigate these bold political economic changes to the way wildlife on private land was managed (see also Chapter 1).

Barnes and Jones describe Namibia's game-ranching sector in some detail. They associate a long-term decline in livestock populations with range degradation and suggest that 15–25 per cent of freehold land is used primarily for game production (although the latest data on wildlife populations are from 1992). Namibia is situated in the southern arid zone, with a narrower range of species than the 'bushveld' of South Africa and Zimbabwe. With fewer valuable game species, the economic advantage of wildlife in Namibia may be less clear cut than in other parts of the region. However, cattle are not very profitable either and wildlife production on private land certainly expanded after livestock subsidies were reduced in the 1990s. Highlighting an important distinction between a financial and economic assessment of wildlife, Barnes and Jones argue the weak positive (financial) incentive for landholders to manage wildlife nevertheless translates into a strong (economic) incentive for the nation to switch to wildlife, because of its economic multipliers. In other words, even where landholders do not make much individual profit from wildlife, the economy of the country benefits because most of the benefits from wildlife accumulate higher up the value chain.

Barnes and Jones demonstrate that wildlife profits increase with scale and as the number and diversity of wildlife increase. Collective action by landholders through the formation of conservancies is a rational financial response, and some of Namibia's 22 private conservancies are larger than state-protected areas. Once the diversity of wildlife (coupled with the natural features of the landscape) provides opportunities for photographic tourism (rather than just hunting), wildlife becomes a better financial and economic option by a factor of ten. Moreover, the terms of trade for wildlife are continuing to improve – by at least 80 per cent over the past decade.

Where Namibia has low densities and species variety of wild animals (which is common on livestock ranches), safari hunting has high returns to capital but low returns to land and is highly profitable only as a subsidiary

enterprise. However, hunting often finances the recovery of wildlife populations and is a stepping stone to the introduction of photographic tourism,
which has excellent financial returns to land.

Zimbabwe

Child's chapter on game ranching in Zimbabwe describes in some detail the
subsidization of livestock that causes the decline of wildlife in the region. The
chapter describes the legislation that tipped land use towards wildlife and how
landholders reacted to this in the 1980s. It describes the important legislative
innovation whereby the regulatory powers necessary to manage the externalities
associated with a fugitive resource such as wildlife were devolved from the state
to groups of landholders (i.e. intensive conservation areas – ICAs) with great
effect. It also summarizes detailed research (Child, 1988; Jansen et al, 1992) to
assess whether wildlife had a financial or economic comparative advantage. An
important inclusion in this chapter is the ecological comparison of wildlife and
livestock on Buffalo Range Ranch, where long-term monitoring confirmed that
the range degradation that anecdotally drove many cattle ranchers into financial
trouble is real. It also suggests that multispecies wildlife systems may well have
an ecological comparative advantage.

Lindsey, du Toit, Pole and Romañach take up the story in Zimbabwe's Savé
Valley that began in the late 1950s (Riney, 1960, 1964). They describe how
white cattle farmers degraded the environment and wildlife resource until livestock undermined its own viability. These problems were hammered home
during the severe droughts in 1984 and 1992, and 18 ranchers collectively
switched to wildlife in 1991, legally constituting the Savé Valley Conservancy
(3442km²). This switch coalesced around the reintroduction of rhinos, which
was facilitated by the Department of National Parks Wild Life Management,
the World Wildlife Fund and the Beit Trust. We believe that the carefully
formulated rules for collective action that characterize this conservancy fed off
many years of experience within an active ICA movement. Technological fixes
such as fencing and grazing systems had failed to help prolong the viability of
the livestock sector. Instead ranchers were saved by institutional innovations
that allowed them to scale up from agro-extractive livestock ranching on single
properties to a much larger wildlife economy on multiple properties based on
hunting and tourism. In addition to Savé Valley, a number of other conservancies developed in Zimbabwe, improving economies of scale as well as habitat
connectivity.

The switch from livestock to wildlife is usually associated with several years
of restricted cash flow as habitats and wildlife recover; without transition
financing, it is easy to see how land can be locked into an unproductive and

degrading state (Walker et al, 2004), even when a superior alternative is possible. Culturally, too, land can be locked into inferior uses. For example, for many years a suspicious political and bureaucratic leadership viewed wildlife holdings as unproductive (pers. obs.); it was difficult for these leaders to accept that economic value added (for example tourism and employment) could be more valuable than the amount of an agricultural commodity produced (meat or grain).

Interestingly, scaling up land use for wildlife may also improve political resilience. Lindsey et al's chapter, for instance, mentions the disproportionate (albeit imperfect) survival of wildlife conservancies in Zimbabwe's current circumstances, which have been more successful at avoiding the disruptive effects of Mugabe's land invasion policy than farms in general, and also than individual wildlife properties. The Savé Valley Conservancy obtained special recognition and protection through wildlife-based land reform administered by the Ministry of Environment and Tourism, a move enhanced by the political abilities of several members of the conservancy. Moreover, long before the land invasions, the Savé Valley Conservancy had articulated a vision that its long-term survival required a responsible relationship with its communal neighbours, and this undoubtedly helped.

Competition and innovation

We turn now to practical and theoretical issues that emerge from these chapters. As noted, Namibia and South Africa devolved the rights to wildlife on the basis of properties being appropriately game fenced, whereas Zimbabwe used a collective action solution. Fences have the advantage that the 'ownership' of wildlife is very clear. However, they are expensive and fragment the landscape causing some of the problems listed in Bothma, Suich and Spenceley's chapter. Fencing is also associated with far more range degradation than unfenced land. For example, in the Zimbabwean Midlands (which was unfenced), valuable species such as sable simply moved off properties that were overgrazed; without fencing, ranchers who wanted good wildlife had to improve their land husbandry.

An important benefit arising from the game ranchers across the region is economic and institutional experimentation. Innovation by landholders is rapidly increasing the number, range and value of wildlife products. In all three countries, wildlife management is being scaled up through collective action as landholders remove fences between farms they have agreed to manage cooperatively, allowing wild animals to range across larger areas of land. In Part IV we see how protected areas, similarly, are scaling up through contractual arrangements with private (and communal) landholders. The devolution of wildlife use rights to thousands of landholders means that competition, exper-imentation and innovation are constantly increasing the value and improving the management of wildlife on private land. By demonstrating new approaches and commercial options, the private sector is also encouraging innovation in state-protected areas as we discuss in some detail in Part IV.

Institutional change and resilience

The following chapters provide examples of the *transformability* of social ecological systems locked in undesirable 'basins of attraction' (Walker et al, 2004). Previously dry savanna systems were characterized by a vicious cycle of rangeland degradation and poverty associated with livestock and the central-ization of wildlife governance. The 'flip' from undesirable (for example unsustainable livestock production) to desirable systems (for example wildlife-based economies) emerged from sequenced reform of institutional and economic rules – attributes that fall within the crucial area (for resilience) of adaptive governance. There was little precedent for this new approach.

The experience that is described for game ranching in the region supports the emerging consensus that economic growth is related more to institutional capital than financial capital (Anon, 2008). Economic institutions, like the new legislation and practices described in the following chapters, can add consid-erable value by allocating resources to higher-value uses and reducing the risk associated with larger and more complex wildlife economies. Better institu-tions allow movement beyond a zero-sum game that assumes a direct trade-off between conservation and development. Indeed, evolutionary economics hypothesizes that organizational energy (for example institutions, property rights and entrepreneurship) converts relatively low-ordered raw materials, such as livestock production and rangeland environments, into more highly ordered products and services (Beinhocker, 2006). This allows the creation of more value with less impact on the environment. Such institutions are particu-larly important for managing resources that are mobile (fugitive) and are therefore an invaluable tool for conservationists. The breakthrough in south-ern Africa was the search for institutional arrangements that built positive incentives, once the limitations of negative incentives such as prohibitive wildlife regulations were recognized.

Conclusions

Part III illustrates how effectively private conservation promotes both economic growth and wildlife recovery, at a very low cost to the state, reminding us of how surprising it is that the sector has been studied so little. The growth of private conservation in the region strengthens the argument for promoting conservation to create jobs and economic growth. The extent and success of wildlife production on private land also challenges the assumption that state-protected areas are the only, or even the best, way of conserving environments. While the natural resources within state-protected areas are often degraded as a consequence of their underfunding, many private reserves are able to pay for habitat rehabilitation. In addition, non-financial values and a passionate commitment to land and wildlife often figure far more prominently in these 'business' choices than landholders are given credit for. In the Agulhas Biosphere Reserve in South Africa, for example, landholders are increasingly conserving rare patches of fynbos habitats for motives other than profit, with simple recognition being an important motivator (see Child, 2007).

The fact that private land often conserves biodiversity at a lower cost to the taxpayer than state-protected areas challenges the assumptions that the latter are superior; yet they are automatically accorded a higher status according to the categories of the IUCN. State-protected areas are normally accorded category I or II status, whereas private land conservation seems to fall into category VI, especially when it entails consumptive use, regardless of how well biodiversity is conserved. It might be wiser if these definitions privileged ecological performance (i.e. the integrity of the biodiversity) over jurisdictional authority (i.e. the fact that the state has set the land aside for conservation regardless of the quality of conservation management), especially given the financial and non-financial incentives associated with classification in certain categories. The importance of private protected areas is increasing rapidly, including in southern Africa, and it may be time to recognize them as such. The sticking point may be that private conservation does not legally entrench a conservation objective 'in perpetuity', although the newness of many state-protected areas and the syndrome of 'paper parks' casts some doubt on their claim of permanency. Scepticism about corporate commitment to the land is a legitimate concern where there is a tight focus on profits and few linkages between investors and the land. However, private conservation is often a family business and a lifestyle choice and associated with high levels of commitment and knowledge about conservation. Given how much biodiversity private landholders are now conserving in southern Africa and elsewhere, and how much diversity they can add to a country's protected area estate, surely it is time that these conservation efforts were formally recognized not as an afterthought in category VI but, where deserving, as category II private protected areas.

References

Anon (2008) 'Order in the jungle', *The Economist*, 13 March

Beinhocker, E. D. (2006) *The Origin of Wealth: Evolution, Complexity and the Radical Remaking of Economics*, Harvard Business School Press, Boston

Bond, I., Child, B., de la Harpe, D., Jones, B., Barnes, J. and Anderson, H. (2004) 'Private land contribution to conservation in South Africa', in Child, B. (ed) *Parks in Transition: Biodiversity, Rural Development and the Bottom Line*, Earthscan, London

Carruthers, J. (1995) *The Kruger National Park: A Social and Political History*, University of Natal Press, Pietermaritzburg

Child, B. (1988) 'The role of wildlife utilization in the sustainable economic development of semi-arid rangelands in Zimbabwe', DPhil thesis, University of Oxford, Oxford

Child, B. (2007) 'Independent mid-term evaluation of the UNDP-GEF CAPE Agulhas Biodiversity Initiative (ABI)', UNDP/GEF, Pretoria, South Africa

Jansen, D. J., Bond, I. and Child, B. (1992) *Cattle, Wildlife, Both or Neither? A Survey of Commercial Ranches in the Semi-arid Regions of Zimbabwe*, WWF Multispecies Animal Production Project, WWF, Harare

Riney, T. (1960) 'Rhodesian wild life is a natural resource', *Wild Life*, vol 2, pp149–154

Riney, T. (1964) 'The economic use of wildlife in terms of productivity and its development as an agricultural activity', FAO African Regional Meeting on Animal Production and Health, Addis Ababa

Riney, T. and Kettlitz, A. (1964) 'Management of large mammals in the Transvaal', *Mammalia*, vol 28, no 2, pp189–248

Walker, B. H. (1979) *Game Ranching in Africa: Management of Semi-arid Ecosystems*, Elsevier Oxford

Walker, B., Holling, C. S., Carpenter, S. R. and Kinzig, A. (2004) 'Resilience, adaptability and transformability in social-ecological systems', *Ecology and Society*, vol 9, no 2, pp2–10

References

8

Game Ranching in Namibia

Jon Barnes and Brian Jones

The development of wildlife-based land uses on freehold farmland in southern Africa is generally well documented. Legislative changes, which bestowed custodial user rights over wildlife to private landholders in South Africa, Zimbabwe and Namibia, and their impacts – which appear to have resulted in investment in wildlife-based land use and increases in wildlife stocks have been described extensively (Joubert, 1974; Luxmoore, 1985; Child, 1988; Cumming, 1990; Jansen et al, 1992; Bond, 1993).

A measurable amount of land in these areas, which is mostly land that receives low rainfall and is marginal for livestock production, has been converted from livestock production to wildlife use. Wildlife in arid lands tends to have high mobility and, partly due to this, some individual landholders have grouped together to share wildlife management activities. The resulting larger land units consisting of several individual properties that are cooperatively managed are known as conservancies. This development followed a similar trend among private landholders in Zimbabwe, a detailed account of which can be found in du Toit (1994).

Some 43 per cent of Namibia's land surface, or 356,886km^2 is occupied by private landholders (Ashley and Barnes, 1996). This land is concentrated in the semi-arid and arid centre and south of the country, and is relatively well served by road and telecommunications infrastructure.

Mean annual rainfall ranges from 550mm in the north-east of Namibia to 50mm in the south-west. About half of the freehold land (i.e. in the northern parts where mean annual rainfall is above about 250mm) is occupied by wooded savanna vegetation of several types, mostly dominated by *Acacia* spp. The other half (i.e. in the drier southern parts where rainfall is below 250mm per annum) is dominated by karroid dwarf shrublands, in which *Rhigozum trichotomum* is common.

Successful crop production is extremely difficult over most of Namibia due to poor soils and insufficient rainfall, and the basic form of land use on freehold land is extensive livestock production. Cattle dominate in the northern savannas and small stock – mostly sheep but also goats – dominate in the southern shrublands. Stock-carrying capacities vary widely with climate, but long-term average capacities range from some 10ha per large stock unit[1] (LSU) in the semi-arid north-east, to some 35ha per LSU in the arid south-west.

Individual ranch sizes average some 8000ha in the northern savannas to some 10,000ha in the southern shrublands. The number of cattle on freehold land has varied between 1 million and 1.5 million since 1972 and has declined by some 55 per cent since the 1950s. Lange et al (1998) investigated this decline and concluded that it was associated with increased herd productivity, although it was probably also accompanied by range degradation. The number of sheep also declined over the period, from about 3.9 million to about 2.2 million (some 624,000 to 352,000 LSU equivalents), but the reasons for this have not been carefully examined. Goat numbers on freehold land have fluctuated around half a million (some 80,000 LSU equivalents) (Lange et al, 1998). The main commercial products from the ranches are beef and mutton, while the production of pelts from karakul sheep in the south declined significantly during the 1980s and 1990s due to poor market conditions, and many farmers switched to mutton production.

Wildlife present on freehold land tends to be species adapted to desert conditions (the Kalahari and Namib deserts). The generally open habitats in the south are dominated by springbok, with associated lesser populations of gemsbok and kudu. Kudu, gemsbok and warthog dominate in the savannas of the north, with associated lesser populations of species such as hartebeest, eland, springbok and dik-dik. Mountain zebra and klipspringer occur in western mountain escarpment habitats, and ostrich, steenbok and duiker tend to occur throughout. Generally, the greatest wildlife diversity is found in the northern savannas, where giraffe, plains zebra and wildebeest also occur, and most game species introductions have taken place here (including that of the black-faced impala subspecies).

Populations of some of the larger predators, for example cheetah, leopard and brown hyaena persist fairly widely, but lion and spotted hyaena have been largely exterminated by farmers. In fact freehold farmland in Namibia is considered to have the largest cheetah population in Africa (Marker, pers. comm.). Small nuclei of elephant, white rhino and black rhino have been established on some northern properties. Some species not natural to the area have also been introduced, such as common impala, sable, waterbuck, blesbok, black wildebeest, nyala, roan, tsessebe and reedbuck. These introductions of exotic species onto private lands were aimed at enhancing the tourism income potential of wildlife on game farms.

Forms of wildlife use

The use of wildlife on private ranches generally developed as a supplementary activity to livestock production, but a small and increasing number of properties are devoted purely to wildlife production. Forms of use include venison production (through formal commercial culling or informal shooting and selling), live game capture and sales, recreational hunting (for either biltong or trophies), specialized semi-intensive ostrich farming and non-consumptive wildlife-viewing tourism. The last of these, in particular, tends to be associated with larger and better-stocked wildlife ranches.

Historically, the basic form of consumptive use of wildlife on private land involved the hunting of common naturally occurring species such as kudu, gemsbok and springbok on a small scale for home consumption either as venison or biltong. The products were also used as rations in part payment for farm labour. (This form of use was technically illegal but largely overlooked by authorities.) Such activities continue to be practised by a very large proportion of livestock producers on private land. With regulatory changes in 1968 followed by legislative change in 1975, which allowed commercial utilization of wildlife, a use category termed 'shoot and sell' allowed landholders to sell venison to local butchers. This requires a licence from the wildlife authority and is an activity that continues to be a common and important use of wildlife.

During the 1970s and 1980s, commercial production of venison for export developed as an industry (Joubert, 1974; Joubert et al, 1983). This was practised by a small number of landholders who hired specialist operators to conduct (night) culls of species such as springbok, gemsbok and kudu, and process these in mobile abattoirs. 'Skin-on' carcasses were then exported to European (especially German) markets by a central meat processing company. A collapse in prices due to oversupply from other countries (primarily Eastern European countries and New Zealand) led to commercial venison culling being discontinued in Namibia in the 1990s. Another form of wildlife use that developed at this time was the capture and sale of live game, which followed developments in South Africa, where a robust market for live game developed and auctions were regularly held. Initially in Namibia, sales were made by the state to landholders, but over time they were increasingly made between landholders (Joubert et al, 1983; Brand, 1984).

An important development in wildlife use on private land was the development of recreational hunting tourism. Private landholders could register to guide hunts for plains game on their own land and on other properties with the permission of the landholder. The most lucrative market was the international trophy hunting market, although some less valuable 'biltong hunting' also developed for the southern African hunting market. Farmers required permits for these activities unless they fenced their properties and registered as hunting farms. Many farms were involved with hunting tourism, with the hunting enterprises tending to be complementary to livestock production taking place on the same farms.

The profitability of trophy hunting resulted in the proactive management and stocking of land with wildlife and the gradual development of non-consumptive guest farm tourism (also commonly in conjunction with livestock production). Where hunting and guest farms had suitable tourism attributes and were of a suitable size, the development of 'pure' game farming developed. Here, livestock production ceased and land was used specifically for tourism, accompanied by middle- and up-market lodge developments. Several hundred such properties have developed and the number is rapidly expanding, and the livestock-dominated land use of the 1960s has been modified to include more wildlife production activities and, in a growing proportion of cases, complete transformation to a purely tourism-based land use.

Private investment in wildlife

Wildlife numbers including species such as the endemic Hartmann's mountain zebra were declining prior to regulatory and legislative change in the 1960s and 1970s, and landholders viewed wildlife as competition for their livestock and therefore as a cost rather than a benefit. Then, the wildlife was owned by the state and farmers had no (legal) control over it. Even if they wanted to maintain wildlife on their land, they were powerless to manage numbers in balance with their livestock, unless through illegal removal of animals. 'Biltong hunting' became popular when, during the declared hunting season, hunters from urban areas and South Africa would indulge in killing sprees for meat to make biltong, as it was the only legal way for landholders to manage and benefit from wildlife on their land.

Regulations enacted by the South African colonial administration in 1968 gave white freehold farmers a limited form of proprietorship over wildlife. In 1975, the rights given to freehold farmers were incorporated into new wildlife legislation, The Nature Conservation Ordinance (No. 4 of 1975). The proprietorship over wildlife granted to landholders was partial, with the state retaining some control through the permit system. Farmers meeting certain conditions (i.e. land of more than 1000ha surrounded by a certain type of fencing) could receive 'ownership' of certain wildlife species designated as huntable game. These species were an eclectic mix of kudu, gemsbok, springbok, warthog, buffalo and bushpig and the farmer was free to use them for personal use, at his discretion.[2] Farmers were also able to use other species designated as protected or specially protected game by successfully applying for a permit (thus these species remained under the control of the state).

Permits were issued by the conservation authorities based on a farm inspection and a farm population estimate for the relevant species. Landholders could also apply for status as a hunting farm, with rights to carry out trophy hunting. Farmers without fenced properties could also use game species through the permit system, but their rights were more restricted.

The motivation behind these changes was to reduce the decline of wildlife on private land and to allow diversification of land use, and appears to have resulted in the growth of the wildlife industry on freehold land (van der Walt, 1987; de Jager, 1996; Barnard, 1998). According to van der Walt: 'A significant change in attitude towards game amongst farmers became noticeable since 1968 when they received by ordinance ownership of game on their farm. Their sense of economic value of this natural asset then lead [sic] to the present booming industry' (1987, p4).

Similar changes occurred at around the same time in Zimbabwe and South Africa. In Namibia, the change in approach from state to conditional private ownership, and use rights over certain species of wildlife, had a major impact on trends in wildlife numbers. Barnes and de Jager (1996) investigated the effect of the legislative changes on wildlife stocks and the impacts on financial and economic characteristics of investment in wildlife on Namibian private land. Table 8.1 shows a general increase in Namibian wildlife stocks, based on successive surveys two decades apart.

While no national monitoring of wildlife numbers on private land has been undertaken, the historical data in Table 8.1 confirm commonly made assertions that since the 1960s, wildlife numbers and diversity on private land have increased. Increase in numbers and biomass appear to have been some 3 per cent per annum over the period, with the effects of recruitment and stock importation outweighing the effects of consumptive offtake. Unpublished evidence indicates that these increases continue to the present. The increase in total biomass has primarily involved the more valuable plains game species, and one can assume that it has had the effect of enhancing wildlife use values for trophy hunting and wildlife viewing. In as much as the increase in wildlife stocks and diversity has restored historically and naturally appropriate compositions, it will also have enhanced conservation values. However, some private landholders have sought to increase diversity as well as biomass, to the point that several species not locally or nationally indigenous have been introduced. Thus the introduction of aliens such as blesbok and black wildebeest and some likely genetic pollution caused by introduction of alien subspecies, may to some extent have jeopardized these values.

Lange et al (1998) showed that between the late 1940s and the mid-1990s, livestock biomass declined by some 55 per cent on Namibian private land. As noted above, this decline tended to be compensated for by enhanced livestock productivity, but was perhaps also accompanied by range degradation, so that

Table 8.1 *Estimates of game numbers and biomass by species for all private land in Namibia, 1972 and 1992**

Year Species	1972		1992	
	No. head	No. LSU**	No. Head	No. LSU**
Black wildebeest	0	0	7177	2009
Black-faced impala	0	0	2144	300
Blue wildebeest	326	130	4935	1974
Dik-dik	13,011	520	15,783	631
Duiker	84,419	6753	75,518	6041
Eland	10,338	10,338	29,150	29,150
Gemsbok	55,406	22,163	164,306	65,722
Giraffe	3760	5039	4552	6099
Hartebeest	16,302	4076	50,804	12,701
Impala	1006	141	4919	689
Klipspringer	29,509	1770	22,879	1373
Kudu	148,211	59,285	203,087	81,235
Mountain zebra	22,531	13,519	34,398	20,639
Nyala	0	0	96	19
Plains zebra	1214	765	4170	2627
Reedbuck	0	0	2303	322
Roan	0	0	633	380
Sable	0	0	6804	2722
Springbok	221,955	22,195	286,113	28,611
Steenbok	18,741	1124	138,941	8336
Tsessebe	0	0	1564	422
Warthog	67,207	12,097	121,250	21,825
Miscellaneous***	5293	1164	12,514	2753
Total	699,227	161,080	1,194,042	296,583
Number of species	16		23	

Note: * Conservative estimates, assuming that densities of game on land of non-respondents were half those of respondents. ** LSU equivalents calculated using method of Meissner (1982). *** Other species, dominated primarily by ostrich and blesbok

Source: Barnes and de Jager (1996)

on balance meat production per hectare was maintained. It had probably also created space for the expansion of wildlife stocks. Berry (1990) suggests that the move to game ranching began in areas that were marginal for livestock, such as on farms in the pro-Namib desert (with rainfall less than 200mm per annum) and in some of the bush-encroached farms of the higher rainfall areas. By 2001, about 15–25 per cent of total freehold farmland was used primarily for commercial game production, including game ranching, safari hunting, live game capture and non-consumptive wildlife viewing (Krug, 2001). By 2003 there were some 510 registered commercial hunting farms (up from some 200 in 1994) varying in size from 3000ha to 10,000 hectares (Erb, 2003). Table 8.2

shows some recent trends relating to livestock and wildlife on private land, confirming that the decline of livestock and the expansion of wildlife described above continues.

Table 8.2 *Some trends relating to livestock and wildlife on private land in Namibia*

	Changes over time			
Livestock	**1971**	**1981**	**1991**	**2001**
Cattle ('000)	1800	1400	1300	910
Percentage change (1971–2001)				–49%
Small stock ('000)	4550	4350	3500	2700
Decline (1971–2001)				–41%
Wildlife				
Huntable game populations	**1972**	**1982**	**1992**	**1997**
Huntable game ('000)*	565	701	1489	1161
Percentage change (1972–1997)				105%
Game auctions**		**2000**	**2001**	**2002**
Number of game sold		400	1954	3122
Percentage change (2000–2002)				681%
Trophy hunting clients***		**1994**	**1999**	**2004**
Number of hunters		1918	4030	5363
Percentage change (1994–2004)				180%

Note: * Numbers for 1972 to 1992 based on questionnaire survey, that for 1997 based on aerial survey. ** All of game auctioned sold to private landholders in Namibia and South Africa. *** More than 85 per cent of hunting clients hosted on private land

Source: Lange et al (1998); Erb (pers. comm.); unpublished Ministry of Agriculture, Water and Forestry statistics (2006)

Successful game and livestock ranching in arid unpredictable environments requires large areas of land on which animals can take opportunistic advantage of pasture growth and water supply. Mobility and flexibility are the key to survival of wildlife. In the late 1980s, a number of individual freehold farmers realized that their individual farm units were not large enough for successful game farming in Namibia's arid environment and began to discuss the possibility of pooling the management of their land, human, financial and wildlife resources to improve wildlife production (de Jager, 1996). After long and arduous negotiations, facilitated by conservation personnel, a group of farmers agreed in 1992 to adopt a model of resource management based on the 'conservancy' approach developed in KwaZulu-Natal in South Africa (de Jager, 1996). These farmers agreed on a constitution, a set of operating rules and criteria for the distribution of income derived from joint wildlife management. Although the government had not then legislated to give formal recognition to freehold area conservancies, the then Directorate of Nature Conservation and

Recreational Resorts developed a policy acknowledging such conservancies, giving conservancies some preference in granting permits (MWCT, 1992).

Supported by the Ministry of Environment and Tourism, the conservancy approach has increased in popularity among freehold farmers.[3] In 2006, at least 22 conservancies existed on freehold land in Namibia, double the number in 1998, covering an area of close to 4 million hectares (Kolberg, pers. comm.) The size of the conservancies ranges from 65,000ha to 387,000ha, and even the smallest freehold conservancy is larger than some state-protected areas.

Financial analysis in Namibia generally confirmed the relatively low financial profitability of ranching on private land in southern Africa (Brand, 1984; Behr and Groenewald, 1990; Conybeare and Rozemeijer, 1991; Jansen et al, 1992; Bond, 1993; Barnes and Kalikawe, 1994; Drew and Schwarting, 1994; Barnes and de Jager, 1996). This low profitability applies to both livestock and wildlife, and is due to the disproportionately large capital investments required for these land uses. This would indicate that profit alone is not the only motivation for Namibian freehold farmers turning to wildlife as a major form of land use. Indeed, Ashley and Barnes conclude that 'part of the value of game to farmers lies in the diversification of risk, and aesthetic (non-use) benefits (which are not captured in the economic analysis). Diversification is particularly important when farming in such a variable environment with relatively low profits' (1996, p7).

Reductions in subsidies during the 1990s substantially reduced the financial profitability of livestock production to levels generally below those for wildlife systems. However, the comparative financial advantage of wildlife over livestock is not always clear, since livestock farms commonly have sunk capital costs, while conversion to wildlife typically involves significant new capital costs.

The results of a comparative economic analysis by Barnes and de Jager (1996) suggest that all the wildlife use activities were economically efficient and thus deserving of consideration of support in policy. The economic analysis measured the contribution of the activities to the national income – i.e. in terms of the returns to the nation – where the financial analysis measured returns in terms of profit to the private investor. (The economic value provides an indication of the true value of an activity, and taxes and subsidies may be used to create (dis)incentives for private landholders to invest in an activity – policies such as taxation and subsidization can make financial profitability lower or higher respectively, relative to economic viability for the same activities.)

In the economic analysis, the effects of subsidies, taxes and the main price distortions, and the worth of investments was examined in terms of the opportunity costs to Namibian society as a whole. The economic analysis showed that where landholders cooperated to form larger management units or conservancies, all land uses become more efficient both financially and economically (Barnes and de Jager, 1996). At this scale, the study showed that there was a weak financial incentive for investors and a strong economic incentive for the nation for the conversion away from mixed livestock and game production toward conservancies producing wildlife only. The study found that livestock production was generally economically less efficient than wildlife production systems. As mentioned above, however, the comparative advantage was not always evident in practical terms due to the considerable capital costs already sunk in the livestock sector (Barnes and de Jager, 1996). These findings provided average figures and are likely to hold for much of Namibia's private land.

More recent detailed empirical work has highlighted that, when tourism potential on private land is high, investment in wildlife-based tourism can result in much higher financial and economic returns than those possible for livestock. Barnes and Humavindu (2003) examined the returns to (non-consumptive wildlife) tourism, commercial livestock production and small-scale traditional livestock production on 80,000ha of private land in the arid shrublands of the south, near to the Ai-Ais/Fish River Canyon protected area. Similar conditions pertain over perhaps a fifth of private farmland, where scenery is especially attractive, where rich and diverse wildlife stocks have been established, where livestock potential is marginal, or where the site can capture the traffic of tourists using nearby state-run protected areas. Table 8.3 shows some comparisons between the two livestock production systems and nature-based tourism. Tourism as a land use was determined to be some ten times more valuable than the alternatives, in terms of both financial and economic measures. The annual financial net profit per hectare, inflated to 2006 prices was N$27 (US$2.97) for tourism, compared with N$0.60 and N$1.10 (US$0.07 and US$0.12) for the commercial and small-scale livestock systems respectively. The direct annual contribution of these activities to the gross national income per hectare, also inflated to 2006 prices, was N$66 (US$7.26) for tourism, N$4.30 (US$0.47) for commercial livestock and N$1.40 (US$0.15) for small-scale livestock (Barnes and Humavindu, 2003).

Clearly there are enormous comparative advantages to wildlife use through tourism in areas such as this – close to main tourist routes, where scenery is of high quality, and where livestock production is marginal. Although localized, the physical potential for expansion in such settings is considerable, and the past and predicted rapid growth in tourism demand (WTTC, 2006), as well as the generally low price elasticity of demand for tourism (Barnes, 1996; Barnes et al, 2004) indicate that the economic expansion potential is also considerable.

Returning to the findings of Barnes and de Jager (1996), the results in Table 8.4 suggest that the aggregate economic value of wildlife use on private land rose by some 80 per cent in real terms between 1972 and 1992. From this study it emerged that the proportion of the total contribution of private

Table 8.3 *Comparative financial and economic values for three land uses on private land at a site in Karas Region, Namibia*

Item	(a) Livestock (traditional)	(b) Livestock (commercial)	(c) Tourism
Financial values*			
Initial capital investment per hectare (N$)	33	56	271
Financial turnover per hectare (N$/annum)	4.71	15	165
Net profit per hectare (N$/annum)	0.93	0.48	23
Financial rate of return over 10 years	5.5%	9.8%	12.9%
Economic values**			
Economic gross output per hectare (N$/annum)	5.45	16	160
Gross national income per hectare (N$/annum)	1.19	3.74	57
Net national income per hectare (N$/annum)	0.10	2.30	46
Economic rate of return over 10 years	11.1%	23.1%	29.4%
Direct employment created (full-time jobs)	24	29	153

Note: * Measured from the perspective of the landholder. ** Measured from the perspective of society (contribution to national income); Prices in 2001 prices, where N$1=ZAR 1= US$ 0.11

Source: Barnes and Humavindu (2003)

rangeland use to national income due to wildlife rose from 5 per cent to 11 per cent over that period. The results of work by Erb (2003) and Barnes and Humavindu (2003) since then, suggest that declines in livestock numbers and increases in wildlife numbers and use have continued unabated.

The direct annual economic contribution of wildlife on private land to the national income in 1992 was estimated to be N$30–56 million (US$8.1–15.1 million) or N$89–167 million in 2006 prices (US$24–45 million). Thus the government's capital and recurrent expenditures in managing this – unlikely to have exceeded N$8 million (US$1.2 million) per annum or N$24 million (US$3.6 million) per annum in 2006 prices – seem to have been a sound investment (Barnes and de Jager, 1996). Policy in the Ministry of Environment and Tourism has supported the use of wildlife and the development of conservancies on private land, and this appears to be economically sound. Other ministries such as those responsible for agriculture and lands have been less ready to recognize the land use value of tourism alongside traditional agricultural land use values, but this seems to be changing slowly.

Conclusions

Changes in government policy and legislation that encouraged sustainable use have resulted in improved conservation of wildlife on Namibia's freehold land. Providing proprietorship, even in a limited form, to freehold farmers has

Table 8.4 *Estimated contribution to the economy of wildlife use on Namibia's private land in 1972 and 1992*

	1972	1992
Northern, predominantly cattle-producing land*		
Total number of properties	2757	2757
Total extent (km²)	192,237	192,237
No. of wildlife LSU equivalents	129,980	242,318
Net national income generated per LSU (N$/annum)	170	170
Total net national income due to wildlife use (N$/annum)	22,096,600	41,194,060
Wildlife net national income per km² (N$/annum)	115	214
Southern, predominantly sheep-producing land**		
Total number of properties	1703	1703
Total extent (km²)	164,650	164,650
No. of wildlife LSU equivalents	31,100	54,265
Net national income generated per LSU (N$/annum)	275	275
Total net national income due to wildlife use (N$/annum)	8,552,500	14,922,875
Wildlife net national income per km² (N$/annum)	52	91
All private land		
Total number of properties	4460	4460
Total extent (km²)	356,886	356,886
No. of wildlife LSU equivalents	161,080	296,583
Net national income generated per LSU (N$/annum)	190	190
Total net national income due to wildlife use (N$/annum)	30,649,100	56,116,935
Wildlife net national income per km² (N$/annum)	85	157

Note: * Land in following districts: Tsumeb, Grootfontein, Outjo, Otjiwarongo, Omaruru, Karibib, Okahandja, Windhoek and Gobabis. ** Land in following districts: Mariental, Maltahöhe, Lüderitz, Bethanie, Keetmanshoop and Karasburg. In constant 1994 prices, where N$1 = ZAR1 = US$0.27

Source: Barnes and de Jager (1996)

opened up the possibility of the commercial use of wildlife, turning it into an asset rather than a liability. The results of research investigating game ranching in Namibia suggest that there are strong financial and economic incentives in selected parts of Namibia's private land for conversion from predominantly livestock to predominantly wildlife production systems. In other parts of the country this incentive is weaker – generally where the scenery is poor, where wildlife stocks are insignificant and where easy access to nearby state-protected areas is not possible. The incentive is also weaker in higher rainfall areas, where livestock production potential is best.

In spite of this, throughout most private land, economic incentives for conversion to wildlife do exist and financial incentives are also increasing. The 1990s saw the virtual elimination of government subsidies favouring livestock production in the commercial farming sector, and the comparative advantage of wildlife – which never benefited from subsidies – has become stronger. The comparative advantage that wildlife-based land use has over livestock production varies spatially, and is extremely high in certain areas where scenery and wildlife resources are of high quality and tourism routes are established. Very clear comparative advantages of wildlife are likely to prevail over about one third of private land.

There are economies of scale for wildlife land uses in Namibia and large properties have clear advantages financially and economically. However, the formation of conservancies involves shared (instead of individual) decision making regarding wildlife and resource management. This can result in high transaction costs (for example, time spent achieving consensus among participants). Transaction costs have not been included in the analyses above and can reduce the general worth of conservancies relative to that of individual ranches. The most successful large properties appear to be those with single, rather than numerous, owners/investors.

Safari hunting on private land is interesting because it appears to provide high returns to capital but low returns to land (due to the low sustainable off-takes possible for trophy animals). It is thus highly profitable, but only as a supplementary enterprise alongside livestock or other wildlife land uses, confirming a similar observation made in South Africa (Scriven and Eloff, 2003). The high profitability of safari hunting appears to have provided the financial incentive for much of the investment in wildlife stocks, which in turn appears to have led to conditions where ranching with only wildlife (and no livestock) and large returns per hectare from non-consumptive, lodge-based tourism are possible (Humavindu and Barnes, 2003).

Since Independence, Namibia has been undergoing a land reform process where private land is being redistributed to previously disadvantaged Namibians. It involves an affirmative action loan scheme for those with some wealth, and a resettlement scheme for poorer citizens (Schuh et al, 2006). So far this process has been orderly and relatively small areas have been involved. However, the planning and execution of these schemes has been suboptimal and it is vitally important that it does not jeopardize the economically efficient land use conversion processes discussed in this chapter. Nature-based tourism also needs to be incorporated into the land reform process.

Notes

1   An LSU is the biomass equivalent of on average one head of cattle and approximately 6.25 goats or sheep.
2   Interestingly, buffalo and bushpig do not occur in Namibia's freehold land, and it is not known why they were included.
3   The conservancy approach, based on collective management of wildlife, also served as a model for the development of community-based wildlife management on Namibia's communal land (see Chapter 14 in this volume).

References

Ashley, C. and Barnes, J. (1996) 'Wildlife use for economic gain: The potential for wildlife to contribute to development in Namibia', *Research Discussion Paper No. 12*, Directorate of Environmental Affairs, Windhoek

Barnard, P. (ed) (1998) *Biological Diversity in Namibia: A Country Study*, Namibian National Biodiversity Task Force, Windhoek

Barnes, J. I. (1996) 'Economic characteristics of the demand for wildlife viewing tourism in Botswana', *Development Southern Africa*, vol 13, no 3, pp377–397

Barnes, J. I. and de Jager, J. L. V. (1996) 'Economic and financial incentives for wildlife use on private land in Namibia and the implications for policy', *South African Journal of Wildlife Research*, vol 26, no 2, pp37–46

Barnes, J. I. and Humavindu, M. N. (2003) 'Economic returns to land use options in Gondwana Cañon Park, Karas, Namibia', unpublished report prepared for Nature Investments (Pty) Ltd, Namibia

Barnes, J. I. and Kalikawe, M. C. (1994) 'Game ranching in Botswana: Constraints and prospects', in Proceedings, 3rd International Wildlife Ranching Symposium: Wildlife ranching: A celebration of diversity, Pretoria, South Africa, 27–30 October, 1992

Barnes, J. I., Zeybrandt, F., Kirchner, C. H., Sakko, A. L. and MacGregor, J. (2004) 'Economic valuation of the recreational shore fishery: A comparison of techniques', in Sumaila, U. R., Steinshamn, S. I., Skogen, M. D. and Boyer, D. (eds) *Ecological, Economic and Social Aspects of Namibian Fisheries*, Eburon Academic Publishers, Delft

Behr, J. and Groenewald, J. A. (1990) 'Profit potentials in game farming', *Agrekon*, vol 29, no 1, pp59–65

Berry, H. (1990) 'Large scale commercial wildlife utilization: Hunting, tourism, animal production – Namibia', unpublished report for the Directorate of Nature Conservation, Namibia

Bond, I. (1993) 'The economics of wildlife and land use in Zimbabwe: An examination of current knowledge and issues', *Project Paper No. 36*, WWF Multispecies Animal Production Systems Project, Harare

Brand, P. A. J. (1984) 'Inskakeling van wild by die boerdery', *Agricola*, vol 1, pp31–6

Child, B. (1988) 'The role of wildlife utilization in the sustainable economic development of semi-arid rangelands in Zimbabwe', DPhil thesis, University of Oxford, Oxford

Conybeare, A. and Rozemeijer, N. (1991) 'Game ranching in Botswana: An assessment of the game ranching potential of eight Controlled Hunting Areas', paper prepared for the USAID Natural Resources Management Programme/Department of Wildlife and National Parks, Gaborone

Cumming, D. H. M. (1990) 'Developments in game ranching and wildlife utilization in East and southern Africa', *Project Paper No. 13*, WWF Multispecies Animal Production Systems Project, Harare

de Jager, J. L. V. (1996) 'The establishment and operation of conservancies in Namibia', in Tarr, P. (ed) *Namibia Environment, Vol. 1*, Ministry of Environment and Tourism, Windhoek

Drew, C. and Schwarting, G. (1994) 'Export opportunities for Namibian venison', paper prepared for the World Wildlife Fund Living in a Finite Environment Programme, Namibia

du Toit, R. F. (1994) 'Large-scale wildlife conservancies in Zimbabwe: Opportunities for commercial conservation of endangered species', in Proceedings 3rd

International Wildlife Ranching Symposium: Wildlife ranching: A celebration of diversity, Pretoria, South Africa, 27–30 October, 1992

Erb, K. P. (2003) 'Consumptive wildlife utilization as a land use form in Namibia', MBA thesis, University of Stellenbosch, Stellenbosch

Humavindu, M. N. and Barnes, J. I. (2003) 'Trophy hunting in the Namibian economy: An assessment', *South African Journal of Wildlife Research*, vol 33, no 2, pp65–70

Jansen, D. J., Bond, I. and Child, B. (1992) 'Cattle, wildlife, both or neither: Results of a financial and economic survey of commercial ranches in southern Zimbabwe', *Project Paper No. 27*, WWF Multispecies Animal Production Systems Project, Harare

Joubert, E. (1974) 'The development of wildlife utilization in South West Africa', *Journal of the Southern African Wildlife Management Association*, vol 4, no 1, pp35–42

Joubert, E., Brand, P. A. J. and Visagie, G. P. (1983) 'An appraisal of the utilization of game on private land in South West Africa', *Madoqua*, vol 3, no 3, pp197–219

Krug, W. (2001) 'Private supply of protected land in southern Africa: A review of markets, approaches, barriers and issues', paper presented at the World Bank/OECD International Workshop on Market Creation for Biodiversity Products and Services, Paris, 25–26 January 2001

Lange, G.-M., Barnes, J. I. and Motinga, D. J. (1998) 'Cattle numbers, biomass, productivity, and land degradation in the commercial farming sector of Namibia, 1915 to 1994', *Development Southern Africa*, vol 15, no 4, pp555–572

Luxmoore, R. (1985) 'Game farming in South Africa as a force for conservation', *Oryx*, vol 19, no 4, pp225–234

Meissner, H. H. (1982) 'Theory and application of a method to calculate forage intake of wild southern African ungulates for purposes of estimating carrying capacity', *South African Journal of Wildlife Research*, vol 12, no 2, pp41–47

MWCT (1992) *Policy Document: The Establishment of Conservancies in Namibia*, Ministry of Wildlife, Conservation and Tourism, Windhoek

Schuh, C., Conroy, C., Grimm, J., Humavindu, M., Kwala, C., Werner, W. and Ströbel, H. (2006) 'Economics of land use: Financial and economic analysis of land-based development schemes in Namibia', report for the GTZ, Windhoek

Scriven, L. and Eloff, T. (2003) 'Markets derived from nature tourism in South Africa and KwaZulu-Natal: A survey of the sale of live game', in Aylward, B. and Lutz, E. (eds) *Nature Tourism, Conservation and Development in KwaZulu-Natal, South Africa*, World Bank, Washington DC

van der Walt, P. T. (1987) 'The status of wildlife conservation and utilization in Namibia', unpublished report for the Directorate of Nature Conservation, Windhoek

WTTC (2006) *Namibia: The Impact of Travel and Tourism on Jobs and the Economy*, Namibia Tourism Board/Ministry of Environment and Tourism, Windhoek

9

Game Ranching in Zimbabwe

Brian Child

This chapter describes the 'flipping' of rangelands in Zimbabwe from a live-stock to a wildlife-dominated system in the mid-1980s, and the economic and institutional factors that made this possible. It addresses the problem posed by resilience theory that social ecological systems get locked into undesirable basins of attraction (for example vicious cycles of rangeland degradation and poverty associated with livestock production) and the question of how society can develop the 'transformability' to avoid such lock-ins (Walker et al, 2004). We observe the importance of leadership and institutional reform in the transformation of Zimbabwe's rangelands towards wildlife-based economic systems, suggesting that transformability falls in the realm of institutions and adaptive governance. Indeed, it is argued that economic growth is related more to institutional than financial capital (Anon, 2008) because organizational energy (for example institutions, property rights and entrepreneurship) is necessary to turn relatively low-ordered raw materials, such as livestock and rangeland environments, into more highly ordered products and services (Beinhocker, 2006). The chapter uses historical data from Zimbabwe's Midlands and Lowveld ranching areas from the late 1980s (when this transformation was occurring) to illustrate how devolving rights to wildlife (proprietorship) to landholders, coupled with improving terms of trade (price), affected land use in Zimbabwe and indeed wildlife on private land throughout southern Africa.

The livestock-dominated rangelands of southern Africa are an example of a social-ecological system locked in a downward spiral of weakening economics and degrading environments. This chapter describes how careful institutional design provided more order to rangeland systems, which in many areas 'flipped' into a wildlife-based economy that was more profitable and more ecologically friendly. These institutional experiments used the 'invisible hand' to allocate resources to high-valued uses, but also strengthened local collective action to manage the externalities associated with the fugitive nature of wildlife, deliberately avoiding bureaucratic regulation. The remarkable recovery of wildlife on private land in southern Africa illustrates the subtle power of what we might call 'institutional ecology', ideas that go largely unremarked but underlie the emerging paradigm of sustainable use in southern Africa (for example SASUSG, 1996).

The policy response to the excesses of the white man's frontier economy in the late 1800s was heavy regulation and centralized state control (or nationalization) of wildlife. However by the 1960s, untrammelled overuse was no longer the primary threat to wildlife. Replacement by other land uses for economic reasons was. This needed a completely new policy response that, in responding to the new challenges facing wildlife conservation, radically altered its political economy so that it was 'owned' by, managed and benefited thousands of landholders, instead of being controlled by government agencies.

There was little precedent for this experimental approach, which stretched conservation biology into the fields of politics and economics. Consequently, wildlife policy used an adaptive management approach to institutional reform, and it was only later that gut instinct behind the reforms was formulated into more scholarly language, for example the price–proprietorship–subsidiarity hypothesis: wildlife would be conserved if it was valuable (price), if landholders were legislatively enabled to capture this value through property rights (proprietorship), and if authority and responsibility was devolved to hundreds of entrepreneurial landholders (see Handy, 1994).

Overview of land use on private ranchland

Only four decades ago, white landholders proudly pushed their cattle through dip tanks and admired their muscling, their hybrid breeding or their calving rates. It was hard to find a property that was not heavily grazed, with grass damage and bush encroachment reducing productivity. Most ranchers thought wildlife was a pest, or at best, a nuisance, and the few pioneer game farmers were often viewed with humour and sometimes as a threat. Paradoxically, it turns out that wildlife rescued many livestock ranchers from ecological degradation and bankruptcy. Today, private land in southern Africa is teeming with wildlife. Tourism and hunting activities in their ever-expanding variety are widely accepted enterprises, creating jobs and growing national economies through biodiversity-friendly uses. Carefully crafted institutional reform was the key to this transformation, especially the devolution of proprietorship of wildlife to landholders, which unlocked the innovative capacity of thousands of landholders at a time that coincided with the rising global value of outdoor recreation and tourism (Booth, 2002).

The subsidized rise of the beef industry

African rangelands evolved under indigenous multispecies systems typically carrying 15 to 25 ungulate species (Cumming, 1995). Domestic livestock entered southern Africa less than 2000 years ago, replacing wildlife only in the last 150 years, yet contributing 90 per cent of southern Africa's biomass (Cumming and Bond, 1991). Livestock had almost squeezed wildlife off freehold land in Zimbabwe by 1980, largely through simple neglect and competition for food and water, notwithstanding dramatic stories of ranchers shooting wildebeest and zebra to protect their grazing, and leaving hundreds of carcasses to rot in the veld. Starting in 1919, the Rhodesian government also made serious efforts to eliminate wildlife on the assumption that it was an important vector for trypanosomiasis and food for unwanted tsetse fly; hundreds of thousands of wild animals were shot, the veld was burned, natural waterholes were destroyed and magnificent shade trees were bulldozed to remove the resting places of tsetse flies (Child, 1995). Later military-type operations opened roads into wilderness areas, which were also sprayed with DDT, all in the name of the beef industry.

In a story with global parallels (see Rifkin, 1992), government support to white settlers was the overriding factor in the expansion of the Zimbabwe's beef herd. This story is illustrated by different historical trajectories in the numbers of cattle owned by price-responsive commercial ranchers and 'native-owned' cattle (see Figure 9.1). The latter have many uses that eclipse beef sales, including storing wealth and providing draught power. Economic signals are highly distorted because privately owned animals get free (open-access) grazing from communal rangelands.

In 1896, a rinderpest pandemic from East Africa killed 90 per cent of cattle and most wildlife in Zimbabwe (Bond and Cumming, 2004). On abundant grazing, both herds recovered at their maximum biological rates until the mid-1920s when low beef prices caused the commercial herd to stagnate, despite cheap land, labour and environmental subsidies. Single channel marketing was introduced in 1937 through a parastatal marketing board – the Cold Storage Commission – that played an important political and economic role in the industry for the next 50 years.

Compulsory destocking of the communal herd in the tribal trust lands (initiated by dedicated soil conservationists for environmental reasons following the experience of the American Dust Bowl) slowed down the growth of the communal herd between 1940 and 1960. At the same time, pricing policies effectively transferred capital from the communal to the commercial herd (Southern Rhodesia, 1952; Turner Report, 1956). The forced destocking caused enormous resentment against the settler government and fed the discontent that led to the liberation struggle as well as a long-lasting suspicion of government conservationists.

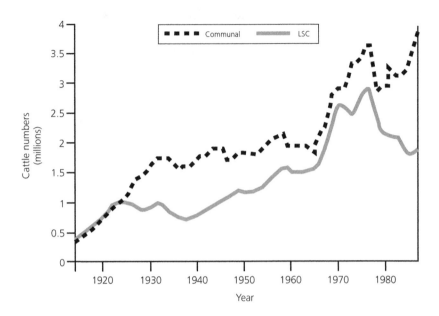

Source: Adapted from Child (1988)

Note: LSC = large-scale commercial

Figure 9.1 *Growth of communal and commercial cattle herds in Zimbabwe, 1914–1985*

The effects of cross-subsidization are readily apparent in Figure 9.1. In the late 1930s, compulsory destocking of the communal herd led to rapid growth in the commercial sector. When compulsory destocking was terminated by parliamentary commissions in the 1960s, the growth of communal livestock resumed immediately. However, without cross-subsidies, the commercial herd stagnated until 1965 when new subsidies and diversification incentives associated with Ian Smith's Unilateral Declaration of Independence were initiated. After 1965, technological innovation in the form of nitrogen feeds allowed a fourfold increase in cattle in arable agro-ecological zones, compared to an increase of 40 per cent in traditional ranching areas (Child, 1988). The war of Independence (1975–1980) affected both sectors severely, after which the communal herd recovered very rapidly (helped by the migration of people into the lightly settled Sebungwe Region and Zambezi Valley). By contrast, the commercial herd stagnated despite heavy price subsidization (government set prices on a production-cost-plus basis), and prices remained 40 per cent above world levels. These fell from US$4.50 in the early 1970 to US$3.00 by the mid-1980s (Rodriguez, 1985) in response to feedlot technology.

Table 9.1 *The (non-)viability of Zimbabwe's beef sector, 1990*

	ZW$ million
Income (additional income from preferential access to EU markets)	149 (42)
Costs	186
Variable costs (i.e. production costs of ranchers and farmers)	74
Fixed costs	??
Marketing board (Cold Storage Commission) subsidy	50
Cost of government services to the livestock sector	62
Capital investment in land and livestock	??
Environmental costs	??
Beef sector financial profit (with EU market)	−36 (6)

Source: Child (1988, p161)

A simple analysis showed that the industry cost more to manage (ZW$186m) than it earned (ZW$149m) even ignoring the costs of land, land degradation and asset depreciation, including abattoirs, fencing and water-points (see Table 9.1).[1] These losses only made government more determined to protect the sector, investing in new abattoirs and courting the European Union to add 8100 tonnes to the subsidized European beef mountain. The single-sector mentality and rigidly technical attitude to disease and ecology damaged the nascent wildlife sector. Buffalo were eradicated (on the scientifically tentative grounds that buffalo were a reservoir of foot-and-mouth disease in cattle) regardless of their emerging value. Marketing of game meat and the movement of live wildlife was severely regulated at exactly the time ranchers were cropping impala on over-stocked properties, and restocking species that had disappeared.

The decline of wildlife

On southern Africa's European frontier, some 20 million wild animals were eliminated between 1780 and 1880, including local and complete extinctions (i.e. bloubok and quagga) (www.kruger2canyons.com). In Part II we saw that the response to this was to centralize control of wildlife in the state, to reject commercial utilization (but not sport hunting), and to establish state-protected areas. Colonial officials became responsible for managing wildlife even on farm lands. The net result was an economic framework harmful to wildlife: killing animals for sport (and for low-value uses like food) was acceptable, but using wildlife profitably was not.

Hunting was controlled though permits issued on behalf of the British monarch (Parker, 2004), and species popular with trophy hunters were deliberately conserved (IUCN, 2004). Hunting for sport was valued by the ruling classes and maintained, but the law generally disenfranchised landholders, both white and black, from using wildlife sensibly.

In Zimbabwe (and elsewhere) the view that you cannot 'farm in a zoo' became the norm. In the 1950s, some ranchers killed wildebeest and zebra to reduce grazing competition; a cheap landholder's licence permitted them to shoot as many animals as they liked (apart from Royal Game) to 'protect their livelihoods'. The meat could be used to feed labour but the sale of all products was illegal. As a result, many zebra and wildebeest were left to rot in the veld.

However, in the 1950s, ecologists were also discovering that savannas were highly variable, stochastic, patchy systems. They suggested that wildlife had a production advantage because of its multiple feeding strategies and superior behavioural and physiological adaptations to harsh savanna environments (Walker, 1979). Making wildlife more productive than livestock, they hoped, was the solution to its disappearance. Three Fulbright Scholars (Thane Riney, Ray Dasmann and Archie Mossman) from Berkeley were invited to Rhodesia by Reay Smithers to experiment with game cropping in the late 1950s, while similar experiments occurred in East Africa (Talbot et al, 1965). Scientific work on diet separation, herbivore community structure, adaptations to dry environments and disease, growth rates and carcass quality, were inconclusive about wildlife's advantages that were, in any case, overshadowed by technical constraints to harvesting and restricted markets (Child, 1988). Decades ahead of its time, this cohort of wildlife managers began to think about wildlife in terms of land use incentives (Buechner, 1964; Riney, 1967; Riney and Hill, 1967).

Legislation, rights and collective regulation

In Zimbabwe, the 1960 Wild Life Conservation Act experimented with commercial use of wildlife under a permit system. Initially, permits followed a survey by the wildlife department, but by 1973 some 160 landholders had permits to crop and sell animals; the department's capacity to conduct surveys was exceeded long before this and, as it turned out, ranchers who used wildlife conserved it, much against the common understanding of the time.

This experience emboldened policymakers to propose entrusting land-holders with wildlife proprietorship. So controversial was this idea that the emerging act went through five years of drafting, protracted meetings with members of parliament and public hearings. The seemingly unending discussions were truncated when a landholder took the government to court over the ownership of wildlife in 1973. The court upheld his argument that the state's claim to ownership of wildlife made it liable to pay him for grazing and other damages, and rendered the previous legislation unworkable.

Soon afterwards, the landmark Parks and Wild Life Act of 1975 radically altered the political economy of wildlife. It replaced state ownership with *res nullius*, or ownership by no-one, but emphatically devolved rights to use wildlife to landholders. The new act 'recognized that most species are best protected by landholders and landholder communities' but without government abrogating its ultimate responsibilities for the resource (Child, 1995). Technically, ownership of wildlife was claimed by controlling it (for example shooting it), but if it moved to the neighbour's property he, similarly, could utilize it.

This created a common property challenge that Zimbabwe ingeniously resolved with a system of devolved community regulation. Since the early 1900s, Zimbabweans had been obsessed with soil conservation (Beinart, 1984), the awareness of which was heightened by the American dustbowl. Avoiding bureaucratic solutions to soil erosion and seeing conservation holistically as wise land use (rather than single species 'preservation'), the Natural Resources Act of 1941 empowered landholder communities with collective responsibility for regulating and restoring the environment through the voluntary formation of catchment-based intensive conservation areas (ICAs), to which government officials provided technical support. So effective was the combination of peer pressure and technical support that ICAs protected 91.5 per cent of arable land in European areas by 1949/1950 (Showers, 1994). Indeed, this combination of private ownership and local collective regulation proved so effective that legislatively backed local collective action became the backbone of Zimbabwe's conservation administration on private land, including soil erosion, grazing, deforestation, fire and, later, wildlife. Speaking economically, ICAs were a collective-action solution that reduced the transaction costs (and increased the social legitimacy) of measures to manage the spatial and temporal externalities associated with natural resource uses. Consequently, the 1975 wildlife legislation was grafted onto them.

Thus, the power to protect species, set quotas and hunting seasons, and place prohibitions on the activities of individual landholders, was legislatively devolved to landholder communities through their ICAs. Unlike national regulatory solutions, locally designed controls had greater social legitimacy and situation specificity. Local rules were usually enforced through informal peer pressure, but both parties had recourse to ministerial or court decisions within two weeks. Controls were placed on only a few species, in specific circumstances and in innovative ways, and varied considerably according to the specific character of each area's problems, much in line with principles of jurisdictional parsimony and delegated aggregation developed by Murphree (2000). For example, the Chiredzi ICA introduced quotas for high-value sable after the community spent a lot of money capturing and feeding these valuable animals in the 1984 drought; the Munyati–Sebakwe ICA only set hunting quotas for high-value mobile species such as sable and waterbuck; and the Bubi ICA used capture quotas to control a landholder who was enticing animals onto his land before catching them for live sale (pers. obs.). Central government retained the power of unilateral action if an ICA neglected its responsibilities, but only used this power once a decade; the mere hint that action was contemplated usually galvanized a committee to resolve the issue itself. This system was economically efficient because it used incentives such as market prices (rather than regulations) to allocate resources to best use while externalities were controlled locally by the community.[2]

However, the system also required that the historical underpricing of wildlife be dealt with. Thus, landholders were encouraged to use wildlife in as many and as profitable ways as possible (provided this was humane). Further, the wildlife department recognized that subsidizing hunting or tourism on public land undercut the incentives for private conservation. Consequently, it steadily increased the price of wildlife by replacing administrative pricing and public allocation of hunting with competitive market mechanisms (Child, 1995).

The de-nationalization of wildlife placed its future in the entrepreneurial hands of hundreds of innovative landholders. It also de-politicized the sector by shifting decisions away from officials and special interests operating in the political-administrative marketplace into the economic marketplace. This worked well because most properties were family businesses with a strong interest in conservation and a rural lifestyle.

The emergence of game ranching

The Midlands

The effects of the new wildlife policy are illustrated by the example of ranchers in the Zimbabwean Midlands, which lies at the interface between rangelands and crop–agriculture production – i.e. Natural Region IIIb, rainfall of 600–700mm (Vincent and Thomas, 1961). Surveys showed that attitudes of ranchers to wildlife improved rapidly in response to the new wildlife policies, and that making wildlife a business magnified positive perceptions about its aesthetic and ecological values (see Table 9.2).

The commercial utilization of wild animals led to a rapid expansion of their geographic ranges and populations, especially for higher-value animals (for example wildebeest, zebra, waterbuck, tsessebe, eland and sable) that were susceptible to competition from livestock (see Table 9.3). Interestingly, the only species to decline – cheetah – was the only species that received special protection from the state.[3]

Table 9.2 *Changes in landholder attitudes towards, and use of, wildlife in the Zimbabwe Midlands*

	% of interviewed farmers		
	1975	1984	Change
Attitudes toward wildlife			
Beneficial	77	100	23
Neutral	18	0	−18
Problematic	4	0	−4
Uses of wildlife			
Aesthetic	82	100	18
Business	55	74	19
Ecologically important	54	74	20
Personal recreation	48	82	34

Source: Child (1988); unpublished data (Vaughan-Evans)

Table 9.3 *Changes in occurrence and utilization of wildlife on properties in the Zimbabwe Midlands, 1975–1984*

	Trophy price (ZW$)	Occurrence (% of properties)			Properties reporting increasing populations (%)	Utilization (% of properties)		
		1975	1984	Change		1975	1984	Change
Leopard	2000	58	91	33	stable	14	20	6
Cheetah		25	100	75	83%	13	0	−13
Duiker	70	100	100	0	stable	16	36	20
Steenbok	70	100	100	0	stable	22	27	5
Bushpig	100	100	100	0	57%	3	55	52
Warthog	100	100	100	0	68%	32	68	36
Impala	100	100	100	0	90%	48	91	43
Klipspringer	175	36	23	−13	stable	14	60	46
Reedbuck	275	88	100	12	decline	17	55	38
Bushbuck	275	45	55	10	0%	30	42	12
Wildebeest	400	28	45	17	100%	100	100	0
Zebra	550	65	86	21	65%	64	68	4
Waterbuck	550	50	91	41	47%	47	50	3
Tsessebe	550	25	68	43	71%	20	53	33
Kudu	550	100	100	0	50%	52	91	39
Eland	700	50	86	36	69%	23	63	40
Sable	1000	52	73	21	81%	58	75	17

Source: Child (1988); unpublished data (Vaughan Evans)

Allowing commercial use of wildlife encouraged ranchers to actively manage it (replacing previous neglect) by controlling poaching and dogs, providing water and restocking. They soon began to integrate their wildlife businesses vertically; the number of ranchers outfitting safaris increasing from 42 to 74 per cent within four years of the end of the Bush War, and they stopped selling animals casually to outfitters (11 per cent in 1984 compared to 32 per cent in 1975). The main enterprise was trophy hunting, which generates cash with low fixed and overhead costs, and also allows rapid population recovery. On Iwaba Ranch, which switched entirely to wildlife, wildlife biomass increased tenfold from 27 tonnes to 226 tonnes between 1963 and 1984 (farm records provided by Seymour-Smith).

These data coincided with the time that wildlife became more profitable than livestock. The value of wildlife rose from ZW$0.57/ha in 1975 to ZW$2.52/ha in 1984 (in real terms), as both wildlife populations and prices increased. At the time of these surveys (1984), cattle were still the primary land use (comprising 83 per cent of 14.1 million kg livemass[4]). However, net margin analyses showed that for each kilogram of cattle replaced by wildlife, profits increased from 7 cents to 17 cents (see Table 9.4). Although we have no specific data for the Midlands, wildlife numbers quadrupled on private land in the next ten years (Booth, 2002) as the number of species, populations, the number of properties involved, and the number and variety of enterprises increased (interviews with landholders). Zimbabwe's tourism also expanded tenfold in this period before the political and economic turmoil that began in 2000.

Table 9.4 *Cattle and wildlife profitability in the Zimbabwe Midlands, 1984*

Profit	Units	Cattle	Wildlife
Income	ha	$13.57	$3.36
Profit	ha	$4.52	$2.93
Income	kg livemass	20c	20c
Profit	kg livemess	7c	17c

Source: Child (1988)

While scientists were still experimenting with meat production and ecological issues, landholders demonstrated that trophy hunting increased profits from lightly stocked land and was ecologically and economically robust. Landholders who mismanage harvesting and trophy quality are quickly punished in the marketplace. They can generate good profits from relatively few animals while simultaneously expanding their stocks of wildlife because trophy hunting harvests only 2–3 per cent of antelope populations that expand at 10–20 per cent. Moreover, in times of political turmoil the demand for hunting does not dry up, unlike tourism.

Another important observation is that the devolved structure of the sector encouraged innovation, with many landholders experimenting with new products and learning from each other. In the early 1980s, landholders had limited wildlife and few high-value species. A few had enough to sell full bags of animals on the five to ten day hunts demanded by international markets, but most had to combine their properties to do so. Landholders also introduced bow hunting because it uses fewer and lower value animals. Over time, hunting improved as wildlife populations increased and live animals were purchased for restocking. Hunting also provided the crucial stepping stone for the later introduction of tourism by generating enough income to finance the recovery of overstocked land and wildlife, a process that usually took at least ten years (pers. obs.; Booth, 2002).

In southern Africa nearly 10,000 private landholders are experimenting with new wildlife products and values in ways that are impossible in a state-dominated sector. Thus, by 2000, Zimbabwe offered some 80–100 different wildlife experiences ranging from bow hunting to elephant-back safaris and walking with lions. Countries where the wildlife sector is centrally controlled and heavily regulated, such as Kenya, offer a far narrower range of products and therefore add less value to wildlife.

The south-eastern Lowveld

In the dry south-eastern Zimbabwe in 1984, a casual observer would have been convinced that landholders were indeed cattle ranchers. Cattle were everywhere and the range was obviously heavily grazed. However, a more careful assessment showed that wildlife earned 64 per cent of the profits from only 32 per cent of the biomass.

Table 9.5 *A comparison of cattle and wildlife profitability in the south-eastern Lowveld, 1985*

	Livemass (kg)		Turnover (ZW$)		Profit (ZW$)	
Cattle	10.1m	68%	$2.0m	60%	$0.7m	36%
Wildlife	4.7m	32%	$1.4m	40%	$1.2m	64%

Source: Child (1988)

Even though most cattle ranchers used wildlife, they were inefficient (net profit ZW$1.16/ha) because they relied on selling meat from common species such as impala (a third of their wildlife income) and selling a few trophies to outfitters. The emerging game ranchers, by contrast, diversified their wildlife populations and destocked cattle. They earned ZW$4.47/ha (net profit) mainly by outfitting trophy hunting themselves (Child, 1988). By the mid-1980s, ranchers reported properties could not carry as many cattle as when they were settled in the 1950s, and were at best breaking even (interviews). They recognized the economic benefits of switching scarce grazing resources from cattle to livestock (see Table 9.6), but this transition was complicated by its cash flow implications. However, by the time the 1992 drought decimated cattle and wildlife (albeit the latter to a lesser extent), many landholders recognized wildlife's financial advantages (see Table 9.7), many expanded their wildlife enterprises, and some had switched entirely to wildlife, such as the 342,123ha Savé Valley Conservancy (see Chapter 11).

Table 9.6 *Comparison of efficiency of cattle and wildlife enterprises in the Lowveld, 1995*

Source of income	Cattle-wildlife Ranches 239,559ha		Cattle ranches 131,484ha	
	$/ha	$/kg	$/ha	$/kg
Wildlife	$4.47	32c	$1.16	11c
Cattle	$1.23	7c	$2.93	7c
Total	$5.71	18c	$4.09	8c

Source: Child (1988)

Buffalo Range Ranch and the question of environmental sustainability

The 20,000ha Buffalo Range Ranch was established by the Style family in 1955. In 1960 George Style set aside 8000ha along the Chiredzi River as a game-only section because heavy past use of the river for watering livestock had left it degraded. He developed 12,000ha away from the river for cattle by investing in fencing and an alternative water source.

The Style's kept fastidious data that tracked cattle and wildlife productivity from 1960 to 1986, a rare data set that is invaluable for understanding savanna sustainability. In 1973, the cattle section was in better ecological condition than the more lightly used (but previously degraded) game section, despite a similar biomass (Taylor and Walker, 1978). However, the reverse was true when transects were repeated in 1986 and 1990 (Child, 1988; Taylor and Child, 1991). The Style's records show that cattle productivity increased steadily until the mid-1970s, but then declined dramatically, despite above average rainfall (see Figure 9.2). By contrast, wildlife productivity increased steadily, reflecting the steady recovery of the range.

While research into the ecological advantages of wildlife has been inconclusive (Cumming and Bond, 1991; Prins, 2000) most of these studies

Table 9.7 *A summary of the relative viability of cattle and wildlife in Zimbabwe, 1984–1986 .*

Farming area or ranch	Natural region (rainfall)	Wildlife (profit/kg biomass)	Cattle (profit/kg biomass)
Iwaba Ranch (long-standing game ranch in Midlands	III (550–700mm)	17–25c	7–10c
Midlands region (cattle ranches building up remnant wildlife populations)	III (550–700mm)	17c	7c
Rosslyn Ranch (managed by long-term proponent of wildlife in an excellent wildlife area that was expropriated in the parks estate in 1975 after a parliamentary commission concluded it was useless for agriculture)	IV (450–500 mm)	ZW$5.29/hectare	losses
Matetsi Region (good wildlife areas adjacent to parks estate)	IV (450–500 mm)	ZW$4.18–8.93/ha	loss of ZW$1.18/ha
Matetsi Safari Area (failed agricultural area that had been converted into a state-owned safari area)	IV (450–500 mm)	ZW$5.14–11.54/ha	large loss
Lowveld (cattle ranches building up remnant wildlife populations)	V (<500mm)	32c	7c
Nuanetsi ranch (million-acre cattle ranch with a new safari enterprise)	V (<500mm)	2.6c	1.7c
Buffalo Range Ranch (owned by a dedicated record keeper with separate wildlife and cattle sections since 1960; well studied)	V (<500mm)		
• financial		0–3c	2c
• economic		13–18c	–4c
• economic analysis including environmental costs		ZW$4.90–6.21/ha	–ZW$8.00/ha

Source: Child (1988)

have been short term. This 25-year data set suggests that wildlife is, indeed, slightly more productive than cattle, while simultaneously allowing range recovery (see the 1973, 1986 and 1990 vegetation surveys: Taylor and Walker, 1978; Child, 1988; Taylor and Child, 1991). This implies that wildlife's diversity of browsing and grazing feeding strategies may be advantageous over the longer term. Moreover, the collapse of cattle productivity supports Campbell et al's (2000) warning that the 'new rangeland scientists' are establishing a false narrative that range degradation is not as serious as once supposed. As noted, the degradation measured on Buffalo Range was reported by other Lowveld ranchers who switched into wildlife. Within 15 years, perennial grasses again appeared abundant, although this needs to be validated more carefully.

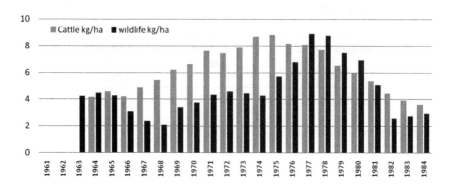

Source: Adapted from Child (1988)

Note: * The decline in wildlife productivity after 1982 reflects the massive die-off in the 1983–1984 drought, noting that the data are smoothed using five-year moving averages

Figure 9.2 *A comparison of the productivity of cattle and wildlife on Buffalo Range Ranch, 1960–1984**

More detailed economic comparisons of wildlife and livestock

In 1992, a detailed investigation of the relative financial and economic profitability of 184 wildlife and cattle enterprises was conducted using a policy analysis matrix (Jansen, 1989), in part to combat agricultural and veterinary policies that continued to disadvantage wildlife and subsidize agriculture. Although many wildlife enterprises were new and understocked, almost half were profitable and the average return on investment (ROI) was 9 per cent (Bond, 1993; Jansen et al, 1992). Cattle ranchers were struggling to make a living out of cattle and less than 5 per cent of cattle enterprises were viable (i.e. with an ROI of 10 per cent or more). Using economic analysis to strip out policy distortions on prices revealed that government policy undervalued wildlife. Surprisingly, it also undervalued livestock (Bond, 1993), although we believe this calculation does not fully account for indirect subsidies, the depreciation of fencing and other assets, and range degradation.

Understanding wildlife's advantages

Ecology, limiting factors and the structure of production systems

The ecological differences between producing meat from wildlife and livestock are small; the advantages offered by wildlife's diversity, mobility and hardiness

Table 9.8 *Comparison of wildlife and cattle profitability in Zimbabwe, 1990*

	Profitable (ROI >10%)	Marginal (10%>ROI>0%)	Loss (ROI<0%)	Average ROI
Financial analysis				
Wildlife	48%	38%	14%	9%
Cattle	5%	59%	36%	3%
Economic analysis				
Wildlife	78%	17%	5%	20%
Cattle	54%	35%	11%	17%

Source: Jansen et al (1992)

are probably exceeded by the cost of harvesting wild animals and the under-developed markets for game meat. Wildlife's advantages lie in its multiple usage, with tourism and hunting adding value through outfitting services, the aesthetic appeal of wild areas and wild animals, hospitality and suchlike. The reduced reliance on commodity values fundamentally alters the production system. A simple, extractive commodity economy is replaced by a complex economy that is less biologically extractive and has many horizontal and verti-cal linkages. Indeed, recent data from South Africa confirm that switching to wildlife increased employment five times, the total wage-bill by 30 times, created numerous upstream and downstream economic multipliers and doubled land values (Langholz and Kerley, 2006). Consequently, private land in South Africa, Namibia and Zimbabwe (before land expropriation) is now protecting as much wildlife as state-protected areas by generating profits and jobs (Bond et al, 2004).

Conclusions

The data presented in this chapter were used by wildlife officials in Zimbabwe to improve the policy environment for wildlife. As late as the 1990s, govern-ment officials were sceptical about the relative profitability of wildlife. However, landholders were not. In 1984, game ranching was a hobby practised by a few dozen ranchers, but by 2000 some 1000 properties in Zimbabwe were conserving wildlife. Indeed, the footprint of the wildlife sector was apparent everywhere from tourism services to hunting vehicles on the roads.[5] By this time, most ranchers in semi-arid savannas were utilizing wildlife to a greater or lesser extent. The booming tourism sector also allowed an increasing minority of farmers in agricultural zones to use uncultivated land, hillsides, farm dams and proximity to tourism hubs to introduce wildlife and wildlife businesses.

There is little doubt that the ecological condition of many properties improved significantly under wildlife, or that the combination of individual property rights and collective regulation was a highly effective conservation strategy. Indeed, despite the unfortunate bureaucratization of the ICA movement in the 1980s, in a number of districts, landholders developed legal agreements to manage wildlife collectively, removing internal fences over large areas of land, but generally managing the commercial aspects of their business as individuals. The bad news is that this experiment in wildlife management was decimated by the invasion of private land in Zimbabwe: without effective institutions of property rights and the rule of law it becomes difficult to manage wildlife. This wildlife conservation model requires institutional stability for the collective action required to manage such a mobile resource and because scaling-up is important for both ecological sustainability and commercial investment – it is easy to sell one cow, but a single wild animal can seldom support a tourism market.

However, it is interesting that several wildlife conservancies have survived Zimbabwe's current lawlessness relatively intact. In at least one case, the collective action developed to manage wildlife also provided political resilience. At the same time, Zimbabwe's elite has made a massive cognitive transition. In the 1980s, land under wildlife was considered to be underutilized. Now wildlife is a sought-after business in which the emerging black elite is financially invested, albeit sometimes through subsidized (even coerced) entry and via state-controlled assets.

It is clear that the survival of wildlife outside state-protected areas can be enhanced by combining clear property rights and a commercial approach, and is particularly effective where institutional scaling-up is facilitated. We have also invested heavily in transferring this model to communal lands to ensure that communal area residents benefit from and conserve wildlife through programmes such as CAMPFIRE. However, wildlife is unlikely to prosper unless the emerging elites in African states genuinely invest in the sector. Achieving this may well be the next conservation challenge.

Notes

1   At this time the ZW$ was worth a little less than the US$ (itself unusually strong) but had equivalent or higher purchasing power, since products in Zimbabwe were generally significantly cheaper than on world markets. For instance, a graduate-qualified junior professional could live on ZW$400 a month.

2   Zimbabwean legislation managed the externalities associated with a fugitive wildlife resource by legally empowering local collective regulation through ICAs. By contrast, use rights in Namibia and South Africa often required a 'certificate of enclosure' and tall game fences. This works but has disadvantages. Fences are

expensive and they also fragment the ecological and social landscape. Moreover, fencing negates many of wildlife's behavioural adaptations to variable environments, and can be associated with overstocking. While fences can delineate ownership in the absence of institutional solutions, it is noteworthy that in all three countries groups of property owners are forming 'conservancies', tearing down the internal fences and working out ways to collectively scale up environmental management.

3   Child (1995) is critical of the retention of a category of 'Specially Protected Species' in the Parks and Wildlife Act of 1975, and of the reliance on negative incentives by international conventions such as the Convention on the International Trade in Endangered Species. He argues that simply listing wildlife 'does nothing to offset ecological constraints to a species' survival' (Child, 1995, pp84–87). An internal analysis by the Zimbabwean wildlife department in the mid-1980s seriously questioned the value of taking species out of the marketplace by giving them special protection status because the primary threat in all cases, except rhino, was habitat change rather than hunting. In no cases did preventing commercial use of these species add to their chances of survival. Indeed, the opposite was usually true: the probability of survival was enhanced by making the species more valuable to landholders, even for poached species such as rhino. Child (1995) also provides an informative comparison of the legally unprotected leopard that fared far better than the 'protected' cheetah on private land.

4   A note about methodology is relevant here – most studies compare the profitability of wildlife and livestock on a per area basis ($/ha). This is misleading because it does not account for relative stocking densities. A much better measure is '$/kg livemass', which controls for differential stocking rates and relates output ($) to the limiting factor (i.e. grazing and browsing, for which livemass or metabolic biomass are good proxies).

5   The fact that private wildlife conservation is expanding so rapidly challenges the assumption that savanna parks require state subsidization.

References

Anon (2008) 'Order in the jungle', *The Economist*, 13 March

Beinart, W. (1984) 'Soil erosion, conservationism and ideas about development: A Southern African exploration 1900–1960', *Institute of South African Studies*, vol 2, no 1, pp52–84

Beinhocker, E. D. (2006) *The Origin of Wealth: Evolution, Complexity and the Radical Remaking of Economics*, Harvard Business School Press, Boston

Bond, I. (1993) 'The economics of wildlife and landuse in Zimbabwe: An examination of current knowledge and issues', WWF Multispecies Animal Production Systems Project, WWF, Harare

Bond, I., Child, B., Harpe, D. D. L., Jones, B., Barnes, J. and Anderson, H. (2004) 'Private land contribution to conservation in southern Africa', in Child, B. (ed) *Parks in Transition: Biodiversity, Rural Development and the Bottom Line*, Earthscan, London

Bond, I. and Cumming, D. (2004) 'Wildlife research and development', draft paper

Booth, V. (2002) 'Analysis of wildlife markets (sport hunting and tourism)' WWF Programme Office, WWF, Harare

Buechner, H. K. (1964) 'Conservation of African wildlife', *Ecology*, vol 45, no 2, pp420–421

Campbell, B. M., Dore, D., Luckert, M., Mukamuri, B. and Gambiza, J. (2000) 'Economic comparisons of livestock production in communal grazing lands in Zimbabwe', *Ecological Economics*, vol 33, pp413–438

Child, B. (1988) 'The role of wildlife utilization in the sustainable economic development of semi-arid rangelands in Zimbabwe', DPhil thesis, University of Oxford, Oxford

Child, G. (1995) *Wildlife and People: The Zimbabwean Success: How the Conflict between Animals and People Became Progress for Both*, Wisdom Foundation, Harare

Cumming, D. (1995) 'Are multispecies systems a viable land use option for southern African rangelands?', in Hofmann, R. R. and Schartz, H. J. (eds) *Wild and Domestic Ruminants in Extensive Land Use Systems*, Humbolt University, Berlin

Cumming, D. H. M. and Bond, I. (1991) 'Animal production in southern Africa: Present practices and opportunities for peasant farmers in arid lands', Multispecies Animal Production Systems Project Project Paper No. 22, WWF Multispecies Project, Harare

Handy, C. (1994) *The Empty Raincoat. Making Sense of the Future*, Arrow Books, London

IUCN (International Union for Conservation of Nature) (2004) *An Introduction to the African Convention on the Conservation of Nature and Natural Resources*, IUCN Environmental Policy and Law Paper No. 56, IUCN, Gland, Switzerland, p60

Jansen, D. J. (1989) 'Cattle, wildlife, both or neither? A guide to an appropriate response', WWF Multispecies Animal Production Project, WWF, Harare

Jansen, D. J., Bond, I. and Child, B. (1992) 'Cattle, wildlife, both or neither? A survey of commercial ranches in the semi-arid regions of Zimbabwe', WWF Multispecies Animal Production Project, WWF, Harare

Langholz, J. A. and Kerley, G. I. H. (2006) *Combining Conservation and Development on Private Lands: An Assessment of Ecotourism-Based Private Game Reserves in the Eastern Cape*, Report No. 56, Center for African Conservation Ecology, Nelson Mandela Metropolitan University, Port Elizabeth

Murphree, M. (2000) 'Constituting the commons: Crafting sustainable commons in the new millennium', 'Multiple Boundaries, Borders and Scale', Eighth Biennial Conference of the International Association for the Study of Common Property (IASCP), Bloomington, IN, 31 May–4 June

Parker, I. (2004) *What I Tell You Three Times Is True: Conservation, Ivory, History and Politics*, Librario, Elgin

Prins, H. T. (2000) 'Competition between wildlife and livestock in Africa', in Prins, H. T., Grootenhuis, J. G. and Dolan, T. T. (eds) *Wildlife Conservation by Sustainable Use*, Kluwer Academic Publishers, London

Rifkin, J. (1992) *Beyond Beef: The Rise and Fall of the Cattle Culture*, Dulton, New York, NY

Riney, T. (1967) *Conservation and Management of African Wildlife*, FAO, Rome

Riney, T. and Hill, P. (1967) *Conservation and Management of African Wildlife*, FAO, Rome

Rodriguez, G. (1985) 'The economic implications of the beef pricing policy in Zimbabwe', LPU Working Paper No. 7, International Livestock Centre for Africa

SASUSG (Southern Africa Sustainable Use Specialist Group) (1996) *Sustainable Use Issues and Principles*, IUCN Species Survival Commission, SASUSG

Showers, K. B. (1994) *Early Experiences of Soil Conservation in Southern Africa: Segregated Programs and Rural Resistance*, African Studies Center, Boston University, Boston, MA

Southern Rhodesia (1952) 'Report of a commission of inquiry into the Cold Storage

Commission of Southern Rhodesia', Government Printer, Salisbury

Talbot, L. M., Payne, W. A., Ledger, H. P. and Talbot, M. H. (1965) *The Meat Production Potential of Wild Animals in Africa*, Commonwealth Agriculture Bureau Technical Communication No. 16, Commonwealth Agriculture Bureau, Farnham Royal, Buckinghamshire

Taylor, R. D. and Child, B. (1991) 'Does wildlife offer comparative advantages over cattle? The Buffalo Range case study: Ecological considerations', 'Meeting Rangelands Challenges in Southern Africa in the 1990s', CSIR Conference Center, Pretoria, 6–11 May

Taylor, R. D. and Walker, B. H. (1978) 'Comparison of vegetation use and herbivore biomass on a Rhodesian game and cattle ranch', *Journal of Applied Ecology*, vol 15, pp565–581

Turner Report (1956) 'Report of a commission of inquiry on the marketing of cattle for slaughter and distribution and sale of beef in Southern Rhodesia', Government Printer, Salisbury

Vincent, V. and Thomas, R. G. (1961) *An Agricultural Survey of Southern Rhodesia*, Federation of Rhodesia and Nyasaland, Salisbury

Walker, B. H. (1979) 'Game ranching in Africa', in Walker, B. H. (ed) *Management of Semi-arid Ecosystems*, Elsevier, Oxford

Walker, B., Holling, C. S., Carpenter, S. R. and Kinzig, A. (2004) 'Resilience, adaptability and transformability in social-ecological systems', *Ecology and Society*, vol 9, no 1

10

Extensive Wildlife Production on Private Land in South Africa

J. du P. Bothma and Helen Suich with Anna Spenceley

South Africa is the scene of one of the greatest reversals of fortune ever seen in wildlife conservation. Wildlife was devastated when millions of animals were killed by the early settlers in the 19th century, and was then almost eliminated by crop and livestock agriculture in the 20th century when it was perceived to conflict with livestock, competing for water and grazing resources and supposedly posing health risks (Cumming, 1990; Flack, 2002a, 2002b). By 1950, wildlife had been severely decimated outside of South Africa's national parks.

The rights to use wildlife depended largely on the presence of wildlife-proof fences, which were first erected on private land in the 1950s (Hearne and McKenzie, 2000). As late as the 1960s, there were only around ten wildlife ranches in South Africa (Van Hoven and Zietsman, 1998). However, by 1993 around 8500 ranchers were involved in some form of wildlife utilization (Benson, 1986; Pauw, 1989 and Skinner, 1989, cited in Cumming, 1990; Bothma and du Toit, 2009). Today wildlife is integral to South Africa's animal production, with more than 10,000 wildlife ranches and some 4000 mixed wildlife/livestock enterprises (Hearne and McKenzie, 2000; Bothma, 2002, 2004, 2005a; Absa, 2003), most situated in semi-arid savannas in the Limpopo and Northern Cape provinces (see Table 10.1) (Bothma, 2005a).

The income-generating activities that occur on a single property usually depend on a variety of factors, including their location relative to tourist routes and large cities, the size of the ranch and its endowment of wildlife (especially the presence or absence of the 'Big Five'). These activities are of three main types: hunting (for trophies and/or meat); photographic or wildlife-viewing tourism; and live sales and breeding (Hearne and McKenzie, 2000; Absa, 2003; Higginbottom and King, 2006).

Much of the initial impetus for wildlife ranching reflected the personal desire of private landowners and investors to provide a 'retreat' for their own enjoyment. Over time, motivations within the industry have diversified – many

Table 10.1 *Mean size and provincial distribution of 'exempted' wildlife ranches in South Africa, 2000**

Province	Mean size	Number of ranches	% of total ranches
Limpopo	1339.9	2480	49.0
Northern Cape	4921.0	986	19.5
Eastern Cape	1412.9	624	12.3
North West	1073.7	340	6.7
Mpumalanga	1346.4	205	4.1
Free State	820.8	180	3.6
KwaZulu-Natal	1876.0	92	1.8
Western Cape	3234.2	82	1.6
Gauteng	1139.9	72	1.4
South Africa	2047.8	5061	100.0

Note: * Exemption was given to properties that were suitably fenced, and entitled the landholder to hunt, capture and sell wildlife all year round (the original exemption was from provincial ordinances that governed wildlife utilization).

Source: Eloff (2002)

landholders care deeply about threatened species, subspecies, natural land-scapes and habitats; profits are important to others; and the sustainability of wildlife production compared to conventional agricultural is of primacy to others (Benson, 1986; Porter et al, 2003; Sims-Castley et al, 2005; Higginbottom and King, 2006).

Wildlife property rights

Until 1991, all wildlife in South Africa was still *res nullius* and belonged to the state. However, in 1988, wildlife producers, as represented by the National Game Organization (now Wildlife Ranching South Africa), requested that the national government examine the possibility of legislation dealing with the theft of wildlife (separate to that dealing with the theft of domestic livestock). The former President, P. W. Botha, immediately requested the South African Law Commission to look into the matter. The Commission subsequently submitted a report to the National Game Organization executive in which it also dealt with the property rights to wildlife on ranches that were suitably fenced, according to the specifications of the various provincial conservation authorities. This in turn led to the promulgation of the Game Theft Act of 1991 (Act 105 of 1991), the aim of which was to, in certain instances, regulate the ownership of game and to combat its theft, illegal hunting and capture. The Act gave the rights of ownership of wildlife, on suitably fenced land, to those landholders who could prove ownership in cases of dispute and exempted them from certain provincial ordinances regarding the use of wildlife on their properties, such as hunting seasons. The exemption permit is valid for three years at a time, and has recently been augmented by several regulations that allow wildlife ranchers to qualify for exemption (under strict guidelines) when hunting threatened or protected species or subspecies. These regulations now provide uniform protection at the national level to listed wildlife species or subspecies by regulating activities such as hunting and selling them.

The growth of private wildlife conservation in South Africa has come at almost no cost to the state, yet has provided enormous biological and economic benefits. Perhaps the most surprising facet of this conservation success is how little documentation is associated with it. The Game Theft Act itself only has nine clauses and is a little more than two pages long.

Impact on biodiversity

There is no systematic documentation of the effect of the massive growth of the wildlife sector on biodiversity conservation. Yet there is more wildlife in South Africa now than at any time over the last 100 years (Bothma, 2004, 2005a; Bothma et al, 2009), and South Africa is one of the few countries where the classifications of rare or threatened species is decreasing (Lindberg et al, 2003). There are now more rhinoceros on private land in South Africa than in the rest of Africa (Absa, 2003), and the bontebok, black wildebeest, Cape mountain zebra and geometric tortoise (MINTECH, 2004, cited in Higginbottom and King, 2006) have recovered from the brink of extinction. Moreover, private landholders also conserve biota and habitats that are poorly represented in formal protected areas, and those properties adjacent to protected areas increase the extent of land under effective conservation management (Absa, 2003; Goodman et al, 2002, cited in Higginbottom and King, 2006). In all, wildlife utilization enterprises were estimated to cover 16.8 per cent of South Africa, compared to 6.1 per cent in provincial and national protected areas in 2007 (Bothma and Von Bach, 2009).

However, wildlife ranching is not without its problems. The necessity of fencing properties in South Africa can lead to fragmented landscapes and genetic isolation, and can encourage overstocking and range degradation – especially on small fenced properties where trophy and/or meat hunting takes place (Absa, 2003). In a survey of wildlife ranches in KwaZulu-Natal province, for instance, up to half were overstocked (Lindberg et al, 2003), although this problem is said to be decreasing over time, as better information about ecological sustainability is becoming available (Higginbottom and King, 2006). A quantifiable stocking density technique was recently developed for herbivores in the African savannas to address overstocking (Bothma et al, 2004) and its implementation should help to alleviate the problem.

Extra-limital introductions, hybridization and disease transmission are also concerns (Bothma, 2005b; Bothma and du Toit, 2009). Some private land-holders 'enhance' tourist wildlife-viewing by clearing woody vegetation and/or manipulating the landscape, sometimes with negative environmental impacts (Sims-Castley et al, 2005). A disadvantage (compared to wildlife populations within protected areas) can be the non-natural wildlife mix – the tendency to understock animals for which there is little hunting interest, as well as those which are potentially damage-causing, such as large predators. The latter contribute to healthy ecosystems, gene pools and biological control, but impact negatively on the commercial viability of the enterprise (Féron, 1995; Absa, 2003; Higginbottom and King, 2006). Moreover, many of the current wildlife ranches are too small to contain large enough predator populations to keep their populations genetically fit.

Notwithstanding such issues, the impact of wildlife production activities on private land on biodiversity has been broadly positive – and certainly better than alternative land uses (see Aylward and Lutz, 2003). Furthermore, the industry is improving as knowledge spreads, supported by some (albeit insuf-ficient) research and monitoring (Higginbottom and King, 2006).

Economic and financial impacts

Economic data on the wildlife-ranching industry is fragmented and often unreliable but indicates the diverse range of value-adding products – both consumptive and non-consumptive – that can generate revenue for wildlife producers (Absa, 2003). Income is generally derived from four sources: live wildlife sales at auctions, hunting, wildlife-viewing tourism and meat production.

The source and composition of income of wildlife producers differs widely between the provinces of South Africa, and as noted above, is particularly affected by distance from the cities, the type of habitat and the wildlife present on the property, particularly the presence of the 'Big Five'. In the remote Bushveld areas, 60–65 per cent of income is from hunting for trophies or meat, 5–10 per cent is from wildlife-viewing and the balance is from the sale of live animals at auctions. In contrast, in Gauteng province, ecotourism generates 40 per cent of the total revenue (Absa, 2002; Reilly et al, 2003).

Table 10.2 presents a range of data sources estimating the size of various components of the wildlife production industry, across provinces and nation-ally, and illustrates the difficulty of attempting to determine the 'real' size of the industry from such fragmented data.

Table 10.2 *Estimates of the value of extensive wildlife production*

Year	Estimated value of extensive wildlife production	Source
2000	The direct gross income to wildlife ranchers from extensive wildlife production was R843 million (US$122 million). Of this income 53.4% was generated by local hunters, followed by live animal sales, foreign trophy hunters, ecotourism and wildlife meat production (see Figure 10.1)	Bothma, 2005a
2000–2001	Foreign trophy hunters contributed almost US$10 million to the economy of the Eastern Cape province alone	Radder et al, 2003
2001–2002	Total investment in wildlife ranching (including land and infrastructure) of about R15.5 billion (US$1.8 billion) in 2001 and R16.5 billion (US$1.6 billion) in 2002, at a time when many wildlife producers were still building up stock populations and were not yet earning much income from their enterprises.	Eloff, 2002; Flack, 2002b

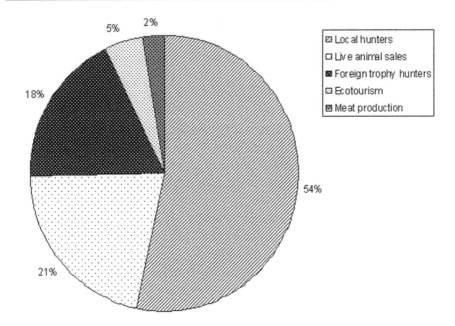

Note: While tourism is a major source of income in national parks, only a small proportion of wildlife ranches cater for (wildlife-viewing) tourism.

Source: Eloff (2002); Bothma (2005a)

Figure 10.1 *Proportions of direct gross income from wildlife ranching, South Africa, 2000*

Wildlife production enterprises are costly to establish, requiring about R6 of investment for every R1 of annual revenue, with this ratio increasing to 20:1 in the Lowveld areas where land prices are high. Consequently, the capital cost for a small ranch that can support 150 large animal units is currently around R2.5 million (US$334,000) rising to R15 million (US$2 million) for a large ranch that can support at least 1000 large animal units, and in excess of R85 million (approximately US$11.4 million) for a property with the 'Big Five', though such an enterprise would struggle to be profitable (Absa, 2003). The size of such ranches varies with habitat quality and productivity and a small ranch ranges from 1200ha in the high-rainfall grassland areas to 16,000ha in the semi-arid Karoo regions, a medium-sized one from 4000ha to 55,000ha and a large one typically exceeds 55,000ha in all cases.

Live animal sales and breeding

The first live animal auction in South Africa took place at Hoedspruit in the Limpopo district in 1975. It was privately run, selling 128 animals and turning over R20,362 (US$20,000 at that time) (Conroy, 1993; Bothma et al, 2009). Such auctions initially faced resistance from most provincial conservation agencies who believed that wildlife conservation should be the exclusive domain of the government (Scriven and Eloff, 2003), and were concerned about the lack of skills in the industry and the possible detrimental effects of managing wildlife for profit. Against the trend of most provincial wildlife agencies, the KwaZulu-Natal provincial conservation agency (now known as the Ezemvelo KwaZulu-Natal Wildlife Service) played a critical role in the development of the live wildlife trade in South Africa, accounting for a considerable proportion of the turnover of South Africa's total annual wildlife sales (Scriven and Eloff, 2003). The value of their sales has since declined as, rather than selling high-priced black rhinoceros, they are restocking new areas for breeding purposes. Over time, the resistance of other provincial agencies has disappeared and they have begun to present their own auctions to augment their allotted budgets (Higginbottom and King, 2006; Bothma and Von Bach, 2009).

Trends in live sales are an important indicator of the growth of the sector. The contribution of animals from conservation agencies declined from 68 per cent in 1991 to 17 per cent in 2002, reflecting the growth of the private sector. It increased to 38 per cent in 2004 as these agencies started to fight for economic survival (Bothma, 2002; Eloff, 2005). In 2001, for example, revenue from live wildlife sales covered approximately 15 per cent of the management costs of Ezemvelo KwaZulu-Natal Wildlife (Higginbottom and King, 2006), while in 2004 the Limpopo provincial government generated a turnover of R19.5 million (US$3 million) mainly from the sale of rare wildlife (Eloff, 2005).

The number of animals sold privately is estimated to exceed the number sold in auctions threefold, and the mean price per animal has increased by approximately 2 per cent per annum in real terms since 1991 (see Figure 10.2) (Absa, 2003; Bothma et al, 2009).

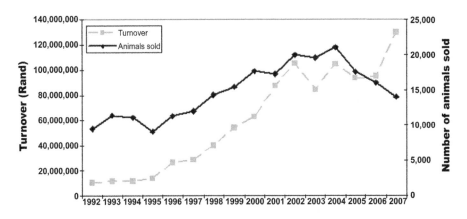

Source: Bothma et al (2009)

Figure 10.2 *Number of live animals sold and turnover
on live wildlife auctions, 1991–2007*

Some analysts believe the live wildlife market is beginning to mature and that prices will stabilize for common wildlife, but record prices are still being paid for more rare ones (Higginbottom and King, 2006; Bothma et al, 2009). Females are still more expensive than males, indicating that most animals are for breeding rather than hunting.

Table 10.3 compares prices from live sales in 2006 and 2007 across South Africa. Rare wildlife (for example disease-free African savanna buffalo, roan, sable and white rhinoceros) comprised only 3.5 per cent of the number of animals sold, but generated 52 per cent of turnover, while low-value wildlife formed 60 per cent of the sales volume but only generated 12 per cent of total turnover (Bothma et al, 2009).

The live animal market is indicative of the rate of innovation in the sector. Wildlife is increasingly sold through catalogue auctions, and internet sales are beginning because of the high costs of capture, holding and commission associated with the traditional methods of live sales (at the boma) and the reduction of stress of the animals. Higginbottom and King (2006) note that these capture and handling costs were approximately two-thirds of the gross sales value of the 2004 Ezemvelo KwaZulu-Natal Wildlife Service's auction, suggesting that gross values give a misleading impression of economic benefits.

Meat production

Since the 1970s, the potential for the production of wildlife meat to be a substantial income generator has been recognized. However, the export

Table 10.3 *Provincial trends for live wildlife auctions, South Africa, 2006 and 2007*

Province	Estimated turnover (R)				No. of animals				Mean price/ animal (R)	
	2006		2007		2006		2007		2006	2007
	Total	%	Total	%	No.	%	No.	%	2006	2007
Limpopo	34,097,884	36	45,527,284	35	6494	40	5072	36	2521	8976
North West	2,171,417	2	2,601,559	2	988	6	846	6	2198	3075
Mpumalanga	4,513,513	5	9,105,456	7	927	6	846	6	4869	10,762
Northern Cape	815,467	1	1,300,779	1	412	3	423	3	1979	3075
Eastern Cape	10,022,654	11	9,105,452	7	1079	7	564	4	9289	16,144
KwaZulu-Natal	8,069,327	9	18,210,913	14	1900	12	1409	10	4247	12,923
Free State	15,389,562	16	19,511,692	15	2607	16	3664	26	4903	5342
Gauteng	19,741,879	21	24,714,810	19	1629	10	1268	9	12,119	19,491
Western Cape	0	0	0	0	0	0	0	0	0	0
Total	94,821,703	100	130,077,945	100	16,036	100	14,092	100	5913	9231

Source: Taljaard and Cloete (2007); Bothma et al (2009); Cloete and Taljaard (2008)

industry has stuttered along rather than grown smoothly and actively. In 2000, wildlife meat generated approximately R20 million (US$2.85 million) (Van der Merwe and Saayman, 2005). Aggressive marketing and strict quality control are essential for marketing wildlife meat in the European Union, but only ten wildlife abattoirs meet the European Union's standards, and some of them only produce ostrich meat.

Hunting

South Africa's trophy hunting industry is relatively large, with 5000 to 6000 hunters having visited the country during the 2003–2004 hunting season. These hunters shot 53,453 animals with a combined value of R265 million (US$40.7 million). In addition, South Africa has an estimated 200,000 resident hunters (who hunt predominantly for meat rather than trophies) and this sector is worth around R2.9 billion (US$450 million). Up to 6000 jobs are provided directly by the hunting industry and estimates suggest in excess of 60,000 jobs are provided by secondary industries such as tourism, taxidermy, professional hunters and skinners. Other benefits include tips, hospitality, levies for conservation authorities and communal landowners, education, training and capacity-building, meat and tourism opportunities (De Wet, 2005). The Eastern Cape and Limpopo provinces seem to draw the most hunters, though the number of animals hunted per client was the highest in the Northern and Eastern Cape provinces (Anon, 2001; Radder et al, 2003).

Trophy hunters are estimated to spend more than four times the amount a non-hunting tourist spends during a visit (on hunting licences, accommodation, professional hunters, trackers, skinners and taxidermists). Hunting operations are also said to be a lower business risk than photographic tourism operations, since hunters tend to book their trips long in advance and hunting associations often conclude package deals with wildlife ranches for extended periods. A wildlife ranch that can successfully combine hunting and non-consumptive tourism (either side by side on a large ranch, or by stipulating different time periods on a smaller-sized ranch) is thought to be able to generate a higher profitability than a ranch that offers only one activity (Absa, 2003).

Wildlife-viewing tourism

In South Africa, the tourism industry is the fourth largest contributor to the national economy following the automotive, chemical and metal industries (DEAT and DTI, 2004), and South Africa attracted almost 2 per cent of total world tourism arrivals internationally in 2005 (ranking it 32nd in the world) (Tourism KwaZulu-Natal, 2006). From 2004 to 2005, tourism's total contribution to the South African economy grew by 13 per cent to R124 billion (US$19 billion), or 8.2 per cent of the gross domestic product of South Africa (South African Tourism, 2006). Despite the importance of tourism to the economy, the relative contribution of wildlife-related tourism is not known, nor is the total contribution of nature-based tourism on private land.

A number of studies have been undertaken in various provinces to estimate the returns to wildlife tourism on private land, the results of which are summarized in Table 10.4. There is considerable variability across the country and particularly between enterprises. Revenues depend primarily on the type of facilities offered, the number of beds and the rates charged rather than the area of land that is devoted to wildlife, and several authors suggest that landholders generally struggle to generate operating profits from these enterprises (Absa, 2003; Porter et al, 2003).

Socio-economic impacts

The most important socio-economic impact of the wildlife production industry is employment. The industry is said to be able to employ more people than an equivalent area of livestock (cattle) ranching (Hearne and McKenzie, 2000), and to rely far less on unskilled labour while creating more opportunities for semi-skilled and skilled labour (Absa, 2003). Across the industry, an estimated 7000 wildlife ranches employ approximately 63,000 staff members at a mean

Table 10.4 *Estimated returns to private nature reserves and wildlife ranches**

Area	Type of operation (number in sample)	Turnover/hectare	Source
KwaZulu-Natal (Zululand)	Private nature reserves and wildlife ranches (27)	Large: R826 (US$103) Medium: R528 (US$66) Small: R718 (US$90)	Porter et al, 2003
Eastern Cape	Private game reserves (7)	US$259	Sims-Castley et al, 2005
KwaZulu-Natal (southern Maputaland)	Tourism accommodation enterprises – large and exclusive luxury lodges, smaller lodges, self-catering facilities, wildlife ranches, hunting lodges and bush camps (12)	Large: R929 (US$134) Smaller enterprises: R1450 (US$208)	Oldham et al, 2000

Note: * Revenue-generating activities considered by these studies include non-wildlife-viewing tourism activities (for example hunting and live sales)

of approximately nine employees per property, up from six employees per property in 1998 (Hearne and McKenzie, 2000; Bothma, 2005a; Van der Merwe and Saayman, 2005). In the Eastern Cape, the transformation of properties from commercial livestock to wildlife-based production enterprises increased employment by a factor of 3.5 to 4.5 and farm labourers also benefited from retraining during the conversion process (Sims-Castley et al, 2005).

Complementing the increase in employment, the mean wage per employee has increased markedly, with the mean annual salary per full-time employee increasing by a factor of approximately five in the Eastern Cape province. Additional non-income benefits, including accommodation, food/provisions and training, are also said to have improved (Sims-Castley et al, 2005). When luxury and budget establishments were compared in Zululand, it was found that wildlife tourism operations in the lower price ranges of R50 (US$7) per bed generated twice as much non-management employment as those priced at R150 (US$22) or above (Aylward and Lutz, 2003).

While few studies have extrapolated information on local salaries to establish the impact of wildlife ranching and tourism on poverty, there are data available for the Sabi Sabi Game Reserve adjacent to the Kruger National Park. Cumulatively, local employment at the reserve was estimated to have risen, with 1120 dependents in the local community above the poverty line, approximately 4 per cent of the local population (Spenceley and Seif, 2003).

The indirect impact of tourism expenditures can also be significant generators of employment and revenue, where purchases of goods and services by tourism enterprises are made locally, though there is virtually no information available with respect to these multiplier effects.

Institutional experimentation on private land

Approximately half of all South African wildlife ranches are owned on a part-time basis by professional people, and the majority of these are less than 500ha in extent and are generally unprofitable. Cooperation among neighbours through the creation of conservancies (Bothma, 2004; Bothma and du Toit, 2009) is often the only possible way to increase profitability to acceptable levels. A conservancy is a group of adjoining wildlife ranches that operate under a cooperative management agreement based on a shared common goal such as wildlife conservation (Absa, 2003; Bothma and du Toit, 2009). The removal of internal fences between the properties involved in conservancies has bio-diversity advantages, allowing owners to stock wildlife (such as elephant) that are not viable on smaller parcels of land. Conservancies also provide a wider pool of skills, experience and knowledge to contribute to the joint management of the operation.

Some government agencies actively encourage conservancies, for instance in KwaZulu-Natal, conservancies obtain preferential access to and discounts of 25 per cent on live wildlife obtained from the conservation authority (Higginbottom and King, 2006), and it is thought that some 3 million ha of land now fall within conservancies in South Africa (Absa, 2003; Bothma, 2004).

'Shareblocking' legislation has also been crucial to the conversion of land to wildlife ranches. The legislation allows a group of people to buy blocks of shares in a company that has purchased land; the shareholders then become co-owners and acquire rights to make use of the land and certain assets according to a 'user agreement'. Any income generated is used to pay for the land and cover the operating costs of the company, with shortfalls being made up by levies on the owners. The legislation has enabled significant investment in the wildlife production sector because it enables large sums of capital to be raised relatively quickly to cover the high capital costs usually associated with the conversion of land from livestock to wildlife ranching (Hearne and McKenzie, 2000).

Other forms of institutional innovation involving the private sector include their incorporation into biosphere reserves and contractual national parks (see also Chapters 19, 20 and 22). Landowners in contractual national parks retain the private ownership of their land, but all fences between their properties and the adjacent protected area can be removed to facilitate the free movement of wildlife. Formal agreements are made, usually regarding access to neighbouring properties, management objectives, the means of achieving them and revenue distribution.

Conclusions

Many of the factors that have made the establishment of private wildlife utilization enterprises attractive have their roots in political and legal developments. The deregulation of the agricultural industry, particularly the phasing out of subsidies and reduced returns from conventional farming, enable wildlife production and other wildlife utilization enterprises to compete economically. The recent decline in the political power base of the farming lobby, changes to the Shareblock Act, the Game Theft Act, legislation regarding conservancies, biospheres and other protected areas and new labour laws have all encouraged the transformation from conventional agriculture to wildlife utilization. A further stimulus has been the growth in international tourism.

Support for wildlife conservation on private land has been particularly strong in KwaZulu-Natal, with protected area managers in official conservation areas initially providing encouragement to landowners by providing excess wildlife free, and later subsidizing wildlife sales and providing free wildlife management extension services – measures that were discontinued once the industry was established in the province. Despite early resistance to the wildlife-ranching industry by some provincial conservation agencies, all of them now see part of their role as supporting the development of commercial wildlife-ranching industries on private land (Higginbottom and King, 2006). However, many private participants in the wildlife-ranching industry still believe that their contribution to conservation and to regional socio-economic development is not adequately recognized by the government. Many industry participants also believe that the establishment of wildlife-ranching enterprises and the further development of the industry are being obstructed by too many government regulations, excessive bureaucracy and the insufficient resources of provincial conservation agencies (Sims-Castley et al, 2005; Higginbottom and King, 2006). The differences in legislation and regulations between provinces are also problematic, and the current development of a single national set of regulations, norms and standards is supported by many within the industry (Higginbottom and King, 2006).

Currently there are various actions under way to improve the management and regulation of the wildlife industry in South Africa. A nationally representative body, Wildlife Ranching South Africa, now represents the wildlife-ranching industry in a formal capacity, and is in continual discussions with national and provincial authorities with respect to the regulation and accreditation of wildlife ranches, the impact and enforcement of new legislation, such as the National Environmental Management: Protected Areas Act, the National Environmental Management: Biodiversity Act and the Threatened or Protected Species Regulations. These regulations will have a notable impact on 'green darting' and canned lion hunting in South Africa as both practices are likely to be banned in the future. However, it may be prudent to allow the 'green darting' and harvesting of rhinoceros horn in the future, provided that animals are not harmed by repeated darting. Harvesting from live darted animals and establishing a legitimate trade in rhinoceros horn may reduce the current high level of rhinoceros poaching (Bothma and du Toit, 2009). One source of concern is whether the new regulations might encourage some owners to remove their land from conservation uses through its restrictions on what they perceive to be desirable wildlife utilization activities.

More stringent veterinary regulations are also gradually being introduced to reduce the risk of disease transmission that is associated with wildlife movements (Higginbottom and King, 2006) and others will require the countrywide removal of extra-limital plants and animals, although climate change may eventually influence this legislation as habitats change over time. This may have a future impact on private conservation areas that seek legal status for their land (Langholz and Krug, 2004, cited in Sims-Castley et al, 2005).

There is currently an excellent economic climate for wildlife ranching in South Africa and it is likely that more land will be converted to wildlife utilization enterprises than would occur for conservation reasons alone. As a form of land use, particularly in marginal agricultural regions, wildlife utilization can generate better returns than conventional agriculture in South Africa, as well as providing improved biodiversity outcomes.

References

Absa (2002) *Game Ranch Profitability in Southern Africa*, Absa, Johannesburg

Absa (2003) *Game Ranch Profitability in Southern Africa*, Absa, Johannesburg

Anon (2001) 'Overseas clients', *Game and Hunt*, vol 7, no 5, p16

Aylward, B. and Lutz, E. (2003) 'The actual and potential contribution of nature tourism in Zululand, considerations for development, equity and conservation', in Aylward, B. and Lutz, E. (eds) *Nature Tourism, Conservation, and Development in KwaZulu-Natal, South Africa*, World Bank, Washington DC

Benson, D. E. (1986) 'Sources in income', *Farmer's Weekly*, 11 April 1986, pp10–11

Bothma, J. du P. (2002) 'Some economics of wildlife ranching', in Penzhorn, B. L. (ed) *Game Ranch Planning and Management,* Wildlife Group, S.A. Veterinary Foundation, Onderstepoort, South Africa

Bothma, J. du P. (ed) (2004) *Game Ranch Management*, 4th edition, Van Schaik, Pretoria

Bothma, J. du P. (2005a) 'Extensive wildlife production in South Africa', keynote address to the Wildlife Seminar of the Northern Game Farmers Organisation, 11 March 2005, Pretoria

Bothma, J. du P. (2005b) 'It is time to end the debate on introducing exotic wildlife', *South African Journal of Wildlife Research*, vol 35, no 1, pp97–102

Bothma, J. du P. and du Toit, J. G. (eds) (2009) *Game Ranch Management*, 5th edition, Van Schaik, Pretoria

Bothma, J. du P. and Von Bach, S. H. (2009) 'Economic aspects of wildlife production in southern Africa', in Bothma, J. du P. and du Toit J. G. (eds) (2009) *Game Ranch Management*, 5th edition, Van Schaik, Pretoria

Bothma, J. du P., Van Rooyen, N. and Van Rooyen, M. W. (2004) 'Using diet and plant resources to set wildlife stocking densities in African savannas', *Wildlife Society Bulletin*, vol 32, no 3, pp840–851

Bothma, J. du P., du Toit, J. G. and Van Rooyen, J. (2009) 'Buying and selling wild animals', in Bothma, J. du P. and du Toit, J. G. (eds) (2009) *Game Ranch Management*, 5th edition, Van Schaik, Pretoria

Cloete, F. and Taljaard, P. (2008) 'Tendense van lewendige wildverkope op wildveilings', *Game and Hunt*, vol 14, no 4, pp21–25

Conroy, A. M. (1993) 'An introduction to the South African game industry', in Woodhouse, I. (ed) *A Salute to World Deer Farming*, New Zealand Deer Farmers Association, Wellington

Cumming, D. H. M. (1990) 'Developments in game ranching and wildlife utilization in East and southern Africa', Project Paper No. 13, WWF Multispecies Animal Production Systems Project, Harare

DEAT and DTI (2004) *Global Competitiveness Project: Summary of Key Findings of Phase 1, South African Tourism*, Department of Environmental Affairs and Tourism/Department of Trade and Industry, Pretoria

De Wet, A. (2005) 'Professional Hunting Association of South Africa Conference, 21 November 2005', www.polity.org.za/article.php?a_id=78017 (accessed 21 August 2007)

Eloff, T. (2002) 'The economic realities of the game industry in South Africa', in Ebedes, H., Reilly, B., Van Hoven, W. and Penzhorn, B. (eds) *Sustainable Utilization: Conservation in Practice*, South African Game Ranchers' Organisation, Pretoria

Eloff, T. (2005) 'Ontleding van alle wildveilings in 2004', *Game and Hunt*, vol 11, no 2, p21

Féron, E. M. (1995) 'New food sources, conservation of biodiversity and sustainable development: Can unconventional animal species contribute to feeding the world?', *Biodiversity and Conservation*, vol 4, pp233–240

Flack, P. H. (2002a) 'Exotic game: Catching up with Texas?', *Magnum,* October 2002, pp76–80

Flack, P. H. (2002b) 'The conservation revolution', *Game and Hunt*, vol 8, no 10, pp29–33

Hearne, J. and McKenzie, M. (2000) 'Compelling reasons for game ranching in Maputoland', in Prins, H. H. T., Grootenhuis, J. G. and Dolan, T. T. (eds) *Wildlife Conservation by Sustainable Use*, Kluwer Academic Publishers, Dordrecht

Higginbottom, K. and King, N. (2006) *The Live Trade in Free-Ranging Wildlife within South Africa, and Implications for Australia*, Rural Industries Research and Development Corporation, Canberra

Lindberg, K., James, B. and Goodman, P. (2003) 'Business and the environment in nature tourism, supply studies', in Aylward, B. and Lutz, E. (eds) *Nature Tourism, Conservation and Development in KwaZulu-Natal, South Africa,* World Bank, Washington DC

Oldham, G., Creemers, G. and Rebeck, T. (2000) 'An economic evaluation of tourism, a case study of accommodation facilities in southern Maputaland', *Development Southern Africa*, vol 17, no 2, pp175–188

Porter, S., Ferrer, S. and Aylward, B. (2003) 'The profitability of nature tourism in Zululand: A survey of private reserves and public protected areas', in Aylward, B. and Lutz, E. (eds) *Nature Tourism, Conservation, and Development in KwaZulu-Natal, South Africa*, World Bank, Washington DC

Radder, L., Van Niekerk, P. and Nagel, A. (2003) 'A profile of the international hunter', unpublished report, Nelson Mandela Metropolitan University, Port Elizabeth

Reilly, B. K., Sutherland, E. A. and Harley, V. (2003) 'The nature and extent of wildlife ranching in Gauteng Province, South Africa', *South African Journal of Wildlife Research*, vol 33, no 2, pp141–144

Scriven, L. and Eloff, T. (2003) 'Markets derived from nature tourism in South Africa and KwaZulu-Natal: A survey of the sale of live game', in Aylward, B. and Lutz, E. (eds) *Nature Tourism, Conservation and Development in KwaZulu-Natal, South Africa*, World Bank, Washington DC

Sims-Castley, R., Kerley, G. I. H., Geach, B. and Langholz, J. (2005) 'Socio-economic significance of ecotourism-based private game reserves in South Africa's Eastern Cape Province', *Parks*, vol 15, no 2, pp6–17

South African Tourism (2006) *Indaba 2006 Fact sheet: 2005 Tourism Arrivals, May 2006*, South African Tourism, Pretoria

Spenceley, A. and Seif, J. (2003) 'Strategies, impacts and costs of pro-poor tourism approaches in South Africa', Pro-Poor Tourism Working Paper No. 10, Overseas Development Institute, London

Taljaard, P. and Cloete, F. (2007) 'Tendense van lewendige wildverkope op wildveilings', *Game and Hunt*, vol 14, no 2, pp24–26

Tourism KwaZulu-Natal (2006) *Some Useful Statistics 2006*, Tourism KwaZulu Natal, Pietermaritzburg

Van der Merwe, P. and Saayman, M. (2005) *National Profile and Economic Impact of Biltong Hunters in South Africa*, Institute for Tourism and Leisure Studies, North West University, Potchefstroom

Van Hoven, W. and Zietsman, M. (1998) 'Game: SA's Hidden Asset', *Farmers Weekly*, 6 November 1998

11

Savé Valley Conservancy: A Large-Scale African Experiment in Cooperative Wildlife Management

Peter Lindsey, Raoul du Toit, Alistair Pole and Stephanie Romañach

The Savé Valley Conservancy (SVC) is a large (3442km²) cooperatively managed wildlife area, comprising multiple properties held by private ranchers, local councils, government and one community. The conservancy is located in the semi-arid South East Lowveld of Zimbabwe, occurring at an elevation of 480–620m, with deciduous woodland savanna, low and variable rainfall (474–540mm per annum) and poor-quality soils. The conservancy is bordered primarily by high-density communal land (of between 11 and 82 people per km²), with some commercial agriculture to the south and east (Pole, 2006).

Settlement and the development of the cattle industry

The area that is now SVC was originally inhabited by San (Bushmen), as indicated by the presence of San rock paintings (Pole, 2006). Bantu people settled in the area in approximately AD500 and ousted the San, though the area was generally sparsely populated because of low rainfall, lack of permanent water and the danger to people and crops from wild animals (Pole, 2006). European hunters and explorers first passed through the area in the 1870s to 1890s, and the settlers in the 'Moodie Trek' in 1892, named the area 'Hell's Wood' in response to the heat, malaria and thick bush. Wildlife was abundant, with significant populations of buffalo, lion, spotted hyaena, wild dogs and many other species.

During the 1920s, the area was settled by European farmers and three large-scale private cattle ranches were developed: Devuli, Angus and Humani. The remainder of what is now SVC was 'Crown Land'. By 1925, there were already 23,000 cattle on ranches in the area (Pole, 2006). The availability of cheap labour, abundant land and access to inexpensive livestock from stocks of indigenous cattle encouraged efforts to expand the beef industry, despite repeated outbreaks of foot-and-mouth disease (Phimister, 1978). In 1972, the remaining Crown Land in the SVC area was sold to individuals who planned to develop cattle ranches. The then Rhodesian government supported the cattle industry with direct financial assistance via subsidies, soft loans, tax concessions and support services. Further assistance to the livestock industry had been rendered in the 1970s in the Savé Valley where the Department of National Parks and Wild Life Management (DNPWLM) eradicated buffalo and most elephant to reduce the risk of foot-and-mouth disease transmission and damage to fencing respectively (Pole, 2006).

During the civil war of the 1970s, the cattle industry in the Savé Valley was impacted significantly by cattle rustling and the herd was reduced from 24,000 in 1975 to 5000 in 1979. After Independence in 1980, economic incentives from the Zimbabwean government for cattle ranching were reduced and stringent conditions were imposed by the European Community on beef exports. Concurrently, the South East Lowveld experienced a prolonged period of below-average rainfall. During those years, the ecological impacts of cattle ranching became apparent – cattle ranchers in Zimbabwe had traditionally established stocking rates based on 'average' rainfall years, which did not account for variable precipitation or competition from indigenous wild herbivores (du Toit, 2004). As a result, cattle were overstocked for decades, resulting in gully erosion, soil capping, increased run-off and the development of lower-productivity grass communities. Sensitive grazers such as reedbuck, Lichtenstein's hartebeest, tsessebe, roan and sable disappeared from the Savé Valley area completely, while other wildlife species suffered from benign neglect and subsistence poaching. Predators were actively persecuted by cattle ranchers, with the effect that wild dogs were extirpated and cheetah, spotted hyaena and lion persisted only at low densities. Protectionist policies at the time prevented landowners from utilizing wild animals occurring on their land, which effectively devalued wildlife and exacerbated population declines (Bond and Cumming, 2006).

Emerging potential for wildlife production

In response to declining stocks of wildlife outside the state-protected areas, the Parks and Wildlife Act of 1975 conferred 'appropriate authority' status on landowners for wildlife that occurred on their land, replacing the earlier protectionist policies (Bond and Cumming, 2006). This law effectively meant that ranchers could utilize wildlife consumptively for profit, such as through hunting or live capture and sale. The right to generate income from wildlife coincided with an increasing realization by some ranchers of the ecological problems associated with livestock ranching. Shortly thereafter, empirical evidence of the competitive advantage of wildlife over livestock began to emerge (Taylor and Walker, 1978; Child, 1988; Bond, 1993), particularly in arid areas (Jansen et al, 1992; Cumming, 1993).

The game-ranching industry began to develop in Zimbabwe in the 1960s, initially in the form of mixed livestock and wildlife production systems, and gained momentum in the 1970s and 1980s (Cumming, 1991). In the Savé Valley, two ranches started to experiment with wildlife as a land use and one rancher (Roger Whittall, Humani Ranch) introduced some waterbuck, giraffe, nyala and white rhinoceros to complement remaining populations of bush-buck, bushpig, duiker, eland, grysbok, impala, klipspringer, kudu, warthog and zebra. In addition to plains game, small populations of cheetah, elephant, hippo, spotted and brown hyaena, leopard and lion remained in the area.

Between 1986 and 1988, 20 black rhinoceros were introduced onto Humani Ranch as part of the Government of Zimbabwe's black rhinoceros conservation strategy, under a custodianship scheme whereby ownership was retained by the state. Meetings were held between Savé Valley landowners, WWF (formerly the World Wildlife Fund) and the DNPWLM to discuss the need for cooperative management of the reintroduced rhinoceros populations. These negotiations, along with leveraged funding provided by the Beit Trust for rhinoceros conservation, led to the development of a cooperative wildlife area or conservancy. Black rhinoceros were thus the 'flagship' species that catalysed the formation of the SVC. Shortly thereafter, a constitution for the nascent SVC was developed, which enshrined the need for cooperative management of wildlife resources while ensuring the sovereignty of individual ranches (see Figure 11.1). In June 1991, 18 ranchers signed the constitution and SVC was formed. A further 13 black rhinoceros were introduced shortly thereafter, and with a high rate of reproduction this population grew to over 100 by 2004.

The Beit Trust provided funds and technical support for the construction of the perimeter game fence, on the agreement that conservancy members would remove internal game fences (thus creating extensive range for the rhinoceros), and provide match funding for wildlife restocking within a stipulated period (du Toit, 1998). This agreement ensured that all members contributed to restocking (pro rata to their ranch areas) and avoided a situation where some ranchers retained fencing around their properties to prevent the loss of wildlife to non-contributing neighbours. The agreement was made with the belief that a concerted effort at restocking would move the conservancy more quickly to a situation where viable tourism operations could be established, and where the rhinoceros would become economic assets rather than liabilities.

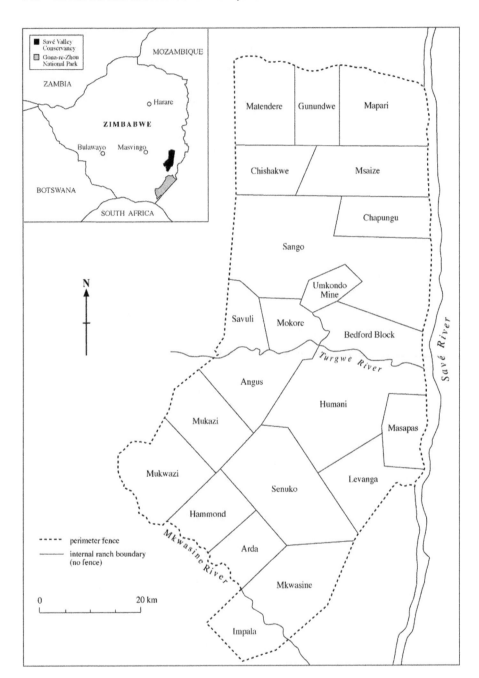

Source: Adapted from Pole et al (2004)

Figure 11.1 *Savé Valley Conservancy*

Following the formation of SVC, some ranchers decided to retain live-stock, pursuing a mixed species production system. However, in 1991–1992, the South East Lowveld experienced the worst drought on record, forcing ranchers to sell cattle at greatly reduced prices. A major cooperative effort was made by the landowners to save grazing wildlife species by importing hay from farms on the Central Plateau. During the drought, a strategic planning meeting was held by conservancy members and a decision was taken to completely remove cattle from SVC and to develop a multi-use wildlife production system for high-quality wildlife tourism.

Development of the conservancy

During the 1990s, a series of steps were taken to foster increases in the diversity and abundance of wildlife within SVC. With further catalytic funding from the Beit Trust, a security system (including personnel) was established to protect the black rhinoceros and control bush meat poaching.

A massive wildlife reintroduction programme was initiated, perhaps the most impressive component of which was the mass translocation of elephants from Gonarezhou National Park. During the 1991–1992 drought, DNPWLM had embarked on a elephant culling programme in Gonarezhou. Seeing an opportunity to obtain elephants for reintroduction into SVC, the conservancy chairman, Clive Stockil, proposed that SVC pay DNPWLM the amount they could expect to gain from the sale of skin, meat and tusks from culled elephants for live elephants. DNPWLM accepted the offer and the SVC immediately embarked on a fund-raising campaign. After initial attempts to capture elephants individually, a method was developed that enabled the capture and movement of whole family groups. This operation was the first time that whole family groups had been captured and translocated, and involved far more elephant than any other translocation operation before or since – 533 individuals were relocated to SVC.

Because the SVC then fell into the foot-and-mouth free (or 'green') zone, and because buffalo are long-term carriers of foot-and-mouth disease, a strong case was required to convince the Department of Veterinary Services (DVS) to permit their reintroduction into the conservancy. A case was presented (Price Waterhouse, 1994) that showed objectively that wildlife had a competitive edge over livestock with respect to returns per hectare, foreign currency generation and scope for the development of economic linkages between ranches and neighbouring communities. The report also demonstrated the crucial importance of buffalo to the viability of wildlife operations. The DVS agreed to buffalo reintroductions under stringent conditions – ranchers were required to remove all remaining cattle within the area and to construct a double game fence of set specifications (Foggin and Connear, 2003). By March 1995, the twin 350km game fences were completed, all internal fencing and remaining cattle were removed and buffalo reintroductions began.

During and following the elephant reintroduction, a major restocking programme of other wildlife species was also pursued. Twenty white rhinoceros were introduced through a donor-supported arrangement whereby local communities would receive the receipts of trade in the progeny of the reintroduced animals (du Toit, 2005). In total, 3128 individuals of 13 wildlife species were reintroduced by the conservancy members (see Table 11.1).

Table 11.1 *Wildlife reintroduced into Savé Valley Conservancy, 1992–2006*

Species	By landowners	Through the IFC[a] loan	Total
Wildebeest	688	27	715
Elephant	553	0	553
Eland	478	18	496
Buffalo	179	247	426
Zebra	326	66	392
Waterbuck	283	78	361
Sable	252	104	356
Giraffe	137	6	143
Tsessebe	79	0	79
Nyala	64	1	65
Black rhino[b]	49	0	49
White rhino[c]	30	0	30
Lion[c]	10	0	0
Total	3128	547	3675

Note: [a] International Finance Corporation; [b] black rhino were provided by the Parks and Wildlife Management Authority as part of the national black rhino conservation plan; [c] 20 of the white rhino and all of the lion were introduced in 2005, which was several years after the next most recent wildlife reintroduction

Source: Pole (2006)

Teething troubles

During the early years of the conservancy, SVC members experienced a difficult transition period where income from livestock had ceased and wildlife densities were too low for high-quality ecotourism. Safari hunting was the most important economic activity during these years and enabled the conservancy to function and develop, though the low densities of high-value species (buffalo, in particular) limited revenues. Further restocking was required to increase the viability of wildlife as a land use and to permit high-quality ecotourism and safari hunting operations. With technical assistance from the Southern African Regional Programme Office of the World Wildlife Fund (WWF-SARPO),

SVC applied to the International Finance Corporation (IFC) for a restocking loan. Following completion of an environmental impact assessment and a process of public consultation, a loan of US$1 million was approved. A stocking plan was developed and implemented, with the primary objective of increasing the diversity and the total biomass of indigenous wildlife, and an additional six species were reintroduced into SVC (see Table 11.1). Wildlife populations in SVC increased steadily in abundance and diversity during the 1990s, and revenues from ecotourism and safari hunting climbed correspondingly, facilitated by political stability and the increasing international exposure of SVC as a conservation success story.

The development of other conservancies in Zimbabwe

During the 1990s, the wildlife industry in Zimbabwe boomed, buoyed by favourable government policies and growing numbers of visiting tourists and hunters (Bond and Cumming, 2006). By 2000, an estimated 27,000km^2 of private land was used for wildlife production (Bond et al, 2004; Child, 2005), including several other large conservancies (see Figure 11.4). SVC and (later) Bubye Valley moved most quickly and completely in developing wildlife populations and replacing livestock with wildlife-based industries.

The role of individuals in the development of conservancies

The formation and evolution of SVC and other conservancies depended on several catalytic and enabling factors, and teamwork among various stakeholders. Possibly the biggest catalyst for the formation of SVC was funding from the Beit Trust, via WWF, which re-enforced the rhino conservation programme, provided technical assistance and created incentives for landholders to amalgamate their properties. Significantly, the Beit Trust funding was flexible and provided over almost a decade, ensuring the scope to adapt the support to changing circumstances. Consultants were engaged at the appropriate times in the conservancy's development to tackle emerging needs, for example, the development of a memorandum of understanding (MoU) between the SVC and neighbouring rural district councils as a vehicle for community outreach efforts. The progressive attitude of the DVS, international recognition for the rhino-breeding success and growing interest of external investors in wildlife ranching propelled the conservancy along the route towards large-scale cooperative management. Later, input from advisers helped secure the crucial IFC restocking loan and more recently key guidance during the land reform programme of the Zimbabwean government.

Criticism of the conservancy approach

The development of the game-ranching industry and of conservancies in particular, was criticized in some quarters and sometimes suffered from a lack of political support (Bond et al, 2004; Cumming, 2005). A key criticism has been that conservancies are underutilized and undermine food security by wasting land that could be used to produce crops or domestic stock (Wolmer et al, 2003, 2004). However, such criticism fails to consider that conservancies are located on land that is generally unsuitable for agriculture due to low rainfall and/or poor soils and livestock-ranching enterprises have tended to rely on external (perverse) subsidies, are generally unprofitable in semi-arid areas, and have tended to result in a cycle of overstocking, ecological degradation and declining productivity (Bond et al, 2004). By contrast, wildlife-based land uses rely on the delivery of safari hunting and tourism that generate higher revenues (in the form of foreign currency) than livestock, which can contribute to national food security (in the same way as the non-edible tobacco crop that is grown in areas of greater agricultural potential than the Lowveld). Further added value may be derived from the live sale of animals, an activity that forms the basis for a large industry in South Africa and Namibia. These returns are less tightly coupled to rainfall than meat production, and are less susceptible to drought than those from livestock farming (Bond and Cumming, 2006). Well-managed wildlife production systems also avoid the ecological degradation typical of livestock production and are thus more sustainable (Bond et al, 2004).

Additionally, there is increasing potential for the local or international sale (from foot-and-mouth disease free areas) of venison as a low fat, free range alternative to meat from domestic animals (Radder, 2002). The sale of affordable meat to local communities from trophy-hunted animals and through the annual harvest of overabundant species can also make a contribution to local protein requirements.

Additionally, a review done during the early 1990s indicated that both the volume and quality of employment opportunities is greater for wildlife production systems than for livestock, and there is no reason to believe that this situation does not hold true today (Price Waterhouse, 1994). If current trends and plans to expand SVC to amalgamate community land and to incorporate community involvement through allocation of shareholdings continue, communities will benefit from opportunities to become wildlife producers and tourism operators.

The electric fencing erected around conservancies has been criticized as representing a symbolic exclusion of impoverished communities by wealthy landowners (Wels, 2000; Wolmer et al, 2003; Spierenberg and Wels, 2006). While the concept of 'moving beyond the fences' to involve and benefit communities from wildlife management is important, the practical significance of fencing must also be considered. In the context of 'hard edge' wildlife areas such as SVC where the boundary of the conservancy abuts densely populated communal land, perimeter fencing is vital for preventing human–wildlife conflict and uncontrolled hunting of wildlife that may wander beyond the borders of the conservancy. Furthermore, the fencing around SVC is a statutory requirement to prevent disease transmission to livestock occurring in neighbouring areas.

Ecological and financial benefits of the conservancy approach

SVC illustrates both the financial and ecological advantages of scale associated with cooperative conservancy arrangements relative to isolated fenced game ranches.

Financial advantages of conservancy scale

The large size of SVC permits the reintroduction of large, charismatic species such as buffalo, elephant and lion, which are key revenue drivers for tourism and trophy hunting (see Figure 11.2) and the pooling of land units permits the marketing of a 'wilderness' experience, which is attractive to both photographic and hunting tourists (du Toit, 2004; Lindsey et al, 2007a, 2007b). The large size of SVC also permits economies of scale that reduce management costs. For example, less fencing and fence maintenance is required, fewer artificial water-points are needed, one annual census can be done for the whole area and the larger land area can support the expensive infrastructure required for high-end tourism (du Toit, 2004).

Ecological advantages of conservancy scale

The conservancy permits the effective conservation of a wider diversity of species than would be possible in smaller land units, as no single ranch encompasses the diversity of habitats found in the conservancy as a whole. The scale of SVC can enhance the resilience of the area to ecological shocks such as fires or droughts by enabling herbivores to make use of patchy primary production resulting from sporadic rainfall (Bond et al, 2004; du Toit, 2004), and by enabling the re-establishment of functional predator–prey relationships. SVC is once again home to significant populations of wild dogs, leopard, cheetah and increasing numbers of lion and spotted hyaena. Predators effectively reduce the amplitude of population fluctuations in non-migratory prey species by preventing overpopulation during high-rainfall years, which may prevent

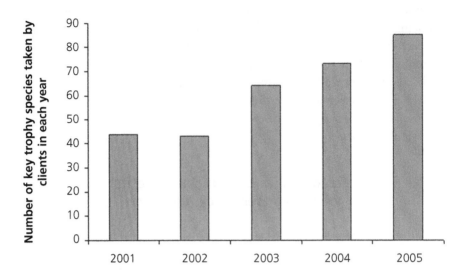

Source: Technical Advisory Committee, SVC

Figure 11.2 *Key trophy species hunted by foreign safari clients, 2001–2005*

population crashes during droughts (Mills and Funston, 2003). Larger areas are more able to support viable populations of wildlife than isolated game ranches, and can host larger populations that are more resilient to stochastic events and are less likely to require augmentation or further reintroductions, and do not require management intervention to prevent inbreeding. Coordinated and cooperative wildlife management has largely prevented issues such as the introduction of extra-limital species, inbreeding or the intentional cross-breeding of closely related species, the breeding of aberrant colour morphs or the persecution of predators, which have reduced the conservation value of game ranching in South Africa (Bond et al, 2004; Lindsey et al, 2007a).

Conservation achievements of SVC

SVC is home to sizeable wildlife populations, including several species of conservation significance (see Table 11.2). There are now nine packs of African wild dogs, occurring at one of the highest densities of that species in the world (Creel and Creel, 2002; Pole, 2006), which had been effectively eradicated from SVC during the cattle production era. Given the high costs and low success rates of wild dog reintroductions in other protected areas, this is one of SVC's more important achievements (Lindsey et al, 2005). SVC also has the largest rhinoceros population in Zimbabwe. Lions recolonized the conservancy from Malilangwe in the south, and the population is increasing rapidly, and the geographic distribution of elephants within Zimbabwe was boosted by approximately 6 per cent following the SVC translocation.

Table 11.2 *Minimum SVC population estimates for wildlife, livestock and human settlements*

Species	Population
Wildlife	
Impala	17,191
Zebra	5075
Wildebeest	4927
Buffalo	1785
Warthog	1426
Eland	1424
Kudu	1150
Elephant	1117
Giraffe	781
Waterbuck	735
Sable	214
Black rhino[a]	120
White rhino[a]	31
Human habitations	4027
Domestic stock	
Cattle	9986
Goats	4715

Note: [a] Known population sizes derived from ongoing monitoring

Source: Joubert and Joubert (2006)

Neighbour relations and political issues

A land use experiment of the size of SVC could not be developed anywhere in the world without raising issues regarding relations with neighbours and official national policies. In Zimbabwe, the further development of the wildlife industry has been constrained by interrelated political and economic tensions. These tensions are associated with a racially inequitable landownership pattern arising from the colonial era. They were present as undercurrents (but not limiting factors) during the early 1990s when the conservancy was established by predominantly white landowners. Many of these landowners had recently bought land in the conservancy, with certificates of approval from the government, and were therefore not inheritors of colonial estates. The architects of the Lowveld conservancies (i.e. the individual ranchers who led the process and the NGO representatives who supported the process technically and financially) were sensitive to these political undercurrents and to the need to make the conservancy relevant to the livelihoods of neighbouring communities. Prior to the Beit Trust funding the development of SVC, political assurances were formally sought and obtained at ministerial level regarding the government's approval for the development of large-scale commercial wildlife projects on semi-arid private land. Clearly, however, these assurances did not reduce the need for community outreach at the local level.

The pace of development of the SVC outreach programme was governed by several factors, of which the most significant (often overlooked by critics of the Zimbabwean conservancy initiatives) is the inevitable inertia involved when a disparate group of landowners is required to make joint decisions on land use transition. The conservancy could not simply spring into existence overnight with all economic, ecological and socio-political aspects being adequately dealt with. Some ranches within the conservancy contributed to community outreach from an early stage by providing direct assistance to small-scale projects such as irrigation schemes, scholarships and community embroidery schemes (Anderson, 2007). Community access was also facilitated on some properties to enable the controlled extraction of renewable resources such as firewood, reeds and edible caterpillars. A significant achievement of SVC community engagement efforts was the establishment of the SVC Trust in 1996. The SVC Trust arose following the signing of a formally negotiated MoU between SVC and the five neighbouring rural district councils (Bikita, Buhera, Chipinge, Chiredzi and Zaka), and has several objectives, including: 'fostering cooperation and communication between the SVC and its neighbouring communities through beneficial and durable economic relationships' (SVC Trust, 2000). The SVC Trust was established as an accountable, not-for-profit body administered by a board of trustees, comprising members from each of the councils, members of parliament, district administrators, traditional leaders and influential citizens. A key initiative of the SVC Trust is to develop a wildlife endowment scheme, where funds are raised to purchase wildlife for reintroduction into the conservancy. The progeny of the restocked wildlife would be regarded as tradable assets of the SVC Trust. In light of current negotiations, these assets are likely to be converted to shareholdings in a new business structure proposed to control the wildlife use rights within SVC and to share earnings between owners and the Trust beneficiaries.

Despite these initiatives, the overall impact of outreach efforts has been limited by the fact that the learning curve regarding community outreach was interrupted by a steep decline in political tolerance for landownership by whites in Zimbabwe. This created a risk that outreach efforts would be derailed, resulting in a lose–lose scenario, rather than a situation where both

SVC and communities could benefit. SVC reflected the problem of racial imbalance at that time – 41 per cent was owned by white Zimbabweans, 54 per cent by white foreign investors and 5 per cent by the government. Landowner-ship was turned into a political flashpoint by the ruling party during a period of challenge by the opposition party in 1998–2000, with the result that private land, including SVC was targeted by the government's 'fast track' land reform programme. By December 2002, five ranches in SVC (Mukwazi, Mukazi, Angus, Chigwete (part of Humani Ranch) and Mkwasine) had been taken over by peasant farmers and three others (Levanga, Masapas and Senuko) were partially settled, comprising a total of 33 per cent of SVC (see Figure 11.3). Other Lowveld conservancies were also affected to varying degrees. Resettlement within SVC and the political and economic instability stemming from the land reform programme in the country as a whole has had several major impacts on SVC.

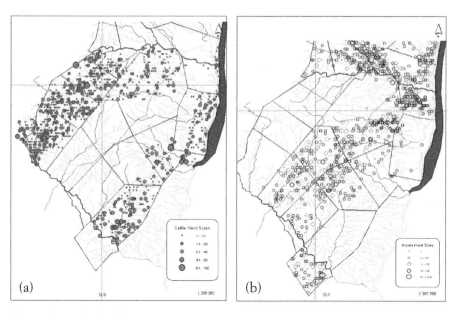

Source: Joubert and Joubert (2006)

Figure 11.3 *Distribution of (a) settlers' cattle and (b) impala in the southern half of Savé Conservancy, 2006*

Disruption of the community outreach programme

With the rights of stakeholders blurred by the radical shift in land tenure, the development of the community outreach programme was stifled. The SVC Trust was derailed because of political allegations that it had been put in place merely as a smokescreen behind which the conservancy hoped to escape land reform.

Increases in bush meat poaching

When the settlers arrived in SVC, the game scouts were quickly prevented from conducting anti-poaching patrols in resettled areas. The last few patrols undertaken on those properties revealed massive levels of snaring for bush meat, often with wire taken from the conservancy fence. During the settlement process, approximately 80km of the double perimeter fence was removed, corresponding to 1280km of wire (enough to produce more than 420,000 snares). Between August 2001 and February 2007, a minimum of 3836 animals were killed on the ranches still run by the original owners (unpublished SVC records). Poaching levels on resettled ranches were undoubtedly much higher – aerial survey data indicates that wildlife populations in those areas have been virtually eradicated (see Figure 11.3b).

Seven years after the onset of settlement within the conservancy, levels of bush meat poaching continue to be extreme. Between August 2005 and February 2007, 2191 incidents of poaching were recorded in SVC resulting in the removal of 13,920 snares and the death of at least 1125 animals. Incidents of poaching are more than twice as frequent on ranches adjacent to resettled properties. Wildlife populations in the south of the conservancy (south of the Turgwe River) are declining, whereas those in the north, away from the resettled areas are stable (Joubert and Joubert, 2006).

Habitat destruction and loss of connectivity

Large areas of woodland have been cleared for subsistence farming within SVC. Due to the poor nutrient content of the soils, resettled farmers practise slash and burn agriculture to boost yields, resulting in the increasing land clearance. The settlement of Chigwete and Masapas ranches in the centre of SVC threatens to prevent wildlife movement between the northern and southern halves of the conservancy, and the occupation in the south-eastern part of SVC (Mkwasine and parts of Senuko and Levanga ranches) jeopardizes connectivity between SVC and the Great Limpopo Transfrontier Area (see Figure 11.4).

Disease outbreaks

By 2006, an estimated 14,700 domestic stock had been moved onto settled ranches within the conservancy (see Table 11.2 above), greatly increasing the risk of transmission of diseases between wildlife and domestic animals. The breakdown of foot-and-mouth disease controls in SVC and other wildlife areas around the country led to a cessation of beef exports from Zimbabwe to the

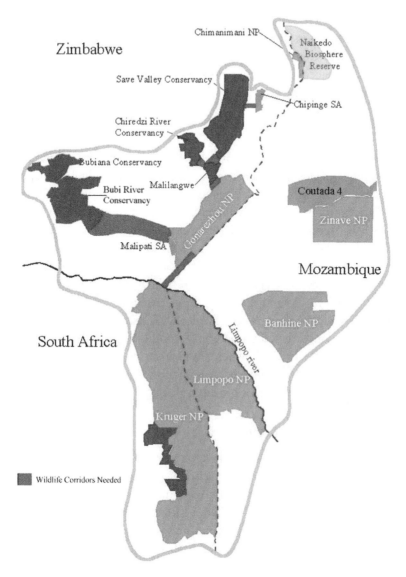

Source: Derek de la Harpe (pers. comm.)

Figure 11.4 *Zimbabwean Lowveld conservancies and the Great Limpopo Transfrontier Conservation Area*

European Union in 2001 (du Toit, 2004). Two outbreaks of foot-and-mouth disease have been recorded in the communal land neighbouring the conservancy since the settlement began (Foggin and Connear, 2003). In addition, in 2005, three wild dogs from a single pack in the southern part of SVC died of symptoms resembling rabies and the pack subsequently disappeared.

Subsequently in 2007, a rabies outbreak in the adjacent Malilangwe Trust resulted in the disappearance of three packs of wild dogs. Finally, a major anthrax outbreak was recorded in SVC/Malilangwe in 2004, related to the breakdown of vaccination programmes in neighbouring communal areas, attributable to the failing economy resulting from the land reform programme.

Human–wildlife conflict

The southern part of SVC now represents a mosaic of natural habitat used for wildlife production and subsistence farming, resulting in conditions conducive to human–wildlife conflict (HWC). The removal of portions of the perimeter fence by the settler farmers has greatly increased HWC in neighbouring communal lands. In SVC, the conflict has been manifested by fatal encounters between humans and wildlife (at least 40 human deaths in or near SVC due to wildlife since 2000), crop damage and livestock depredation (Lindsey, 2007). In response to crop damage, several elephant bulls are killed in problem-animal control operations every year, significantly reducing potential revenues from trophy hunting each year (Lindsey, 2007; Martin, 2007). Settler farmers living in the conservancy no longer employ traditional (conflict-reducing) husbandry techniques employed effectively elsewhere (Woodroffe et al, 2006); and as the lion population increases, complaints of livestock losses appear to be increasing in frequency, resulting in the risk of predators being poisoned by affected farmers.

Major drop in tourism revenues

With the onset of the political turbulence in Zimbabwe, revenues from tourism declined dramatically. In 1999, ecotourism was the primary source of revenue for four ranches in SVC and comprised a significant component of income on another seven properties. By 2004, all of the ranches in SVC relied primarily on income from trophy hunting. The hunting industry has proved to be considerably more resilient to political instability than ecotourism. With the onset of the land reform programme, tourist occupancy in Zimbabwe fell by 75 per cent, whereas trophy hunting revenues dropped by only 12 per cent (Booth, 2002; Bond et al, 2004). In SVC, trophy hunting revenues on the ranches not affected by land reform increased after 2000, due to increasing populations of buffalo, permitting the sale of more high-value 'dangerous game' hunts (see Figure 11.2).

Erosion of property rights dissuading investment

In 2005, the government declared that all agricultural land owned by Zimbabwean nationals belonged to the state, and white farmers continue to be evicted from their properties. Though no further evictions have occurred in

SVC since 2002, the uncertainty of tenure has stifled further investment and reduced enthusiasm for the nascent community outreach initiatives. The resettled farmers in SVC face similar uncertain tenure over the properties they occupy, which, in combination with open access to natural resources has encouraged unsustainable bush meat hunting.

The future

As will be obvious from the issues outlined above, the future of SVC as an experiment in large-scale cooperative wildlife management depends greatly upon the resolution of pressures arising from the land reform programme. Specifically, the future of SVC will depend on the extent to which the conservancy can incorporate and benefit a broader range of stakeholders, especially local communities, through arrangements that retain business viability.

The government has recently reversed its stance on the resettlement of conservancies for small-scale agriculture and has stated that Lowveld conservancies will remain wildlife production zones. After a series of discussions between conservancy representatives, government officials and technical advisers, the Zimbabwean Cabinet approved a short general policy statement on wildlife-based land reform in 2005. The Ministry of Environment and Tourism has been tasked with the development of guidelines to implement this policy statement. Commencing in early 2006, under the auspices of the Ministry of Environment and Tourism, progressively more detailed discussions have been taking place within the group of government officials, conservancy representatives and non-governmental advisers. At present, various models for wildlife-based land reform have been proposed by the government, including a variety of combinations of current landholders, neighbouring communities, the Parks and Wildlife Management Authority (PWMA, formerly the DNPWLM) and indigenous entrepreneur investors.

All of these models require conservancies to engage with neighbouring communities and provide a continued role for those landowners who have not already been evicted from their ranches. A continued role for existing landholders is important, as this would enable the retention of expertise and functioning tourism operations while new entrants are eased into the industry. Since private land has been nationalized, the primary assets that remain as tradable commodities (as identified in the current policy statement) are wildlife populations. Arrangements are thus being developed to form new corporate structures that manage and trade the use rights for these wildlife assets. Shareholdings are envisaged for existing operators, for community trusts (thus the SVC Trust has regained political recognition) and for black entrepreneurs who wish to buy shares. Shares could also be acquired by adding wildlife or land to the conservancy, which is the most feasible route by which PWMA would become a partner (notably, through the inclusion of Chipinge Safari Area, east of SVC). Large shareholders would undertake commercial wildlife operations within the conservancy, based on the wildlife use rights that arise from their shareholding. Practical requirements for the effective functioning of the SVC include the need for rational alignment of wildlife and agriculture and repairing the boundary fence where it has been removed.

These concepts are still evolving and the final form of the arrangements for wildlife-based land reform cannot be stated with certainty. However, irrespective of the composition of shareholders and of where the conservancy boundaries are re-established, SVC will need to renew its community outreach effort through measures such as those outlined below.

Developing durable business partnerships with neighbouring communities

As outlined above, the SVC Trust should be included as a shareholder in any a new corporate structure. A minimum 10 per cent shareholding for the Trust is currently envisaged and would provide dividends related to wildlife populations within the conservancy (providing the incentive for communities to reduce poaching). In addition, there is scope for SVC to be expanded to include portions of community land. Negotiations between SVC and the Nyangambe community (adjacent to Hammond ranch) have already resulted in the decision by that community to incorporate approximately 25km² of land into the conservancy. There are other such opportunities around other parts of SVC, despite generally high human population densities. These areas would have to be restocked by SVC, adding to the community shareholding and providing direct business opportunities for trophy hunting and ultimately ecotourism – especially those linked to cultural tourism possibilities. Finally, joint ventures could be developed, such as out-grower vegetable schemes to supply safari camps, micro-industries such as tanneries and craft production (A. Sithole, pers. comm.).

Other contributions to neighbouring communities

Communities in the South East Lowveld suffer regular food shortages due to low and irregular rainfall and the lack of irrigation (Cumming, 2005). Food shortages, increasing poverty levels and ineffective legal deterrents combine to ensure that SVC experiences severe levels of poaching for bush meat, most of which is sold locally. Bush meat poaching is extremely inefficient: for every animal successfully extracted by poachers using snares, four others are killed and not recovered. If the illegal offtake of bush meat could be replaced with a sustainable legal source of bush meat, SVC could contribute a constant supply of protein to neighbouring communities and the impact on wildlife populations would be lower. For example, SVC could remove 50 or more elephants annually to limit population growth and obtain a supply of meat for distribution to neighbouring communities (Martin, 2007). It has been proposed that elephant meat would be granted to the SVC Trust, which could sell it at a subsidized price to local communities to generate funds for community development projects.

Developing channels of dialogue between SVC members and their neighbours

Ranch/ward subcommittees within the conservancy should be established and regular meetings should be held between ranchers, neighbouring community leaders and local politicians. These forums would enable ranchers and community members to get to know one another and enable discussion of both problems and potential avenues for cooperation.

Developing comprehensive education and awareness programmes

Education programmes are obviously important in raising awareness of conservation issues and of the potential importance of wildlife to local community development. Interviews with local communities revealed misconceptions relating to wildlife and conservation, such as the belief that bush meat poaching has no impact on wildlife populations (S. Matema, pers. comm.). During these interviews, community members have indicated that the opportunity for their children to view wildlife would constitute a valued benefit from SVC.

Conclusions

SVC is an experiment in progress and cannot yet be said to be a tested model for cooperative wildlife management in southern Africa, though it has achieved significant gains for biodiversity conservation and has the potential to act as a key contributor for local community development. These achievements and potentials are, however, dependent on a favourable resolution of the political processes under way at the time of writing in Zimbabwe. If SVC survives the political furnace in which the experiment is currently being forged, it will illustrate important lessons for other countries in southern Africa where similar challenges exist, or can be expected to arise in the near future.

References

Anderson, N. (2007) 'Draft strategy for the Savé Valley Conservancy Outreach Programme', unpublished report, Savé Valley Conservancy, Zimbabwe

Bond, I. (1993) 'The economics of wildlife and landuse in Zimbabwe: An examination of current knowledge and issues', Project Paper No. 36, WWF Multispecies Animal Production Systems Project, Harare

Bond, I. and Cumming, D. H. (2006) 'Wildlife research and development', in Rukuni, M. P., Tawonezi, C., Eicher, M., Munyuki-Hungwe, F. and Matondi, P. (eds) (2006) *Zimbabwe's Agricultural Revolution Revisited*, University of Zimbabwe Publications, Harare

Bond, I. Child, B., de la Harpe, D., Jones, B., Barnes, J. and Anderson, H. (2004) 'Private land contribution to conservation in South Africa', in Child, B. (ed) *Parks in Transition*, Earthscan, London

Booth, V. (2002) 'Analysis of wildlife markets (sport hunting and tourism)', WWF-SARPO Report, Harare

Child, B. (1988) 'The role of wildlife utilization in the sustainable economic development of semi-arid rangelands in Zimbabwe', DPhil thesis, University of Oxford, Oxford

Child, B. (2005) 'Principles, practice and results of CBNRM in southern Africa', report prepared for the Sand County Foundation, Monona, WI

Creel, S. and Creel, N. M. (2002) *The African Wild Dog: Behaviour, Ecology and Conservation*, Princeton University Press, Princeton, NJ

Cumming, D. H. M. (1991) 'Wildlife products and the market place: A view from southern Africa', in Renecker, L. A. and Hudson, R. J. (eds) *Wildlife Production: Conservation and Sustainable Development*, University of Alaska, Fairbanks

Cumming, D. H. M. (1993) 'Multi-species systems: Progress, prospects and challenges in sustaining range animal production and biodiversity in East and southern Africa', in Proceedings VII World Conference on Animal Production, Edmonton, Alberta

Cumming, D. H. M. (2005) 'Wildlife, livestock and food security in the South East Lowveld of Zimbabwe', in Osofsky, S. A., Cleveland, S., Karesh, W. B., Kock, M. D., Nyhus, P. J., Starr, L. and Yang, A. (eds) *Conservation and Development Interventions at the Wildlife/Livestock Interface: Implications for Wildlife, Livestock and Human Health*, IUCN, Gland

du Toit, R. (1998) 'Case study of policies that support sustainable development in Africa: Save Valley Conservancy, Zimbabwe', paper presented at Scandinavian Seminar College Workshop African Experiences with Policies and Practices Supporting Sustainable Development, 28–30 September 1998, Harare

du Toit, R. (2004) 'Review of wildlife issues associated with the land reform programme in Zimbabwe', World Wide Fund for Nature – Southern African Regional Programme Office, Working Paper No. 10, World Wide Fund for Nature, Harare

du Toit, R. (2005) 'Implementation of a rhino endowment model for community participation in rhino conservation, Save Valley Conservancy, Zimbabwe', *Pachyderm*, vol 39, pp114–116

Foggin, C. and Connear, G. (2003) 'Introduction of foot-and-mouth disease-infected buffalo into the Save Valley Conservancy in Zimbabwe: Success or failure?', paper presented at the IUCN World Parks Congress AHEAD Forum, 14 September 2003, Durban

Jansen, D. J., Child B. and Bond, I. (1992) 'Cattle, wildlife, both or neither: Results of a financial and economic survey of commercial ranches in southern Zimbabwe' WWF Multispecies Project Paper No. 27, World Wide Fund for Nature, Harare

Joubert, D. and Joubert, L. (2006) 'Aerial survey of the Savé Valley Conservancy, September 2006', unpublished Savé Valley Conservancy report, Zimbabwe

Lindsey, P. A. (2007) 'Human-elephant conflict in and around Savé Valley Conservancy', unpublished Savé Valley Conservancy report, Zimbabwe

Lindsey, P. A., Alexander, R., du Toit, J. T. and Mills, M. G. L. (2005) 'The cost efficiency of wild dog *Lycaon pictus* conservation in South Africa', *Conservation Biology*, vol 19, pp1205–1214

Lindsey, P. A. Roulet, P. A. and Romañach, S. S. (2007a) 'Economic and conservation significance of the trophy hunting industry in sub-Saharan Africa', *Biological Conservation*, vol 134, pp455–469

Lindsey, P. A., Alexander, R., Mills, M. G. L., Woodroffe, R. and Romañach, S. S. (2007b) 'Wildlife viewing preferences of visitors to protected areas in South Africa: Implications for the role of ecotourism in conservation', *Journal of Ecotourism*, vol 6, pp19–33

Martin, R. B. (2007) 'Savé Valley Conservancy management of the elephant population', unpublished Savé Valley Conservancy report, Zimbabwe

Mills, M. and Funston, P. (2003) 'Large carnivores and savannah heterogeneity', in du Toit, J., Rogers, K. and Biggs, H. (eds) *The Kruger Experience*, Island Press, Washington DC

Phimister, I. R. (1978) 'Meat and monopolies: Beef cattle in southern Rhodesia, 1890–1938', *Journal African History*, vol 29, no 3, pp391–414

Pole, A. (2006) 'Management plan for the Savé Valley Conservancy', unpublished report

Pole, A., Gordon, I., Martyn, L., Gorman, L. and MacAskill, M. (2004) 'Prey selection by African wild dogs (*Lycaon pictus*) in southern Zimbabwe', *Journal of Zoology*, vol 262, pp207–215

Price Waterhouse (1994) *The Lowveld Conservancies: New Opportunities for Productive and Sustainable Land-Use*, Savé Valley, Bubiana and Chiredzi River Conservancies, Harare

Radder, L. (2002) 'Restaurants and venison marketing: A South African experience', *Food Service Technology*, vol 2, pp109–114

SVC Trust (2000) 'Memorandum of understanding between the Save Valley Conservancy and the Rural District Councils of Bikita, Chiredzi, Buhera, Chipinge and Zaka', unpublished document

Spierenburg, M. and Wels, H. (2006) 'Securing space: Mapping and fencing in transfrontier conservation in southern Africa', *Space and Culture*, vol 9, pp294–312

Taylor, R. D. and Walker, B. H. (1978) 'Comparisons of the vegetation use and herbivore biomass on a Rhodesian game and cattle ranch', *Journal of Applied Ecology*, vol 15, pp565–581

Wels, H. (2000) 'Fighting over fences: Organizational co-operation and reciprocal exchange between the Savé Valley Conservancy and its neighbouring communities', PhD thesis, Free University, Amsterdam

Wolmer, W., Chaumba, J. and Scoones, I. (2003) 'Wildlife management and land reform in Southeastern Zimbabwe: A compatible pairing or a contradiction in terms?' Sustainable Livelihoods in Southern Africa Research Paper 1, Institute of Development Studies, Brighton

Wolmer, W., Chaumba, J. and Scoones, I. (2004) 'Wildlife management and land reform in Southeastern Zimbabwe: A compatible pairing or a contradiction in terms?', *Geoforum*, vol 35, pp87–98.

Woodroffe, R., Frank, L. G., Lindsey, P., ole Ranah, S. and Romañach, S. (2006) 'Tools for conserving large carnivores in Africa's community rangelands: A case-control study of livestock husbandry', *Biodiversity and Conservation*, vol 16, pp1245–1260

Part IV

Community-Based Natural Resource Management

12

Community Conservation in Southern Africa: Rights-Based Natural Resource Management

Brian Child

Part IV describes national community-based conservation programmes in Zimbabwe, Namibia, Botswana and Mozambique. While community conservation in southern Africa has generally fallen under the rubric of community-based natural resource management (CBNRM), this term was not coined in the region and nor does it adequately define southern Africa's particular brand of community conservation sufficiently to differentiate it from practices elsewhere. This concerns us because we use CBNRM as shorthand for a set of economic, political and organizational principles within a strongly devolutionary rights-based approach (see Barrow and Murphree, 2001), though the same term is applied to different ideas in different regions of the world. The conceptual integrity of the CBNRM principles are not always followed and weak implementation is undermining the importance of these principles; it is certainly attracting academic criticism and resulting in the dissipation of support from the donor community.

It seems that the 'crisis in CBNRM' (Hutton et al, 2005; Rihoy et al, 2007) has resulted from weak implementation of the principles and not from a weakness in the principles themselves. That this criticism, which does not always recognize the stochastic pathway by which new ideas emerge, is serving to undermine a concept that is vital for the simultaneous conservation, development and political emancipation of poor people is of great concern. For example, in a common response to an increase in negative and, we believe, partial analyses based on limited fieldwork (for example Mashinye, 2007), a Norwegian funding officer recently told me that 'CBNRM does not work' and the United States Agency for International Development (USAID), having provided substantial support to CBNRM for nearly 15 years withdrew just as it was becoming clear that the concept was working. The region is characterized by climatic variability, low productivity and fugitive resources, including mobile herbivores. Rural people in southern Africa need multiple livelihood strategies to survive, and the potential of some of these strategies (such as wildlife management) is highly dependent on collective action. If the search for effective models of collective action were abandoned, the use of these environments will remain unsustainable and economically ineffective.

This is not to say that community conservation in southern Africa has been universally well implemented, or that criticism, so long as it is constructive, is not a vital part of the evolutionary process. Certainly, even within countries, results have been uneven. On the one hand, this makes for a rich learning experience, with the proviso that the monitoring and knowledge networks are in place to ensure that learning occurs. On the other, there are sufficient examples of 'success' to show that community conservation works – and can work consistently, if it is implemented properly.

The CBNRM 'model' as understood in southern Africa is a multifaceted approach that combines economic, political and institutional goals. Economically, its objective is to provide institutions of collective action that improve the productivity of non-agricultural systems by allowing higher-valued land uses, such as wildlife, to be adopted. While the model has concentrated on wildlife since the 1990s, reflecting the economic power and growth of the wildlife-tourism sector, the concept is in no way limited to wildlife, but can take advantage of whichever natural resource is valuable at the time. Practitioners have been trying to diversify community conservation for many years and have learned that developing new products takes skill, investments, marketing and luck – and a lot of time. Given the vast areas of land that community conservation is protecting, the next big market may well be for water or carbon.

Rights-based community conservation

Community conservation in southern Africa emphasizes the centrality of institutions of local collective action for allocating land to its highest valued uses. Community conservation has as its critical ingredient the transfer of the rights to use, manage and allocate resources, especially high-value resources, to grass-roots communities. Stronger communal property rights improve resource allocation by better internalizing the costs and benefits of land use, especially where they are supported by effective information to support decision making. Each of the chapters that follow emphasize the necessity of devolving to communities the proprietorship and control necessary for them to benefit from, manage and control wild resources. Authors such as Adams and Hulme (2001) have categorized the southern African model as typifying high levels of community control and a focus on conservation for use values. Improved institutions are an important means for reducing the ecological over-utilization of natural resources while simultaneously increasing livelihood benefits; hence the claim that community conservation is about both conservation and development. The propensity for institutions to convert low-order commodities into higher-order goods and services lies behind the statement that 'the tradeoff between conservation and development has been exaggerated, and the synergies have been under estimated' (Whande et al, 2003, p1).

The four chapters that follow demonstrate the home-grown nature of community conservation in southern Africa. It took three decades of experimentation before economic models of wildlife utilization were transferred from private land to communal lands, and two decades have already been spent developing institutional and organizational models that still have a way to go. Regional networks have been important for accelerating this learning process, and while this experience may have informed the global narrative on community conservation, the reverse is less true with most ideas being developed locally by empirical trial and error.

Southern Africa's community conservation programmes have been the recipients of substantial amounts of development assistance. Donors have funded the spread of community conservation to other regions, though often with insufficient attention to the principles and practices upon which they are based, the importance of genuine local ownership, or to the long timeframe needed for them to mature; this may be why community conservation is reported to be in crisis.

In a book about innovation, it is useful to analyse how the ideas about CBNRM were developed. They certainly began with some theoretical basis, be these economic ideas about property rights and resource allocation, or political ideas about devolution and empowerment. However, ideas were only refined into operational principles by scholar–practitioners working hand-in-hand with rural communities to implement them. In other words, the principles and practices of community conservation emerged inductively, with scholar–practitioners using observed instances to produce a universal claim or principles. Marshall Murphree has played a prominent role in articulating many of the ideas emerging in sustainable use and community conservation in southern Africa. Murphree's laws,[1] as they are commonly referred to in the region (see Martin, 2007), are interwoven into the dynamic interactions between policy and practice and remain alive as they are being constantly tested and refined. Indeed, this interaction between scholarship and practice has been a critical ingredient in the growth of community conservation in the region.

Southern Africa's scholar–practitioners argue that the only way to develop these ideas is to give communities the right to use wild resources, and to then work with them to iteratively improve institutional, biological, economic and political aspects of management (pers. obs.). Management capacity seldom emerges in the absence of the rights. As articulated by Murphree 'authority is a pre-requisite for responsible management and should not be held out as a reward for it' (Martin, 2007). A key ingredient in this learning process is the tracking of performance, something that still needs to be done better.

We need to improve the understanding of several issues – first, the internal conditions that allow community conservation to work properly, and second, the external conditions that allow the emergence of local collective action. In practice, local failure can often be traced to an external environment that prevents communities from adhering to the CBNRM principles. The hesitation to devolve sufficient discretionary authority and responsibility is particularly problematic.

Consensus about the internal conditions necessary for effective CBNRM is emerging (see chapters in this section; Murphree, 1991; 2000; Jones and Murphree, 2004). Put briefly, CBNRM works best where it is managed by communities small enough to interact face-to-face, and where there are many internal feedback loops of accountability and information that use techniques appropriate to oral societies. In additional, the majority (preferably all) of the benefits of wildlife should by right be controlled by individuals (albeit individuals acting collectively) and they should only flow upwards to implementing committees through a process of upward delegation by the community. In other words, strong communities need to be built by locating discretionary power in the individual and collective will of ordinary people, not the committees that lead them. Only then should the nested hierarchies necessary to manage natural resources begin to be constructed through the upward delegation of authority (Martin, 2007).

The external conditions for the emergence of community conservation are far more variable and they are not yet well understood. The region includes countries with various colonial histories and with very different trajectories of administrative and political change. Consequently, developing and protecting the conditions that allow the emergence of local collective action has been opportunistic and highly variable. Both Zimbabwe and Namibia confidently pursue home-grown national programmes with government blessing and strong, locally staffed NGOs. Botswana was initially more dependent on external expertise and assistance, but in the absence of this, most responsibility now falls on the government, while in Mozambique and Zambia community conservation has 'bubbled up' (Wilson, 2005) in different forms and in different places depending on personalities and aid projects. Perhaps the common feature of community conservation in the region is the cohort of committed individuals finding unique ways of working with country-specific opportunities and constraints towards a vision of CBNRM as closely aligned as possible to the one articulated above.

Zimbabwe

Taylor's chapter provides a multidimensional assessment of the Communal Areas Management Programme for Indigenous Resources (CAMPFIRE) programme 20 years on. As a government ecologist in Matusadona National Park in the Sebungwe region of Zimbabwe, Taylor participated in the 'design' of CAMPFIRE as part of a small cohort of park managers who recognized that parks and wildlife were seriously threatened by land conversion and rapid demographic change, and that the economic potential of wildlife needed to be directed toward improving the livelihoods of rural people living and (increasingly) settling near parks and wildlife if parks were to survive (Taylor, 1978; Martin, 1981).

Many outsiders perceive CBNRM as being concerned primarily with conserving wildlife. Insiders differ – Taylor, for instance, describes how CAMPFIRE quickly graduated from a concern with conservation to a vision based on devolved rural governance, democratization and resource allocation (see also Martin, 1986; Child, 1993). The high value of wildlife was convenient for promoting fiscal devolution, and therefore for building democratic and empowerment processes around the management of revenues by local people (Child and Peterson, 1991). Many of the original champions of CAMPFIRE were not wildlife conservationists (see Metcalfe, 1993; Child et al, 1997), and its birth coincided with recentralization of the Zimbabwe government, and may well have been a response to it. CAMPFIRE was also designed as an institutional mechanism (not a wildlife programme) for improving the allocation of natural resources (Minister of Environment and Tourism, 1992). This was based on the theory of property rights: the devolution of property rights to local people would enable them to allocate resources to higher-value uses, thereby addressing the ineffective and ecologically unsustainable use of natural resources under the prevailing open-access regime in Zimbabwe's communal lands. Thus CAMPFIRE was essentially about the devolution of rights to manage, use, dispose of (i.e. trade or sell) and benefit from natural resources.

CAMPFIRE originated locally and donor funding followed rather than led the programme. Taylor describes how state wildlife officials used legislative changes to transfer their power to communities, ensuring adherence to the 'CAMPFIRE Principles', using simple guidelines (see Maveneke, 1996) and soft yet powerful adherence to the devolutionary 'spirit' of the Parks and Wildlife Act of 1975. CAMPFIRE deliberately avoided blueprint planning; its essence was that local people should be given the right to benefit from and manage wildlife and the space in which to experiment and learn, but within a framework that encouraged high levels of participation.

Taylor outlines a number of national performance metrics that suggest that CAMPFIRE has been broadly successful in the face of challenging demographic and political circumstances, but that this success contains considerable variability. Importantly, he also examines the programme's macro-environment. He ends on a positive note, describing in some detail how, after a period of stagnation in the early 2000s, a devolutionary impulse has reasserted itself when least expected and under extremely difficult conditions. Led by individual communities such as Masoka, direct payments to communities are now spreading nationally.

We are only beginning to understand why CAMPFIRE has been so surprisingly resilient and adaptable. At the micro-level, Taylor associates this with participatory democracy, where the leaders are held accountable to the followers and the linkages between leaders and followers are numerous. At the macro-level, there are several possible explanations. First, while the capture of significant revenues by district councils – because of a legal situation that allowed only partial devolution (Murphree, 1997) – has certainly harmed the programme, this relationship may also have advantages (Hasler, 1995). At least a few district councils reciprocate by providing political and technical support, including protecting local democratic processes and acting against local elite capture (Rihoy et al, 2007). The programme invested significant resources, building the capacity of district councils to support local CAMPFIRE communities and to buy into the devolutionary principles upon which it is based. Second, the CAMPFIRE Association has played an important role in providing the programme with political legitimacy and leadership. Early on, Zimbabwe took the bold step of creating this political organization to represent CAMPFIRE communities. Interestingly, Murombedzi (pers. comm.) suggests that, ultimately, CBNRM will only work if it becomes a political movement which, in Zimbabwe, it may well do. Finally, it is not coincidental that the recovery of CAMPFIRE is linked to new leadership in the wildlife authority that supports its principles; in our frustrations with government agencies, we sometimes forget that they hold the key to devolution and are sometimes more willing to promote devolution and good practice than we give them credit for.

Namibia

The Namibian national CBRNM programme grew from Garth Owen-Smith's community game guard programme in the early 1980s. Like CAMPFIRE, from which it drew many lessons, its strength emerged from a carefully crafted legislative foundation (see Chapters 8 and 14). In common with CAMPFIRE, these legislative changes were driven by skilled professionals within the government who, against the bureaucratic impulse, deliberately gave power to communities and sourced implementation capacity from NGOs staffed largely by local conservationists. One of the strengths of the Namibian programme is the strong programmatic coordination between a number of highly motivated CBNRM support agencies and the relationships they have built with the Ministry of Environment and Tourism.

The Namibian programme is evolving rapidly, suggesting that successful CBNRM is a moving target. In its game guard phase, it sought to undo the complete disenfranchisement of local people and leaders from wildlife (arising from apartheid) by involving them directly in wildlife management. In the early 1990s, it was described as a social empowerment programme because it lacked the economic foundation of CAMPFIRE (Margie Jacobsen, pers. comm.), and was paid for by donor money. By the late 1990s, wildlife had recovered so well in conservancies that Namibia's programme became known for its impressive financial trends. However, as Jones and Weaver explain, this switch from donor-funded wildlife management to self-generated income upended mechanisms of accountability and created new challenges. Top-down mechanisms of accountability were created to manage grants provided for improved natural resource management, and these tended to entrench the authority of committees and managers. Now that many communities have weaned themselves off donor funds and created their own sources of income, new mechanisms of internal (bottom-up) accountability are needed. Overall, CBNRM has created a general awareness of the validity of wildlife as a legitimate land use, and committees are managing wildlife and natural resources with increasing competence. However, these conservation and economic successes mean that CBNRM now faces new challenges: it needs to increase the impact of CBNRM at the household and not just community level, and to promote further devolution and localized forms of participatory accountability.

Botswana

Like the other authors in this section, Nico Rozemeijer played a long and critical role championing CBRNM in Botswana from his position in an NGO. CBNRM in Botswana was initiated externally by funding from USAID between 1989 and 1999, and managed by a firm contracted from the US. The programme was also supported by a few small NGOs with limited funding until the mid-2000s. However, as a middle-income and successful country, Botswana lost donor support for its nascent civil society and associated NGOs. The consequent reduction in civic capacity impacted the emerging CBNRM sector negatively. The programme is now managed primarily by the wildlife department, which has low but improving capacity.

In Botswana's case, CBNRM rights are 'leased' from land boards for 15-year periods, allowing communities to sell hunting concessions and retain all revenues, but giving few other rights – they did not include the right to set quotas or manage wildlife. Despite weak monitoring systems, it is clear that participation, income and employment increased rapidly in areas where people had few alternatives. Learning from Zimbabwe and Namibia, Botswana's CBNRM programme introduced open competitive marketing of hunting and tourism concessions in the early 1990s. This increased the price of wildlife very significantly throughout Botswana. An important but unintended benefit was that, once it was realized how valuable wildlife was, a very inefficient system of uncontrolled citizen hunting was brought under increasing control. Rozemeijer also notes that poaching was reduced in participating communities.

Although theoretically designed as participatory democracies, the mechanisms for operationalizing participation (such as quota setting, revenue distribution and quarterly community meetings) were never put in place, a situation by no means limited to Botswana. Although single village communities such as Sankoyo, Mababe and Khwai appear to be working fairly well, the absence of effective institutions in many of the multi-village communities has made them particularly prone to elite capture by elected management committees who seldom refer back to their constituencies or provide them with benefits. Some of these committees have reputations, deserved or not, for financial inefficiency and misappropriation, which provided ammunition to agencies and individuals uncomfortable with devolution. With their case weakened by these accusations and, without a national-level political organization or strong NGO support to fight for them, communities were unable to politically defend their new-found rights. When the CBNRM policy was finalized in 2007 (Government of Botswana, 2007), it included clauses that may recentralize 65 per cent of wildlife revenues into a national trust fund. While this will greatly undermine the ability of CBNRM to work, these clauses have not yet been invoked; intriguingly, wiggle room was added to this clause at the last minute with a statement that 'the Minister may however, vary these percentages depending on the circumstances and needs of a particular community-based organization' (Government of Botswana 2007, sect 10.3). Interviews suggest that key government officials want CBRNM to work but are extremely concerned that local misappropriation is damaging the programme and that mechanisms are needed to counteract this.

Rozemeijer's chapter brings up two issues: the relationship between the community and the government, and the relationship between the committee and its constituency. Concerned about the low level of political support for CBNRM in Botswana, he suggests that the strategy of by passing local councils may need to be rethought. A more nuanced approach to the relationship between communities and district authorities may be needed so that district councils are more interested in supporting CBNRM in ways that hold both communities and councils more accountable, and that avoid the 'differential taxation' of wildlife.

In a theme common to all four chapters, Rozemeijer suggests that CBNRM in Botswana is too centralized even at local level (see Murombedzi, 1999) where it benefits and empowers committees but with insufficient attention to the communities they represent. The accountability of committees to communities is an area that CBNRM is still addressing. So is the fact that individuals bear the costs of wildlife yet benefits are often managed collectively. Rozemeijer argues that CBNRM in Botswana would benefit greatly by encouraging household cash distribution[2] to compensate people for the costs imposed by wildlife. Also, because it is extremely difficult to set up profitable enterprises (especially community-managed businesses) in remote areas, he insists that wildlife revenues used for such projects are wasteful. For wildlife to become a viable land use it has to compete economically with alternatives in the eyes of individual land users.

Concerns about the success of CBNRM expressed by Rozemeijer originate from structural weaknesses in the design of Botswana's CBNRM programme. First, there has been insufficient investment (including time) in building the structures and capacities of participatory democracy, including accountability mechanisms that address the threat of elite capture or misappropriation, and information systems that lead to informed decision making. Second, the mechanisms for allocating wildlife income are problematic. There is little discussion of trade-offs between alternative uses except within the committee, and benefits almost never get to individuals. In a nutshell, CBNRM in Botswana has generated considerable income and wildlife conservation but (like many of the programmes in the region) it needs to focus on ensuring that individuals within communities participate and benefit, and on improving performance monitoring and governance.

Mozambique

Nhantumbo and Anstey describe a rather different pattern of emergence of community conservation in Mozambique. Many years of war and a socialist outlook left Mozambique with a depleted wildlife resource, a nationalized land tenure system and almost no private sector. Unlike the national programmes in Zimbabwe, Namibia and Botswana, CBNRM in Mozambique has been project based, with some 68 projects testing a diversity of approaches and institutional arrangements for managing forests, wildlife, coastal resources, fisheries and tourism. Despite this diversity of origins, over time and in common with the neighbouring countries, the importance of secure resource tenure and the generation of benefits have emerged.

Some of Mozambique's new policy and legislation is conceptually progressive (albeit contradictory). However, Nhantumbo and Anstey suggest there is a gap between intention and implementation that is undermining the expectations of communities and encouraging a reversion to open-access harvesting of natural resources. Genuine rights to resources have not been devolved and where they have, regulations allow communities to retain only 20 per cent of the government revenue from forest and wildlife resources. To get even this, communities need to jump through difficult administrative hoops. CBNRM has also been restricted largely to resource-poor areas, with the higher-value resources being appropriated by private interests. Consequently, it has been difficult to developing a sound financial base for CBNRM. It has also been difficult to develop democratic grass-roots governance because local co-management structures are dominated by stakeholders and bureaucrats (including appointed chiefs) rather than by communities.

Thus, Nhantumbo and Anstey conclude that a promising start towards community conservation and its considerable potential for conservation and pro-poor development has so far been undermined by a failure to conform to the underlying principles of community conservation. Nevertheless, Mozambique's wild resources may well be recovering. The big question is whether this recovery can be sustained if local people are disenfranchised from the resources they live with. The patterns described for Mozambique are similar to Zambia. They suggest that without strong central leadership to align policy, legislation and practice with the CBNRM principles, progress is uneven and reflects the momentary interests of donor projects or energetic individuals.

Conclusions

Part IV summarizes nearly two decades of experience with CBNRM in four countries. It presents evidence that, provided implementation adheres to a set of principles developed in the region, there is a high likelihood of success. Equally, the many disappointments associated with CBNRM are invariably associated with a failure either to understand these principles or to implement them. All four chapters emphasize the importance that CBNRM places on the devolution of rights to manage, use, dispose of and benefit from natural resources. They highlight the importance of basing CBNRM on working economic models, and the difficulties of simultaneously developing both the economic and organizational components of CBNRM, as well as the dangers of setting up false expectations about benefits (see Kiss, 2004).

Despite its strong association in the region with high-value wildlife resources, there is a collective understanding among CBNRM practitioners that CBNRM is not about any particular resource but is a way of thinking about local collective action and institutions. There have been some successful attempts to diversify CBNRM's sources of benefit on a small scale. Perhaps the next commercial breakthrough lies with marketing the environmental services that many communities are already providing as an adjunct to wildlife management (such as carbon sequestration, water and biodiversity conservation), instead of providing these as a free service to the global community.

There is clear evidence that CBNRM has significantly improved people's attitudes towards wildlife and natural resources. Improving management has rehabilitated and protected wildlife and its habitats and has generated substantial economic benefits. However, CBNRM is also a moving target and the new challenge is to manage expenditure more effectively and democratically to reduce household poverty. All four chapters argue for further devolution, emphasizing the importance of participatory local democracy (whereby the power to make choices and apply accountability is vested in the people rather committees). They also stress that individuals need to benefit and participate far more than is currently the case. A common theme is the importance of adaptive management and performance monitoring, and the stronger programmes are associated with stronger monitoring systems. Within the region, systems for monitoring wildlife and income are well developed, albeit not always adopted. However, the new challenge is to monitor:

- the effectiveness of expenditure (and not just how much income is generated);
- the performance of CBNRM as perceived by individuals; and
- the governance of CBNRM including the participation of and accountability to individual members of communities.

In conclusion, at the level of the community, we have a good idea of how to make CBNRM work with a high probability of success. Indeed, the following four chapters show considerable commonality of principle at the local level. Nevertheless, we need to be more rigorous about applying CBNRM principles at local level, with an emphasis on individual participation and benefit.

However, at the national level circumstances vary considerably so there is substantial variety and opportunism in the ways that scholar–practitioners have manoeuvred to apply these principles. To become more effective at CBNRM, we need to develop a better understanding of this enabling environment. Several points of agreement emerge. First, CBNRM is seldom sustainable outside of a legal environment that adequately devolves rights to local communities; the lower the better. Second, CBNRM is more successful where it is led by an enlightened government agency, and seldom works if the national agency does not support it in practice, as well as in rhetoric (see Ribot, 2004; Nelson, 2007). Ideally government should establish principled policy to guide CBNRM and monitor conformance to these principles (including protecting weaker members of communities), while opening space for NGOs to build community capacity. Third, effective CBNRM is invariably associated with quality light-touch facilitation; money does not substitute for dedicated professionals. Part IV suggests that in practice the champions with the insight and tenacity to develop CBNRM have been local professionals. Fourth, CBNRM has emerged undercover (see Murphree, 1995) as a technical solution to resource management, but ultimately the rights of local people will need to be defended in the political realm. We need to know more about this, with issues of government buy in, the role of political associations such as the CAMPFIRE Association and the financial relationships between communities and local governments being fertile areas for thought. Finally, CBNRM is a complex long-term process that stretches beyond the timeframe and imagination of donor projects. Its ability to adapt will depend on the quality of feedback loops and information, suggesting that ultimately monitoring and information and long-term personal relationships and experience are critical to its success.

Notes

1   It is interesting to note the congruence between Murphree's ideas and the well-known principles underlying common property management developed by, for example, Ostrom (1990). In a process of parallel evolution, Murphree developed principles remarkably similar to, if somewhat more operationalizable, than Ostrom's, although Murphree only met Ostrom and read her work when his key ideas had already been articulated (Murphree, pers. comm.).

2   It is important to point out that 'revenue distribution' differs fundamentally from a cash handout with which it is often confused. It refers to the highly participatory process whereby a community sits together for several days to weigh the advantages and disadvantages of allocating their wildlife revenue among competing uses, one of which is a cash dividend. The trade-off between cash and other investments, and between individual and collective gain, leads to intensely debated decisions. Moreover, the distribution of cash to individuals is powerful and full of symbolism designed to represent people's rights to decide how to allocate resources for themselves (Child, 2006). That the actual cash handout is highly visible should not obscure the fact that it is a symbolic culmination of an intense process of debate and decision making designed to build mechanisms of accountability. Indeed, revenue distribution symbolizes individual choice and empowerment, whereas cash handouts epitomize dependency.

References

Adams, W. and Hulme, D. (2001) 'Conservation and community: Changing narratives, policies and practices in African conservation', in Hulme, D. and Murphree, M. (eds) *African Wildlife and Livelihoods: The Promise and Performance of Community Conservation*, James Currey, Oxford

Barrow, E. and Murphree, M. (2001) 'Community conservation: From concept to practice', in Hulme, D. and Murphree, M. (eds) *African Wildlife and Livelihoods: The Promise and Performance of Community Conservation*, James Currey, Oxford

Child, B. (1993) 'Zimbabwe's CAMPFIRE programme: Using the high value of wildlife recreation to revolutionize natural resource management in communal areas', *Commonwealth Forestry Review*, vol 72, no 4, pp284–296

Child, B. (2006) 'Revenue distribution for empowerment and democratisation', *Participatory Learning and Action*, vol 55, pp20–29

Child, B. and Peterson, J. (1991) *CAMPFIRE in Rural Development: The Beitbridge Experience*, Centre for Applied Social Sciences, University of Zimbabwe, Harare

Child, B., Ward, S. and Tavengwa, T. (1997) *Zimbabwe's CAMPFIRE Programme: Natural Resource Management by the People*, Canon Press, Harare

Government of Botswana (2007) 'Community based natural resource management policy', Ministry of Environment, Botswana Government Printer, Gaborone

Hasler, R. (1995) *Political Ecologies of Scale: The Multi-tiered Co-management of Zimbabwean Wildlife Resources*, IIED Wildlife and Development Series No 7, IIED, London

Hutton, J., Adams, W. and Murombedzi, J. (2005) 'Back to the barriers? Changing narratives in biodiversity conservation', *Forum for Development Studies*, vol 2, pp341–370

Jones, B. and Murphree, M. (2004) 'Community-based natural resources management as a conservation mechanism: Lessons and directions', in Child, B. (ed) *Parks in Transition: Biodiversity, Rural Development and the Bottom Line*, Earthscan, London

Kiss, A. (2004) 'Is community-based ecotourism a good use of biodiversity conservation funds?', *Trends in Ecology and Evolution*, vol 19, no 5, pp232–237

Martin, R. (1981) 'Sebungwe Region: A planning framework', Department of Physical Planning, Joint Report, Department of National Parks and Wildlife Management, Harare

Martin, R. (1986) 'Communal areas management programme for indigenous resources' (April 1986, revised edition), Department of National Parks and Wildlife Management, Branch of Terrestrial Ecology, Harare

Martin, R. (2007) 'Murphree's laws', paper presented at a Symposium to honour Professor Marshall Murphree, Leopard Rock Hotel, Vumba, 22–24 May

Mashinya, J. (2007) 'Participation and devolution in Zimbabwe's CAMPFIRE program: Findings from local projects in Mahenye and Nyaminyami', PhD thesis, School of Public Policy, University of Maryland, Washington DC

Maveneke, T. N. (1996) 'Local participation and benefit sharing in wildlife management: The Zimbabwe CAMPFIRE Programme', World Bank/UNDP Africa Forestry Forum, Nairobi

Metcalfe, S. (1993) 'Rural development and biodiversity: Prospects for wildlife habitat on communal land in Zimbabwe's Zambezi Valley', paper presented at 'Biodiversity in Practice Symposium', South African Wildlife Management Association, Port Elizabeth, June

Minister of Environment and Tourism (1992) 'Policy for wild life', Department of National Parks and Wildlife Management, Ministry of Environment and Tourism, Harare

Murombedzi, J. S. (1999) 'Devolution and stewardship in Zimbabwe's CAMPFIRE Programme', *Journal of International Development*, vol 11, pp287–293

Murphree, M. (1991) *Communities as Institutions for Resource Management*, Centre for Applied Social Sciences, University of Zimbabwe, Harare

Murphree, M. (1995) 'Optimal principles and pragmatic strategies: Creating an enabling politico-legal environment for community based natural resource management (CBNRM)', keynote address to Conference of the Natural Resources Management Programme, SADC Technical Coordination Unit, Malawi, USAID-NRMP Regional, Chobe, Botswana, 3 April

Murphree, M. (1997) *Congruent Objectives, Competing Interests and Strategic Compromise: Concept and Process in the Evolution of Zimbabwe's CAMPFIRE Programme*, Institute for Development Policy and Management (IDPM), University of Manchester, Manchester

Murphree, M. (2000) 'Constituting the commons: Crafting sustainable commons in the new millennium', paper presented at 'Multiple Boundaries, Borders and Scale', Eighth Biennial Conference of the International Association for the Study of Common Property (IASCP), Bloomington, IN, 31 May–4 June

Nelson, F. (2007) *Emergent or Illusory? Community Wildlife Management in Tanzania*, Issue Paper No 146, IIED and Pastoral Civil Society in East Africa, Russell Press, Nottingham, UK

Ostrom, E. (1990) *Governing the Commons: The Evolution of Institutions for Collective Action*, Cambridge University Press, Cambridge

Ribot, J. C. (2004) *Decentralization of Natural Resource Management: Encountering and Countering Resistance*, World Resources Institute, Washington DC

Rihoy, E., Chirozva, C. and Anstey, S. (2007) 'People are not happy: Speaking up for adaptive natural resource governance in Mahenye', Programme for Land and Agrarian Studies, University of the Western Cape, Cape Town

Taylor, R. D. (1978) 'Regional planning for the Sebungwe: An outline description and evaluation of the natural resources with the Sebungwe region in relation to rural development', Department of National Parks and Wildlife Management, Salisbury

Whande, W., Kepe, T. and Murphree, M. (eds) (2003) 'Local communities, equity and conservation in southern Africa: A synthesis of lessons learnt and recommendations from a southern African technical workshop', Proceedings, PLAAS, University of Western Cape, Cape Town

Wilson, K. (2005) 'Of diffusion and context: The bubbling up of community based natural resource management in Mozambique in the 1990s', in Brosius, P., Lowenhauot Tsing, A. and Zerner, C. (eds) *Communities and Conservation: Histories and Politics of Community-Based Natural Resource Management*, Rowman, Altamira

13

The performance of CAMPFIRE
in Zimbabwe: 1989–2006

Russell Taylor

Zimbabwe's Communal Areas Management Programme for Indigenous Resources (CAMPFIRE) pioneered CBNRM in southern Africa. It was conceived some 30 years ago in the northern Sebungwe region where the environmental impacts of tsetse eradication programmes and associated in-migration of people onto marginal agricultural land threatened the potential comparative economic advantage of wildlife, which was still abundant in the region (Child, G., 1995). Although initially driven by a conservation agenda, it was quickly appreciated by those involved in developing the programme that a people-centred approach was needed, as were devolved governance and rural democratization processes. A number of important principles for community conservation emerged, including economic benefit, devolution and collective proprietorship for a common pool resource, wildlife (Jones and Murphree, 2001), which have been put into practice with varying degrees of success and indeed failure, and the programme continues to evolve adaptively and experimentally, with stakeholders. This chapter summarizes the progress of CAMPFIRE since its formal introduction in 1986, providing as complete an assessment as is possible with incomplete data, using several key variables to measure conservation, social benefit and governance outcomes.[1]

Background

Much of Zimbabwe is semi-arid, with a low and variable rainfall, making the country prone to drought. Comprised predominantly of broad-leafed deciduous miombo and mopane woodland and savanna, land use varies from intensive crop production to extensive cattle and wildlife production along a rainfall–altitude gradient. Much of the country is ecologically best suited to extensive rangeland production systems (see Vincent and Thomas, 1960) and economic evidence suggests that wildlife is a highly competitive form of land use in drier regions (Jansen et al, 1992; Price Waterhouse, 1994).

In Zimbabwe, the protected area network comprises some 50,000km² or 13 per cent of the country, situated mostly in the lower-altitude, poor-rainfall areas around the periphery of the country (see Figure 13.1 and also Chapter 5). An equivalent area of 40,000–50,000km² of communally occupied land is either adjacent to or near protected areas where wildlife populations are relatively abundant. Wildlife is typically more abundant especially where human population density is low (less than 10 people per km²) and wildlife habitat (more than 50 per cent of land area) is intact (Taylor, 1999). It is in these less developed, more remote areas, that CAMPFIRE was initially implemented in the late 1980s.

The programme was designed by the then Department of National Parks and Wild Life Management (DNPWLM), now the Parks and Wildlife Management Authority (PWMA) in the mid-1980s (Martin, 1986). It is a long-term programmatic approach to rural development that uses wildlife and other natural resources as a mechanism for promoting devolved rural institutions and improved governance and livelihoods (Child et al, 2003). The cornerstone of CAMPFIRE is the devolution of rights to manage, use, dispose of and benefit from natural resources.

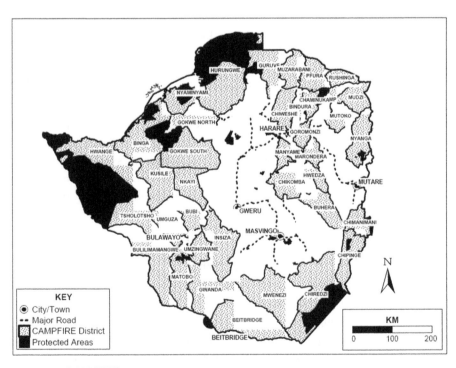

Source: WWF-SARPO (2008)

Figure 13.1 *Zimbabwe's CAMPFIRE districts*

As originally envisaged, CAMPFIRE was to focus on the conservation and exploitation of four natural resources: wildlife, forests, grazing and water. Because wildlife is able to provide direct and immediate tangible financial benefits, the initial success of the programme was premised on the utilization of large mammal wildlife resources, mostly through high-value trophy hunting safaris. Subsequently, CAMPFIRE continued to focus more on the consumptive and non-consumptive use of large mammals rather than other natural resources. However, it has been able to diversify its resource management activities somewhat to include non-consumptive tourism ventures, timber and bamboo harvesting, honey and fruit production, trout fishing, the production of edible mopane caterpillars and the sale of non-renewable resources such as river sand for construction purposes.

Institutional arrangements

The PWMA is the legally mandated authority responsible for wildlife resources in the country. The 1975 Parks and Wild Life Act decentralized state authority and conferred privileges on owners or occupiers of alienated land as custodians of wildlife, fish and plants. Private landowners or land occupiers were designated 'appropriate authority', giving them de facto responsibility for wildlife and making them the beneficiaries of sound wildlife conservation and use. After 1980, similar rights were extended to communal farmers through an amendment to the Act in 1982, which delegated appropriate authority (AA) to rural district councils (RDCs). In practical terms, AA represents the decentralization of authority and control over wildlife to RDCs only (Murombedzi, 2001), and this has had important implications for CAMPFIRE.

Funding and other support

At the time CAMPFIRE was designed, donor aid was uncommon in Zimbabwe and the early establishment of the programme was characterized by a relatively low level of external funding (Child et al, 2003). DNPWLM originally envisaged government funding through a public sector investment programme to initiate CAMPFIRE, with these funds ultimately contributing to the creation of a rural sector economy based on wildlife production. Furthermore a CAMPFIRE agency under an appropriate ministry was also planned, for which short-term donor funding would have been sought (Martin, 1986). Thus funding mostly followed, rather than led, the programme.

Technical and other support was provided by a coalition of support agencies including Zimbabwe's Centre for Applied Social Sciences, Zimbabwe Trust, WWF and the Ministry of Local Government, Rural and Urban Development. The inputs of these organizations were coordinated by the

CAMPFIRE Collaborative Group under the leadership of DNPWLM and, later, the CAMPFIRE Association. The collaborative group was replaced in 1994 by the CAMPFIRE Service Providers. This local support was supplemented by funding primarily from the United States Agency for International Development (USAID), but also the WWF Multispecies Animal Production Systems Project and the Norwegian Agency for Development Cooperation.

Implementing CAMPFIRE

In the process of implementation, a number of strongly interlinked principles embedded in the original design have contributed significantly to the evolution of CAMPFIRE policy and practice (Jones and Murphree, 2001).

Economic benefit

The underlying assumption regarding the motivation for the sustainable utilization of wildlife outside of protected areas in Zimbabwe since the 1960s has been that economics ultimately determines decisions regarding the allocation of land and the resources thereon. The early success of policy promoting the high economic and financial value of wildlife, a key incentive for its sustainable management on freehold land, provided compelling arguments for its wider application in the communal sector of the country, particularly after 1980. In the context of rural development and CAMPFIRE, placing wildlife in the realm of economics and land use, rather than conservation (Jones and Murphree, 2001), provided an important opportunity to complement conventional and subsistence agricultural practices in the communal lands of the country. (This may mean, however, that wildlife could be displaced by other more viable forms of land use at some time in the future.)

Devolution[2]

Devolution of responsibility for the management and use of wildlife (coupled with alternative economic opportunities and incentives for rural development) was intended to better serve wildlife conservation given the inadequate government resources to do so outside of protected areas. It was formalized in the 1975 Parks and Wild Life Act and amended in 1982, and devolved responsibility was initially granted to nine RDCs in 1989 and 1990. Those involved in the development of CAMPFIRE recognized that long-term success depended on further devolution to sub-district levels, even to a community-level institution. However, the absence of any legal persona below the level of RDC obliged DNPWLM to decentralize administrative authority and legal rights to wildlife to RDCs, on condition that rights and benefits were to be further devolved to what were termed 'producer communities'.

Collective proprietorship

While the transfer of proprietorial rights, together with the financial incentives, was highly successful on commercial farmland, replication in communal lands faced numerous legal and institutional impediments. What was required was a communal property regime behaving as a proprietorship unit over land and resources, which comprised a defined group collectively managing and exploiting common pool resources within a defined jurisdiction (Jones and Murphree, 2001).

In the event, ward-level producer communities emerged through the establishment of Ward Wildlife Management Committees (WWMCs) or Ward Wildlife Committees (WWCs).[3] These village-elected committees were formally constituted with a membership comprising a chairperson, secretary and treasurer and others, with or without a specific portfolio or responsibility. The chairperson represented his/her ward on the District Wildlife (or Natural Resources) Committee, a subcommittee of the district council. Although in effect the new committees were subcommittees of local government units (Murombedzi, 2001), one of the difficulties they faced was the perception that they were 'parallel' institutions to already existing ward and/or village development committees, and thus potentially competitive or even subversive.

Participation and implementation

In 1988, two RDCs in the Zambezi valley, Guruve and Nyaminyami, were granted AA status and commenced earning revenue by marketing trophy hunting quotas to international safari hunting clientele. The first two RDCs were rapidly followed by another seven districts requesting AA status. By 1992, 12 RDCs had acquired AA, and by 1996 there were 19 CAMPFIRE districts (see Figure 13.2).

The CAMPFIRE Association was formed in 1991 to lobby for and promote the role of communal land wildlife producers. This gave CAMPFIRE an important level of political legitimacy and an ability to play a proactive advocacy role locally and internationally. Its membership, however, has remained the RDCs and not the true wildlife producer communities at a subdistrict level.

Following major donor inputs for building and strengthening institutional capacity and natural resource management micro-project development, 37 RDCs had AA status by 2001. However, many were not traditional wildlife-producing districts – 19 districts (51 per cent) generated income for

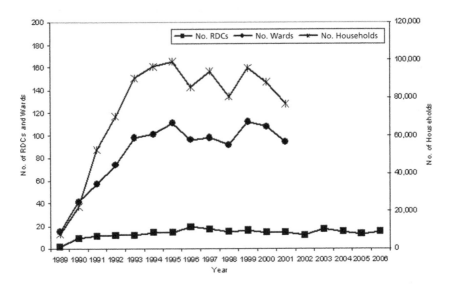

Note: * As reflected by those RDCs granted AA

Source: Unpublished WWF-SARPO data

Figure 13.2 *Numbers of participating wildlife-producing communities**

communities through sustainable natural resource management activities and received benefits in terms of funded projects, training and membership of the CAMPFIRE Association (i.e. were considered to be 'fully participating'). Over 70 per cent of the 271 wards and 1217 villages in these 19 districts could be considered also as fully participating producer communities.

The CAMPFIRE model

Sport hunting and tourism provided the primary economic and financial basis for the implementation of CAMPFIRE between 1989 and 2006. Although there is considerable biophysical and socio-economic variability between RDCs with AA status, Bond (2001) described a general model for the income or revenue earned from the use of wildlife and the subsequent allocation of this income (see Figure 13.3). Consumptive (sport hunting) and/or non-consumptive tourism rights are leased to private sector operators by the RDC, with lease conditions determined primarily by the RDC and negotiated with the safari operator.

Decentralization has encouraged the application of market-based mechanisms by RDCs and greater efficiency of resource use by safari operators (Bond, 1999). Lessees pay all their fees to the RDC but the level of involvement

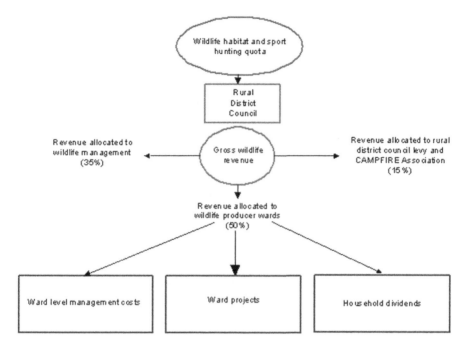

Source: Adapted from Bond (2001)

Figure 13.3 *The CAMPFIRE financial model for revenue generation and allocation*

of sub-district community representatives in the lease allocation process, though variable between districts, generally has been minimal (Bond, 2001; Jones and Murphree, 2001).

Revenue earned is allocated to district council levies, district wildlife management activities and to wildlife producer communities as represented by wards. While the breadth and depth of wildlife management activities varies between districts, most have a small core team of personnel who undertake law enforcement, problem-animal management and wildlife monitoring. Income allocated to communities through WWMCs is intended to provide the financial incentive for households to participate in the collective management of wildlife (Bond, 2001). Wards have chosen to allocate revenue to resource management (salaries for resource monitors, allowances for committee members, fence repairs and maintenance), projects (grinding mills, schools, clinics) and household dividends in the form of drought relief (Taylor and Murphree, 2007) or uncommonly, cash.

Revenues raised and distributed

Between 1989 and 2006 the income to RDCs with AA totalled nearly US$30 million (ZW$390 billion).[4] Some 90 per cent of this income (US$26.8 million) was earned from the lease of sport hunting rights to commercial safari operators. The remaining income came from the lease of tourism rights (2 per cent or US$541,000), ivory sales and sale of hides (4 per cent or US$1.3 million) and other minor resources such as crocodile and ostrich eggs and firewood (4 per cent or US$1.3 million).

In 1989, when only two RDCs had AA and were operational, annual income was US$350,000 (ZW$750,000). In 1999, income exceeded US$2.75 million following a one-off sale of elephant ivory approved under the Convention on International Trade in Endangered Species (see Figure 13.4). In 2002, exchange rate distortions at the time reflect total CAMPFIRE income as US$4.6 million (ZW$255 million), which is unrealistically high. In 2006, income was over US$2 million (ZW$365 billion), at which time only 15 (out of 23) wildlife-producing RDCs were providing financial data, or indeed earning income.

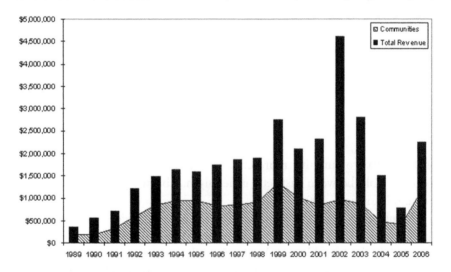

Source: Unpublished WWF-SARPO data

Figure 13.4 *CAMPFIRE wildlife revenue, 1989–2006 (US$)*

Bond (2001) defines five categories for the allocation of income earned from wildlife by RDCs. Councils with AA are not legally obliged to devolve income to sub-district levels but are encouraged to do so by a set of guidelines, originally developed by DNPWLM in 1991. More recently, the guidelines have been revised and endorsed by the CAMPFIRE Association in its financial management manual (Anon, 2003). These seek to ensure that producer communities are the primary beneficiaries of the income earned and recommend that at least 50 per cent of gross wildlife income should be devolved to ward level, up to 35 per cent can be retained for wildlife management purposes at RDC level and no more than 15 per cent should be retained as a council levy.

The allocation of income over the past 18 years has been somewhat variable and not always satisfactory in terms of the revenue guidelines (see Figure 13.5). Even wildlife-rich and well-endowed districts have been unwilling to devolve the recommended 50 per cent of income earned to wards and households (Taylor, 1994). The declines in distribution to sub-district levels can be attributed largely to the declining economic condition of the country, associated exchange rate distortions and the unwillingness of cash-strapped RDCs to disburse scarce financial resources. Following a meeting called by PWMA to remind RDCs of their obligations, the proportion increased again in 2005 and 2006. This subsequently led to agreement in 2005 where safari operators pay the proportion due to the wards directly into ward-level bank accounts.

On average, 52 per cent has been disbursed to community level over the past 18 years (see Figure 13.5). Some 14 per cent of income (US$3 million) remained unallocated between 1989 and 2001, increasing to US$9.1 million (31 per cent) since and matched by a dramatic decline in allocations to district-level wildlife management activities. Unallocated revenue is generally assumed to have been committed to activities not related to wildlife and CAMPFIRE (Bond, 2001).

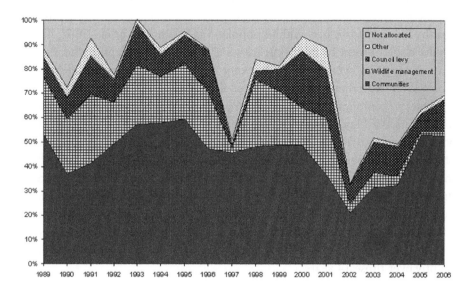

Source: Unpublished WWF-SARPO data

Figure 13.5 *CAMPFIRE revenue allocations, 1989–2006*

While there has been diversification beyond wildlife into tourism and other natural resource products, a number of RDCs have treated income from these activities as general revenue. These income-generating activities may have provided as much as 30 per cent more revenue than is reflected in the CAMP-FIRE accounts (PwC, 2001; Child et al, 2003), and most if not all failed to reach communities. While average household benefits are low there is, nevertheless, site-specific variability around these values, with households in resource-rich areas with relatively low human population densities receiving considerably more income than the long-term national average.

Biodiversity impacts

Impact on wildlife populations

Despite available aerial census data being uneven across CAMPFIRE districts, available data are adequate for the purposes of indicating broad trends in elephant and buffalo populations for 8 of the 12 primary wildlife producing districts (see Figure 13.6). These two species are considered here as they make significant contributions to CAMPFIRE revenue generation (through trophy hunting).

The estimated mean number of elephants was 6840 at a density of 0.43 animals per km^2, ranging from a minimum of 4181 in 1989 to a maximum of 12,707 elephants in 2001. Densities are approximately half or less those found in state-protected areas, but consistent with the maintenance of woodlands and healthy habitats for other wildlife populations (Cumming et al, 1997). These results also conform to increasing national and regional elephant herds that have grown at an average of 4–5 per cent per annum over the past decade or more (Price Waterhouse, 1996; Blanc et al, 2005).

Due to sampling difficulties, estimates of buffalo numbers tend to be highly variable (Taylor and Mackie, 1997) and buffalo are not present in all wildlife districts. The mean number estimated was 12,042 buffalo at a density of 0.96 animals per km^2. This is low for the southern savannas and numbers may have been influenced by a regional rainfall deficit, coupled with periodic droughts in the Zambezi valley and the South East Lowveld, particularly during the early 1990s (Foggin and Taylor, 1996), resulting in substantial wildlife die-offs. As a general observation, elephant numbers have increased in CAMPFIRE areas and buffalo numbers are either stable or have declined slightly over those 14 years.

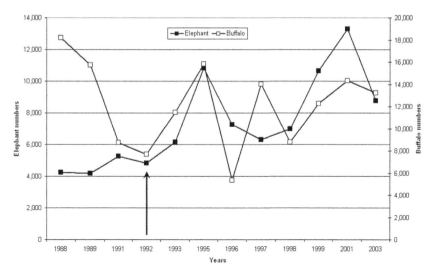

Note: The arrow indicates the severe drought of 1992

Source: Unpublished WWF-SARPO data

Figure 13.6 *Elephant and buffalo numbers for CAMPFIRE areas, 1988–2003*

Quota offtakes and trophy quality

The value of big game trophies has been maintained or increased in CAMP-
FIRE areas, the consequence of improved marketing (Child, B., 1995; Child et
al, 2003) and more effective trophy hunting and monitoring systems. Although
national quotas and actual offtakes for elephant and buffalo increased between
1992 and 2002, trophy quality (as opposed to animal numbers) for these two
species has been declining. For lion and leopard, offtakes have either declined
or are stable, while trophy quality is stable for leopard and increasing for lion.
There is also a strong correlation between increasing quotas, declining trophy
quality and increased 'catch effort' (Grobbelaar and Masulani, 2003).

There has been a noticeable shift in DNPWLM quota-setting policy in the
latter half of the 1990s. Quotas had initially been set to maximize returns, but
around 1996 there was a switch in emphasis to more sustainable trophy quality, resulting in a reduction of most quotas. This shift may be linked to the
growing acceptance and adoption of a participatory approach to quota setting
and monitoring (Taylor, 2001), successfully initiated in CAMPFIRE areas,
which emphasized the adaptive management of quotas in response to indices
of animal abundance, trophy quality, community monitoring, illegal offtakes
and safari operator 'catch effort'.

Maintenance of wild land and wildlife habitat

An assessment of the likely extent of wild land and habitat within CAMPFIRE areas was undertaken that used wildlife-producing wards as a proxy for this land, recognizing that such wards comprised a mosaic of wild and settled land (Taylor, 1999). On average wildlife producer wards made up 36 per cent of the total number of wards in CAMPFIRE districts, with their land area of 39,580 km² constituting 55 per cent of the total area under the programme.

For 12 primary wildlife districts the amount of wild land varied from less than 500 to over 5000km² with an average size of 3300km² (see Figure 13.7). Of these, three districts had wild land in excess of 90 per cent of the district area, six had 50–70 per cent wild land and in three districts only, less than 35 per cent of the district constituted wild land.

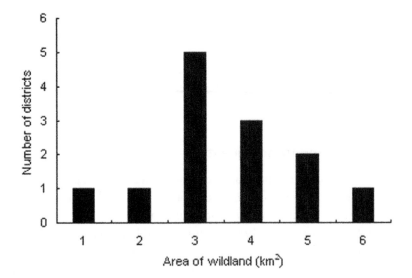

Source: Taylor (1999)

Figure 13.7 *Distribution of wild land in CAMPFIRE areas*

The availability of wild land is negatively correlated with human population density, that is, the maintenance of wild land (greater than 50 per cent of area) is more likely under lower rather than higher population densities (less than 10 persons/km²) (Taylor, 1999). At a coarse scale of resolution, these results suggest that wild land has been maintained in an intact state. Conybeare's (1998) study in three areas of the Zambezi Valley would seem to support this, with just 2 per cent of wild land lost between 1989 and 1997. This also has implications for potential household earnings from wildlife with those areas sparsely populated and with relatively high wildlife densities standing to bene-fit more (Bond, 2001; Murombedzi, 2001).

Examination of habitat quality on a much finer scale, established for three Zambezi Valley districts following the commencement of CAMPFIRE in 1989, shows that the amount of habitat destroyed remains stable, though good qual-ity habitat decreased in two out of three districts. In Nyaminyami District, good quality habitat actually increased; in Gokwe North, habitat has been maintained in the wildlife corridor as show in Table 13.1, even though else-where this district has lost most of its natural vegetation (Dunham et al, 2003).

Table 13.1 *Habitat quality changes over time in three CAMPFIRE districts, 1981–1999**

Habitat quality defined from the perspective of elephant cow herds		District and year								
		Nyaminyami			Binga North			Gokwe North		
		1981	1993	1999	1981	1993	1999	1981	1993	1999
Habitat quality	Defined as		%			%			%	
1. Habitat destroyed	Settlement/ cultivation	5	18	16	16	23	23	27	57	57
2. Poor quality habitat	Natural vegetation inside 2km buffer zone around settlement/cultivation	30	42	27	42	44	55	43	34	40
3. Poor quality habitat	Natural vegetation patch outside buffer zone but <57 km²	1	2	<1	0	<1	2	<1	<1	1
4. Good quality	Natural vegetation	63	38	57	43	33	21	30	8	2

Note: * Values are the percentages of each district that were in each habitat category

Source: Dunham et al (2003)

While causality is difficult to demonstrate, there is some indication that the loss of natural habitat may have been slowed in these wildlife-producing CAMP-FIRE areas. Further work is required on the effects that people have on the spatial distribution of wildlife, especially those species that contribute signifi-cantly to CAMPFIRE revenues, notably elephant and buffalo (Dunham et al, 2003).

Governance

From the inception of CAMPFIRE to the mid-1990s, the amount and proportion of revenues devolved to producer communities increased rapidly, providing the primary impetus for wildlife conservation and for improvements in community institutional development and governance. Subsequently, the rate of devolution levelled off and after 2000 the process reversed itself (Child et al, 2003). Nevertheless, by 2003 the concept and level of devolution in many districts was still relatively strong, with a strong correlation between fiscal devolution and institutional development (Zimbabwe Trust, 2003), despite the high level of taxation imposed on producer communities by RDCs through their various levies. More recently, some wards and village collectives, notably in Chiredzi, Chipinge, Guruve and Nyanga districts, have begun to negotiate directly with safari operators and other private sector partners, resulting in payments of hunting and other tourism revenues direct to ward level.[5] Some RDCs, especially their technical staff, tend to support such innovation, even if only implicitly, recognizing their own limitations and inability to overcome the revenue retention problem, which is largely political in nature.

One of the more notable achievements of CAMPFIRE has been the strength of institutional development at the producer community level (Child et al, 2003). However this does need some qualification. While the use of wildlife dividends has been decided democratically in some instances, this has not always been the case. Likewise, some projects have been implemented properly and finances managed transparently, but not everywhere. Variability, in locality and over time, should not be unexpected in processes of institutional change.

In excess of 100 democratically elected and constituted village and ward CAMPFIRE committees exist in 23 districts, which provide for a high level of community participation and decision making with a transparent flow of information relating to key issues, planning and projects. These committees have been equipped with basic organizational skills and the fundamentals of project and financial management. Community leadership and locally employed natural resource monitors have also gained a number of wildlife management skills including counting wildlife, setting quotas, monitoring hunting, marketing wildlife and undertaking problem-animal mitigation measures.

Assumptions for the success of CAMPFIRE as an innovative CBNRM model

It is pertinent to examine the assumptions and external factors underlying the success or otherwise of CAMPFIRE. A review of the assumptions identified as necessary for the successful implementation of CAMPFIRE (see Table 13.2) indicates that at least three are largely unfavourable for CAMPFIRE (Wright, 1998; WWF-SARPO, 2003).

Recent developments

Cessation of USAID and other funding for CAMPFIRE in 2003 coincided with larger macro-level policy changes in Zimbabwe after 2000, the most significant being the implementation of the government's land reform process (Marongwe, 2002) and the subsequent adverse socio-economic conditions that resulted.

The maintenance of a foreign sport hunting market has been important for CAMPFIRE, though national-level hunting revenues have declined (see Booth, 2002). Small-scale communally managed tourism projects have not fared as well, as many of these initiatives came on line at precisely the time that Zimbabwe's foreign tourism industry began to decline. Domestic tourism has also been negatively affected due to inflation and fuel shortages.

In the absence of major donor funding and in the face of economic hardships, two case studies were made during late 2006 and early 2007 of two CAMPFIRE community projects at Masoka (Guruve District in the mid-Zambezi valley) and Gairezi (Nyanga District in the eastern highlands) (Taylor and Murphree, 2007). Some of the more important observations and unexpected results are provided below.

Masoka

Masoka was the first community in Zimbabwe to inaugurate a natural resource management regime for local development under CAMPFIRE and largely conforms to the CAMPFIRE model as originally envisaged, when external donor aid was relatively uncommon (as it is presently). The community has received no donor support since 1989 (when donated funds helped to construct an electrified game fence around the original settlement area). Since 2003, substantial hunting revenues had been appropriated by the RDC, imposing a grossly unfair tax on Masoka. In 2005, the Masoka WWC, with strong community support, challenged the Guruve RDC, the contracted safari operator and the CAMPFIRE Association over this issue. Following acrimonious exchanges with the RDC, an agreement was made between the community, council and the safari operator whereby all income legitimately due to the community (i.e. no less than 50 per cent) would be paid by the safari operator directly into the Masoka Ward bank account. Consequently an all time low of US$11,437 received in 2004 was followed by a dramatic rise to US$132,522 in 2006, a tenfold increase in income (see Figure 13.8).

Table 13.2 *Assumptions and indicators for CAMPFIRE*

Assumption	Area of indicator	Indicator	Comment
1. Continued government commitment to devolve CBNRM to local levels	Changes in policies for wildlife, forestry and other natural resources	Wildlife policy and legislation. Forestry policy and legislation. Land policy and legislation	No substantive revision of natural resource legislation or defined legal framework for devolved natural resource management in communal areas. Policy remains fragmented and largely centrally controlled.
	DNPWLM/PWMA have personnel and other resources for their CAMPFIRE projects	DNPWLM/PWMA real budget (US$)	DNPWLM/PWMA budgets reduced from about US$200 per km^2 to about US$10 per km^2 over past 20 years
2. Favourable market environment for CAMPFIRE products and/or services exist or are maintained	Demand for CAMPFIRE products and services maintained or expanded	Total number of days of sport hunting per annum. Gross value of sport hunting per annum. Trophy fees paid for key species at the Zambezi Valley auction hunts	Number of tourists, including sport hunters and associated revenue declined dramatically with land reform-related political instability. Declines, followed by an increase in trophy fees since 2001 requires further analysis
3. No extreme climatic conditions	Frequency and distribution of droughts and other extreme climatic events	Mean annual rainfall	No clear links between nine years of below average (five drought years) and five years above average rainfall (three flood years) and CAMPFIRE performance and/or land use change. Nevertheless, food shortages are common in most CAMPFIRE districts on an annual basis
4. Favourable macro-economic environment prevails	Macro-economic performance maintained or increased	Changes in gross domestic product (GDP) (1990–2001). Changes in per capita GDP (1990–2001)	GDP shrank from US$8.4 million in 1996 to US$3.3 million in 2001 with GDP/capita constant from 1991 to 1997 at about US$700. By 2001 GDP/capita had declined to about US$238.
		ZW$:US$ exchange rate (1990–2001)	Between 1990 and 1997 the value of the Zimbabwe dollar declined from ZW$2.47 to ZW$12.44 and to ZW$55 in 2001, when it was fixed.
		Proportion of total population in formal employment (1990–2001). Real wages (1990–2001)	The proportion of the population in formal employment declined from 14% in 1980 to less than 9% by 2001. Real wages are less than 50% of those paid in 1980

Source: Adapted from WWF-SARPO (2003)

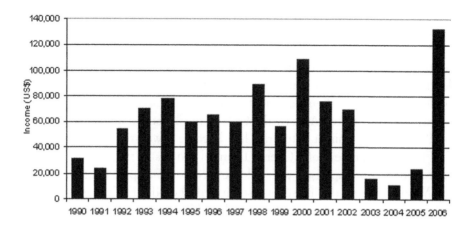

Source: Taylor and Murphree (2007)

Figure 13.8 *Masoka CAMPFIRE revenue, 1990–2006*

The significance of this development cannot be underestimated. Although no legislative change has been made, and although council deductions could still be considered exorbitant, Masoka now has 50 per cent of its earnings paid directly into its bank account at predictable quarterly intervals. Nyaminyami District immediately followed suit and at the 2007 CAMPFIRE Annual General Meeting, all wildlife-producing districts supporting safari hunting agreed to the 'direct payments' system (Jonga, pers. comm.).

It could be argued that a very different picture might have emerged had the more favourable macro-economic conditions of the 1990s persisted (Taylor and Murphree, 2007). The declining performance of the country's economy in the intervening period has had the effect of forcing a level of self-sufficiency, resilience, adaptability and partnership that has enhanced performance beyond expectations. While this is undoubtedly linked to earlier investments in empowerment and capacity-building, local knowledge of what the revenue distribution guidelines required, coupled with strong leadership and representation played a role in forcing the hand of RDCs.

Gairezi

A fly fishing lease runs along a short length of the Gairezi river, within the Gairezi Resettlement Area, the residents of which had always been involved in the protection and management of the river and its immediate environs for both conservation and local benefit. In 2003, following the formation of the

Gairezi Development Trust (the successor of the Gairezi River Protected Area Committee), a Memorandum of Agreement was drawn up between the Trust, the Nyanga RDC and the Nyanga Downs Fly Fishing Club (NDFFC) governing their rights and responsibilities in an ecotourism project that was to be developed, based at least initially, on trout fishing. According to the memorandum, the devolved management authority for natural resources in the Gairezi River Protected Area by the RDC to the Trust is unprecedented.

By 2007, the Gairezi enterprise had become a well-run project with the NDFFC playing a key role. Overall financial returns to the community have been modest by comparison with wildlife income from sport hunting in areas like Masoka. Before the Trust was established, the NDFFC had made annual cash payments to the community around of ZW$4500 per annum, equivalent to ZW$20–25 per member. In 2004 the payment was raised to ZW$10,000,000, which at that time was a substantial improvement. The income earned over the past three years, an annual average of US$4000, suggests that the project is earning in the order of at least US$250 per km^2 or US$1.50–2.50 per ha from this part of the Gairezi river. The financial return to households, albeit modest, is threefold that of CAMPFIRE nationally.

In 2005, the Trust suggested to the NDFFC that members would prefer to receive payment in fertilizer, difficult for Gairezians to obtain and transport, which was subsequently purchased and delivered by the NDFFC. This highly practical mode of revenue distribution clearly made a great impact. Higher yielding fertilized maize fields contrasted strongly to those without treatment. On fertilizer delivery alone, the Gairezi project was perceived as a success in 2006 and 2007 by community members (Taylor and Murphree, 2007).

Conclusions

One of the more important contributions of CAMPFIRE in Zimbabwe has been the lesson that fiscal devolution can lead to improved rural democratization, governance and natural resource management (Child et al, 2003). Democratization takes place in different contexts and on the basis of people's experience and estimates of who can best represent their collective interests. Thus in Masoka for example, people look for a corporate profile of leadership that can deliver effective management and guard their collective interests against internal and external sectional greed (Taylor and Murphree, 2007). However, there is always the danger that rights could be withdrawn or that internal success in corporate solidarity could lead to avarice in the leadership. Thus where devolution has been successful, such as in Masoka, promising results have been achieved and the recent acceptance and implementation of direct payments to communities is probably the most significant development since 2000. That this has happened can be attributed to CAMPFIRE enabling communities to maximize their roles within the existing set of rules, and by so doing, allowing these rules to be challenged.

Institutionally, much emphasis is given to leadership in CBNRM and related development projects, with the 'followers' referred to only infrequently. The importance of the accountability of the leadership to its constituency is emphasized in the cases of Masoka and Gairezi, which illustrate the strong influence on organizational and institutional development of collective will and social capital among a community. Where management committees and their office bearers are accountable, the community can then play a role in providing direction, keeping the direction focused and influencing change when required. Thus the followers help ensure the efficiency, legitimacy (transparency and accountability) and representational ability of their leadership – crucial elements of sound governance. Local communities are now beginning to bring such accountability to bear on the bureaucrats who endeavour to 'run' them.

However, CAMPFIRE has been, and will continue to be embedded in and constrained by a number of fundamental issues, mostly in the policy arena, that are still being addressed. Government policies have mostly not worked well in rural areas and many CAMPFIRE communities have recognized this. In the current socio-economic and political environment in Zimbabwe, some wards provided their RDCs with 50 per cent or more of their revenue, despite punishing taxes either directly through levies and/or indirectly through distorted or misused exchange rates, together with neglect and/or bureaucratic interference. Given this poor administrative history, CAMPFIRE must be one of the best innovations seen in rural Zimbabwe since its Independence.

Notes

1   The analysis in this chapter is based on data sets that are relatively complete between 1989 and 2001, with some data continuing to 2006. Increasing gaps in the data occur following the cessation of major donor funding in 2003. Important site-specific 'snapshots' were made possible through two case studies in 2006/2007, augmented by other qualitative data and information. Together with the long-term data sets, these case studies provide invaluable insights, suggesting the future for CAMPFIRE, in certain contextual settings, is far from bleak.

2   Although the terms 'devolution' and 'decentralization' are used interchangeably here, Murphree (2005) defines 'decentralization' as the delegation of responsibility and limited authority to subordinate or dispersed units of hierarchical jurisdiction, which have a primary accountability upward to their superiors in the hierarchy, and 'devolution' as involving the creation of relatively autonomous realms of authority, responsibility and entitlement, with a primary accountability to their own constituencies.

3   In Zimbabwe, provinces are made up of districts comprised of wards. Wards in

turn comprise a number of villages. These spatially and physically defined group-ings also reflect the lower-level administrative structures of the country, namely ward development committees and village development committees. Members of the CAMPFIRE Collaborative Group (subsequently the Service Providers) and other organizations worked primarily through the WWCs and WWMCs.

4   Between 1989 and 2006 the ZW$ exchange rate moved from ZW$2.126 to ZW$161,909.32 = US$1.

5   See Ingwe Safaris 2005 Year End Report to Guruve Rural District Council and minutes of the Gairezi Development Trust and the Nyanga Downs Fly Fishing Club, 2005/2006.

References

Anon (2003) *CAMPFIRE Revenue Manual*, CAMPFIRE Association/WWF-SARPO, Harare

Blanc, J. J., Barnes, R. F. W., Craig, G. C., Douglas-Hamiliton I., Dublin, H. T., Hart, J. A. and Thouless, C. R. (2005) 'Changes in elephant numbers in major savanna populations in eastern and southern Africa', *Pachyderm*, vol 38, pp19–28

Bond, I. (1999) 'CAMPFIRE as a vehicle for sustainable rural development in the semi-arid communal lands of Zimbabwe: Incentives for institutional change', DPhil thesis, University of Zimbabwe, Harare

Bond, I. (2001) 'CAMPFIRE and the incentives for institutional change', in Hulme, D. and Murphree, M. (eds) *African Wildlife and Livelihoods: The Promise and Performance of Community Conservation*, James Currey, Oxford

Booth, V. (2002) 'Analysis of wildlife markets (sport hunting and tourism)', unpub-lished report, WWF-SARPO, Harare

Child, B. A. (1995) 'A summary of the marketing of trophy quotas in CAMPFIRE areas (1990–1993)', unpublished report, Department of National Parks and Wild Life Management, Harare

Child, B., Jones, B., Mazambani, D., Mlalazi, A. and Moinuddin, H. (2003) 'Final eval-uation report: Zimbabwe Natural Resources Management Programme – USAID/Zimbabwe Strategic Objective No. 1. CAMPFIRE Communal Areas Management for Indigenous Resources', unpublished report, USAID, Harare

Child, G. (1995) *Wildlife and People: The Zimbabwean Success*, WISDOM Foundation, Harare/New York

Conybeare, A. (1998) 'Assessment of habitat maintenance, diversity and productivity under communal management', unpublished report, WWF-SARPO, Harare

Cumming, D. H. M., Fenton, B., Rautenbach, I. L., Taylor, R. D., Cumming, G. S., Cumming, M. S., Dunlop, J. M., Ford, A. G., Horvorka, M. D., Johnston, D. S., Kalcounis, M., Mahlangu, Z. and Portfors, C. V. R. (1997) 'Elephants, woodlands and biodiversity in southern Africa', *South African Journal of Science*, vol 93, no 5, pp231–236

Dunham, K. M., Davies, C. and Muhwandagara, K. (2003) 'Area and quality of wildlife habitat in selected CAMPFIRE districts', unpublished report, WWF-SARPO, Harare

Foggin, C. M. and Taylor R. D. (1996) 'Management and utilization of the African buffalo in Zimbabwe', in Penzhorn, B. L. (ed) *Proceedings of a Symposium of the African Buffalo as a Game Ranch Animal*, South African Veterinary Association, Onderstepoort, South Africa

Grobbelaar, C. and Masulani, R. (2003) 'Review of off take quotas, trophy quality and 'catch effort' across the four main wildlife species, elephant, buffalo, lion and leopard', unpublished report, WWF-SARPO, Harare

Jansen, D., Bond, I. and Child, B. (1992) 'Cattle, wildlife, both or neither: Results of a financial and economic survey of commercial ranches in southern Zimbabwe', WWF MAPS Project Paper No. 27, WWF-SARPO, Harare

Jones, B. and Murphree, M. (2001) 'The evolution of policy on community conservation in Namibia and Zimbabwe', in Hulme, D. and Murphree, M. (eds) *African Wildlife and Livelihoods: The Promise and Performance of Community Conservation*, James Currey, Oxford

Marongwe, N. (2002) *Conflicts over Land and Other Natural Resources in Zimbabwe*, ZERO Regional Environment Organization, Harare

Martin, R. B. (1986) 'Communal Areas Management Programme for Indigenous Resources (CAMPFIRE). Revised version', Department of National Parks and Wild Life Management, Harare

Murombedzi, J. C. (2001) 'Committees, rights, costs and benefits: Natural resource stewardship and community benefits in Zimbabwe's CAMPFIRE programme', in Hulme, D. and Murphree, M. (eds) *African Wildlife and Livelihoods: The Promise and Performance of Community Conservation*, James Currey, Oxford

Murphree, M. W. (2005) 'Congruent objectives, competing interests, and strategic compromise', in Brosius, J. P., Tsing, A. L. and Zerner, C. (eds) *Communities and Conservation*, AltaMira Press, Walnut Creek, CA

Price Waterhouse (1994) 'The Lowveld conservancies: New opportunities for productive and sustainable land use: Savé Valley, Bubiana and Chiredzi River conservancies', Price Waterhouse, Harare

Price Waterhouse (1996) 'Elephant census in Zimbabwe: An analysis and review 1980–1995', Price Waterhouse, Harare

PwC (2001) 'CAMPFIRE revenues. Draft report', Price Waterhouse Coopers, Harare

Taylor, R. D. (1994) 'Wildlife management and utilization in a Zimbabwean communal land: A preliminary evaluation in Nyaminyami District, Kariba', in van Hoven, W., Edebes, H. and Conroy, A. (eds) *Wildlife Ranching: A Celebration of Diversity*, Promedia, Pretoria

Taylor, R. D. (1999) 'Wilderness and the CAMPFIRE programme: The value of wildlands and wildlife to local communities in Zimbabwe', in Cooper, T. G. (ed) *Proceedings of the Wilderness Management Symposium*, Waterberg Plateau Park, Namibia

Taylor, R. D. (2001) 'Participatory natural resource monitoring and management: Implications for conservation', in Hulme, D. and Murphree, M. (eds) *African Wildlife and Livelihoods: The Promise and Performance of Community Conservation*, James Currey, Oxford

Taylor, R. D. and Mackie, C. S. (1997) 'Aerial census results for elephant and buffalo in selected Campfire areas', *CAMPFIRE Association Publication Series*, vol 4, pp4–11

Taylor, R. D. and Murphree, M. W. (2007) 'Case studies on successful southern African NRM initiatives and their impacts on poverty and governance. Zimbabwe: Masoka and Gairesi', www.frameweb.org/iucn_frame_cbnrmstudies (accessed 23 March 2008)

Vincent, V. and Thomas, R. G. (1960) *An Agricultural Survey of Southern Rhodesia. Part I: Agro-ecological Survey*, Government Printer, Salisbury

Wright, A. (1998) 'Monitoring and evaluation plan for CAMPFIRE. Report of the CCG Monitoring and Evaluation Sub-Committee', CAMPFIRE Collaborative Group, Harare

WWF-SARPO (2003) 'Quantifying the assumptions for CAMPFIRE', unpublished report, WWF-SARPO, Harare

WWF-SARPO (2008) 'Map of Zimbabwe showing CAMPFIRE districts', prepared by WWF-SARPO, Harare

Zimbabwe Trust (2003) 'Process oriented monitoring system evaluation of select sub-district CAMPFIRE institutions', unpublished report prepared for WWF-SARPO, Harare

14

CBNRM in Namibia:
Growth, Trends, Lessons and Constraints

Brian Jones and L. Chris Weaver

CBNRM in Namibia has grown from a small pilot project initiated by a non-governmental organization (NGO) and local traditional leaders to a multifaceted national programme that addresses natural resource management, economic development and local governance and involves local communities and 12 Namibian non-governmental support organizations. It has been embraced by government as a means of combating poverty and has been adopted by the Ministry of Environment and Tourism (MET) as an important conservation initiative outside of protected areas.

The four key elements of the conceptual foundation of the programme are sustainable use as a conservation paradigm, economic instrumentalism, devolutionism and collective proprietorship (Jones and Murphree, 2004). These derive from the idea that the main threat to wild habitats and resources in Africa is not overuse but the conversion of land for agriculture and livestock (SASUSG, 1996), which implies that biodiversity conservation depends on the provision of the right incentives for landholders to adopt sustainable land uses (i.e. uses that do not lead to environmental degradation and loss of biodiversity). The highest probability of achieving sustainable use is where the primary beneficiaries of such use are the people living with the resources, as these people have a vested interest in ensuring that they can continue to derive benefit from the resource (Jones and Murphree, 2004). Within Namibia, the success of private conservation on freehold farms demonstrated how a combination of economic incentives and proprietorship could create appropriate conditions for sustainable use of wildlife (see Chapter 8).

Efforts by NGOs in the 1980s to involve rural people in wildlife conservation in the Kunene Region of north-west Namibia demonstrated the viability of community-based approaches to natural resource management on communal lands (Jones, 2001). Conservationists based in Kunene helped local communities establish a network of community game guards who reported to traditional leaders. Later a pilot project to bring tourism revenue to a particular community as an incentive for conservation of local wildlife was established by a local NGO, Integrated Natural Resource Development and Nature Conservation (IRDNC). Significantly, community leaders and many residents agreed to take on some responsibility for conserving wildlife before there was any prospect of economic benefit (Jones, 1999). This conservation commitment of local communities has since played a major role in allowing wildlife numbers in Kunene Region to recover, particularly in the case of the region's black rhino population (Durbin et al, 1997).

In the design of the Namibian initiative, lessons were drawn from the experience of the CAMPFIRE in Zimbabwe: the importance of economic incentives, the necessity that authority should be devolved to the lowest possible unit of management, and that all income generated should go directly to communities instead of being shared with various levels of government. Experience in Zimbabwe showed that where rural communities received income directly related to the use and management of wildlife and perceived that the benefits of wildlife exceeded the costs, they were conserving their wildlife and its habitat (Murphree, 1997).

Namibian government officials were also able to draw upon important advances in management theory and practice about common pool resources (for example Bromley and Cernea, 1989; Ostrom, 1990) when drafting the new policy and legislation. New ideas about the design of common pool resource (CPR) management institutions helped answer some of the questions about how to devolve proprietorship over a common resource (such as wildlife) to a group of individuals on land owned by the state. These new ideas suggested that successful CPR institutions needed to have a defined membership and geographical area (in which the resource is 'owned' and managed), an agreed set of operating and resource use rules, the ability to monitor and enforce these rules, and legitimacy in the eyes of the resource users and the state (Ostrom, 1990).

Thus, Namibian CBNRM policy and legislation provide for rights over wildlife and tourism to be given to communal area residents who form a conservancy. In order to form a conservancy, a community needs to define its membership, define its physical boundaries, elect a representative committee, agree on a plan for the equitable distribution of benefits and adopt a legally recognized constitution. Box 14.1 outlines the evolution and growth of the Namibian CBNRM approach.

Box 14.1 *Major steps in the development of the Namibian CBNRM approach*

1990–1992: A series of participatory 'socio-ecological surveys' is carried out by the then newly created Ministry of Wildlife Conservation and Tourism (MWCT) and NGOs. These identified key issues and problems concerning wildlife, conservation and the MWCT from a community perspective. The surveys lead to the development of several localized community-based conservation projects enabling residents to be involved in wildlife management and to benefit from it, as far as is possible under existing legislation. Government officials and NGO partners realize that policy and legislation must change for these projects to be successful.

1992: MWCT develops the first draft of a new policy providing for rights over wildlife and tourism to be given to communities that form a conservancy.

1993: The Living in a Finite Environment (LIFE) programme brings major donor support (USAID and WWF-US) to CBNRM in Namibia. This investment facilitates the evolution of a national programme, involving a partnership between government, NGOs and rural communities.

1995: Cabinet approves the new policy for communal area conservancies. Work begins on drafting legislation to put the policy into effect.

1996: Parliament passes the new conservancy legislation.

1998: The first communal area conservancy is gazetted on 16 February. Three more communal area conservancies are gazetted by mid-year. The national CBNRM coordinating body is launched to promote synergy and lesson sharing in the development and support of communal area conservancies. President Nujoma officially launches the Namibian Communal Area Conservancy Programme in September. The innovative nature of the programme is recognized by WWF's 'Gift to The Earth' award.

1999: The second phase of LIFE Programme begins, to run for a further five years.

2000: The Namibian Association of CBNRM Support Organizations (NACSO) is constituted in March. (The association was previously known as the CBNRM Association of Namibia.)

2004: 31 communal area conservancies are registered, and more than 50 others are in the process of being formed. The national programme is supported by three major donors (LIFE Plus, the Integrated Community-Based Ecosystem Management (ICEMA) project funded by the Global Environment Facility, and IRDNC activities in Kunene and Caprivi regions funded by WWF, United Kingdom).

2007: 50 communal area conservancies are registered and 20–30 are under development.

2008: The LIFE project ends in April, but the CBNRM programme continues with government and other donor support.

Growth of the Namibian programme

In 1998 the first communal conservancy was gazetted, and by the end of that year, there were four conservancies covering an area of 1.7 million hectares in three regions of the country that had been formally recognized. By March 2007 there were 50 communal area conservancies (see Figure 14.1) in Namibia covering 11.9 million hectares or just in excess of 14 per cent of the country's surface area (see Figure 14.2) (LIFE, 2007). Approximately 230,620 Namibian citizens – almost 13 per cent of the country's population, located in nine of the 13 regions – live within these conservancies (see Figure 14.2). These figures represent significant growth over nine years, especially as forming a conservancy can be a lengthy and time-consuming process (particularly negotiating boundaries with neighbours and registering members).

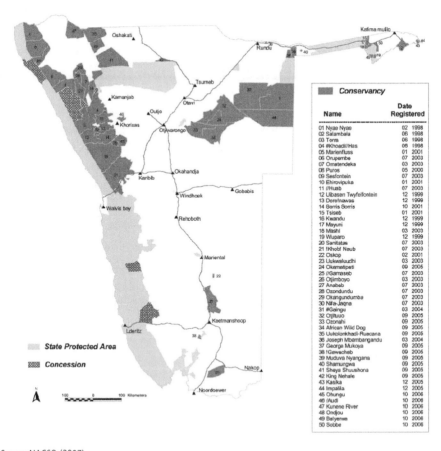

Source: NACSO (2007)

Figure 14.1 *Registered communal area conservancies in relation to protected areas in Namibia, March 2007*

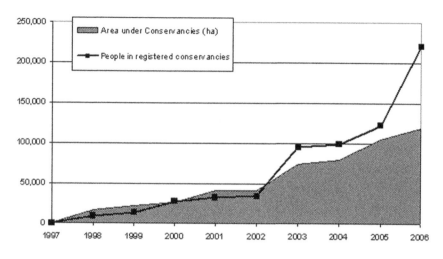

Source: NACSO (2007)

Figure 14.2 *Increased area under, and people living within, communal area conservancies, 1997–2006*

Conservation impacts

The programme has increased community stewardship over wildlife, which is leading to a recovery of wildlife populations across large parts of northern Namibia, in particular the north-west and north-east (see Figures 14.3 and 14.4). For example, populations of endangered species such as black rhino and Hartmann's zebra are growing in north-west Namibia, as are lion numbers (Stander, 2006) and anecdotal evidence based on sightings and reports of live-stock losses suggests that cheetah populations are also increasing. Not only are wildlife numbers increasing, but the distributions of some species are also expanding – in particular elephant ranges are expanding in both the north-west and north-east of Namibia.

Communal conservancies in Namibia cover important habitats that are not represented by the state-protected areas, much of which are desert and proclaimed because there was no other use for them, resulting in several key biomes and vegetation types being poorly represented in national parks and reserves (Barnard, 1998). Communal conservancies now cover much of the northern Namib escarpment and nearby inselbergs of the Kunene and Erongo regions, which together form an important regional centre of endemism (Jones, 2005). Others cover important seasonal wetlands or areas of remnant riverine vegetation, and provide opportunities for improved conservation of these habitats.

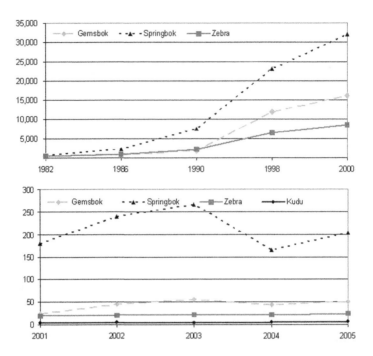

Note: The top graph shows trends from aerial counts (1982–2000); the bottom graph shows trends from road counts (2001–2005)

Source: NACSO (2006)

Figure 14.3 *Growth trends (1982–2005) for some plains game species in north-western Namibian registered and emerging conservancies*

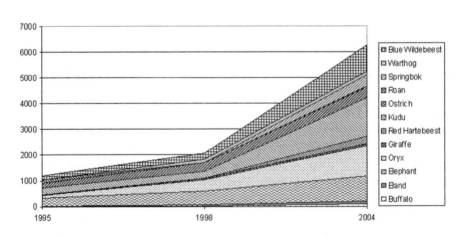

Source: Stander (1995, 2004); Craig (1999)

Figure 14.4 *Game populations trends in the Nyae Nyae Conservancy, 1995–2004*

Of the existing communal conservancies, 31 are adjacent to protected areas, lie across important corridors between protected areas or provide wet season dispersal areas for wildlife that leaves protected areas, increasing land available to wildlife by more than 50 per cent beyond the existing protected area system (LIFE, 2005). The availability of such corridors and dispersal areas is particularly important for large mammals such as elephants whose seasonal movement patterns require extensive tracts of land in Namibia's arid to semi-arid habitats (Jones, 2005).

Escalating demand for the recovery of wildlife populations

Since 1999, more than 3500 mixed plains game animals have been reintroduced into nine communal conservancies, and the tangible financial benefits that conservancies are receiving (as described below) are creating increased demand for wildlife to be reintroduced onto communal lands. The programme has contributed to a change in community attitudes and attendant reduction in poaching, particularly in the north-west (NACSO, 2007), which has created a fertile and safe environment for game reintroductions. A consequence is that the MET (formerly the MWCT) has commenced the reintroduction of black rhino into conservancies, with two conservancies so far being recipients of this important flagship conservation species.

Promotion of wildlife and tourism as legitimate land uses and maintenance of wild habitat

There are a number of reasons that communities form conservancies – not all of which are directly related to wildlife conservation. Although in some cases there might be unrealistic expectations concerning income generation, in others it appears that residents believe conservancies can provide useful institutional arrangements for managing other natural resources, such as grazing, and for gaining a stronger claim over their land. Further, conservancy registration provides a community with government recognition of their claim over land and resources.

The growth in the number of communities forming conservancies, in the number of wildlife-based tourism enterprises and the continued interest in conservancy formation suggests that wildlife and tourism are increasingly being accepted as legitimate land uses on communal land. By the end of 2006 there were 13 formal joint venture tourism enterprises in conservancies (between the conservancy and a private sector operator) and another six conservancies were receiving income from tourism operators for traversing rights or resource utilization. Twenty potential joint venture agreements were under negotiation in 2007, and 21 trophy hunting concessions were in operation across 24 conservancies (NACSO, 2007).

Several of the older conservancies have zoned specific areas of their conservancies as dedicated wildlife management areas in which trophy hunting and/or tourism is promoted. These zones preclude the use of land for settlement and cropping and, in many cases, for the grazing of livestock.

Socio-economic impacts

Since communal area conservancies began to be registered in 1998, the total benefits (cash income, employment and in-kind benefits) generated have increased from almost N$600,000 (US$101,200) to N$26.8 million (US$3.9 million) (see Figure 14.5) (LIFE, 2007). The total estimated 2006 direct income and benefits to conservancies and community members (some beneficiaries of the CBNRM programme reside outside of conservancies) increased by approximately one third over the 2005 figure of N$20 million (US$3.1 million) (LIFE, 2007).

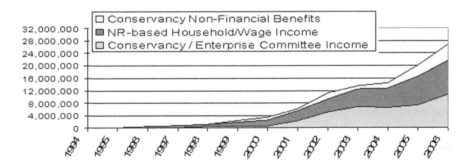

Source: LIFE (2006)

Figure 14.5 *Namibia CBNRM programme benefits (cash income, employment income and in-kind non-financial benefits), 1994–2006*

The diversity of benefits being generated by Namibia CBNRM enterprises is a significant strength, as illustrated in Figure 14.6. Direct cash income to conservancies increased from N$5.2 million (US$814,900) during 2005 to N$7.6 million (US$1.1 million) in 2006. The revenues received from conservancies ranged from a low of N$7200 (US$1060) in the Kunene River Conservancy to a high of N$927,950 (US$138,850) for the Torra Conservancy (LIFE, 2007). Of the 32 conservancies receiving cash income, a total of 17 received more than N$115,000 (US$16,960) during 2006. In contrast, 18 conservancies did not receive any cash income during that year.

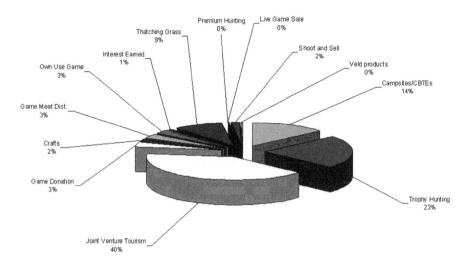

Note: CBTE: community-based tourism enterprise

Source: LIFE (2007)

Figure 14.6 *Source of Namibia CBNRM benefits, 2006*

An estimate of the total revenues generated by CBNRM-supported enterprises (i.e. joint venture lodges, trophy hunting concessions, thatching grass sales, community-based tourism enterprises, crafts and live game sales) shows these enterprises conservatively contributed more than N$93.4 million (US$13.9 million) to private sector stakeholders and approximately N$185 million (US$27.6 million) to the Namibian economy during the 2006 calendar year. These enterprises employ 794 full-time and 5101 part-time conservancy residents (LIFE, 2007).

The 2006 level of income has allowed 13 conservancies to fully cover their annual operating expenses (i.e. they can pay all salaries, vehicle operating and infrastructure maintenance costs), while allowing for benefit distributions to conservancy members. A further four conservancies were poised to reach this important sustainability milestone during 2007. In the case of the 18 non-income-earning conservancies, in some cases, the conservancies are relatively new and so have not yet developed their tourism and safari hunting potential; in other cases, low earnings are linked to a lack of income generation potential because of a lack of wildlife or tourism attractions.

While it is easy to track the income being generated at the conservancy level, it is more difficult to measure the impact this is having at household level, partly because Namibian conservancies (in contrast to communities involved in CAMPFIRE in Zimbabwe) have rarely opted for cash payouts (or 'dividends') to residents. Namibian conservancies have generally opted for job creation and 'social projects' such as adding a classroom to a school, soup kitchens for the elderly and so on. In such circumstances it is necessary to look for other indicators of impact, such as the overall satisfaction of residents with the conservancy.

Namibian researchers have begun to document conservancy expenditures on benefits for residents, to measure levels of satisfaction of residents with conservancies and to assess the contribution that CBNRM is making to combating the multiple non-income dimensions of poverty. Table 14.1 provides data on the types of benefits being provided by four potentially high

Table 14.1 *Benefits provided to members in four high-potential conservancies in Kunene Region*

Benefit	Anabeb	Marienfluss	Puros	Sesfontein
Borehole for domestic water use		✓	✓	Planned
Lifts to clinics and in other emergencies with conservancy vehicle		✓	✓	
Employment in tourism lodges/activities linked to the conservancy such as joint ventures (figures in brackets are part-time jobs)	8	14 (2)	35 (2)	
Conservancy employment (including conservancy owned campsites)	11	13	23	
Employment in hunting (figures in brackets are part-time jobs)	(4)			(5)
Support to local school (food and/or cash)	✓	✓	✓	✓
Own use hunting quota divided among local institutions			✓	
Transport children to and from schools in Sesfontein and Opuwo		✓	✓	
Donation of game meat for meetings and festivals	✓	✓	✓	✓
Meat from own use and other hunting to households	✓	✓	✓	✓
Income from craft sales to individuals			✓	
Loans to individuals			✓	
Social fund to help people in emergencies		✓		
Financial support for school children/prizes, bursaries for tertiary education		✓		✓
Financial support to provide teachers				✓
Acquisition of conservancy property to house members on visits to town		✓		
Contribution to human/wildlife conflict self-insurance scheme payments to members to offset livestock losses	✓		Planned	✓
Provision of infrastructure (e.g. room for mobile clinic, classroom, etc.)		✓	✓	
Payments/cash support to traditional authority	✓			✓
Support to local sports teams/tournaments	✓			✓

Source: Jones and Mazambani (2007)

income-earning conservancies in Kunene Region and Tables 14.2 and 14.3 provide examples of benefit distributions in two conservancies, #Khoadi //hoas Conservancy (Kunene Region) and Nyae Nyae Conservancy (Otjozondjupa Region). #Khoadi //hoas is an example of the conservancies that are spending on social projects, while Nyae Nyae is one of the few that has opted for cash dividends.

Table 14.2 *Community benefit expenditure by #Khoadi //hoas Conservancy, 2004 and 2005*

Benefit	2004*	2005**
Contribution to two schools	25,000	
Support for livestock vaccination	4000	
Diesel for elephant water-points (collected by residents to pump water)	6235	4580
Diesel at subsidized price to pump water for livestock where elephants consume most of water at a settlement	15,645	10,292
Payment to offset livestock losses to elephants	700	5500
Payment to local traditional authority	2500	2500
Loan of breeding animals for improving farmers' small stock	20,000	
Soup kitchen for old people	7824	6000
Contribution to building elephant protection walls around water-points	61,520	116,100
Support to Grootberg Farmers' Union		2000
Implementation of benefit distribution plan (fuel and travel)	14,952	16,391
Total (N$)	158,376	163,363

Note: * Overall income: N$133,932 (US$20,750); accumulated surplus: N$388,599 (US$60,150); ** overall income: N$214,245 (US$33,575); accumulated surplus: N$340,705 (US$53,400)

Source: Jones and Barnes (2007)

Table 14.3 *Community benefit expenditure by Nyae Nyae Conservancy, 2005 and 2006*

Benefit	2005*	2006**
Water protection team transport	96,080	66,000
Water protection equipment	2593	8000
Natural Resource monitoring transport	50,901	67,000
Governance meetings	9453	16,484
Cash dividend distribution	0	197,100[a]
Cash dividend distribution costs	0	11,913
Seed for gardens	0	4876
Total	159,027	359,460

Note: * Overall income: N$742,701 (US$106,100); ** overall income: N$804,032 (US$114,860); [a] 657 members received N$300 each

Source: Nyae Nyae Development Foundation of Namibia-cited, in Jones and Mosimane (2007)

Given the contexts of rural livelihoods and poverty in Namibia's communal areas, it is a reasonable conclusion that the types of benefits indicated in the tables above are helping to combat poverty and improve livelihoods in several ways. In terms of benefits to the community as a whole, conservancies can generate discretionary income at community level for social welfare (for example meat for funerals, transport for sick people to hospital) or other purposes (for example infrastructure development, water-point protection in the north-west, ensuring water is available for domestic and livestock use). The programme can also facilitate better working relationships and formalized partnerships with the government and other support institutions.

Individual benefits are also important, where jobs in tourism and natural resource management activities can provide additional income and/or livelihood diversification options for some residents. More recently, the introduction of a compensation scheme provides cash to offset losses of livestock to predators and for crop losses from elephants. Building skills and capacity (training in financial management, planning, wildlife management, tourism management, exposure to the private sector, business management and to similar activities in other regions in Namibia and other countries) is also important. Devolved decision making and financial management, improved advocacy and access to government and institutional development also play a significant role in empowering marginalized rural people. Strengthening or building local institutions for CPR management can also help to drive local development – conservancies enable collective management of wildlife as common property and act as vehicles for HIV/AIDS awareness in Caprivi; while in Kunene conservancies facilitate access to education and medical services.

However, not all conservancy residents perceive that conservancies are improving their livelihoods. Initial findings from a number of attitudinal surveys conducted in 2007 suggest that conservancy residents have mixed perceptions about the benefits being provided by conservancies (Mosimane, unpublished data; Diez, unpublished data; Davis, unpublished data). A majority of residents in conservancies are able to identify the benefits being provided, and those that have jobs, receive meat and other direct benefits tend to be the most positive about the importance of these benefits. Those that do not receive such benefits tend to be less positive about the overall impact conservancies have on their standard of living, although these perspectives vary from conservancy to conservancy.

At the same time, the costs of living with wildlife are increasing in key areas of the north-west and the north-east. The increased number of elephants and predators in the north-west is leading to increased damage to water installations and more livestock losses. In the north-east, the main problem is crop damage by elephants. The increased costs of living with wildlife, a symptom of conservation success, will also clearly affect perceptions of the benefits brought by conservancies.

Governance impacts

As already noted, the conservancy legislation requires the formation of a local management institution that has a defined membership, a committee that represents the membership and a constitution that sets out the way in which the institution will be governed (NACSO, 2007). This institution takes decisions such as those regarding wildlife management (which can involve closing areas to settlement and livestock grazing) and about how conservancy income is spent. Governance is therefore a critical issue. If the conservancy committee is not accountable to the membership, the potential for misuse of funds, elite capture and corruption increases.

Research on governance issues indicates a dynamic situation of learning and incremental institutional change within conservancies. Long and Jones (2004) found that there were a number of governance issues that needed addressing. These included the overlapping roles of committee members who were also employees of the conservancy, a lack of participation by members in annual general meetings (AGMs), a lack of communication between committees and members, a lack of detailed discussion of financial matters at AGMs and members not being asked to approve conservancy budgets. Further, because conservancy committees were receiving funds from NGOs to help pay operational costs until they could generate sufficient income of their own, committees tended to be accountable upwards to the NGOs rather than downwards to their membership.

The situation has been changing. Within conservancies there has been change in the way that representation and decision making are structured (NACSO, 2007). Instead of electing a committee of individuals from anywhere in the conservancy, there is a trend towards dividing the conservancy into sub-units which then elect their own representatives to the committee. This allows for better accountability as area representatives can give feedback to local residents about the activities of the committee and can represent local interests at the central committee level (NACSO, 2007).

NGOs have also provided increased levels support on governance issues to conservancies over recent years (Jones and Mazambani, 2007). Once donor grants to conservancies were withdrawn and the accompanying financial reporting requirements no longer applied, conservancies did not apply the same accounting standards to their income (Davis and Kasaona, 2007), resulting in conservancies being unable to account for sums of money (although the incidence of fraud was actually low). At the same time, conservancy members started refusing to allow AGMs to be held because committees could not produce proper financial statements. As a result, a number of methodologies and tools for helping conservancies improve internal governance were developed (Jones and Mazambani, 2007), training was provided in implementing financial management systems that meet accepted standards, and guidelines were developed for holding AGMs. Conservancies have also been supported in the implementation of a form of business sustainability planning, while others in the Caprivi Region are piloting a system for monitoring institutional development.

Key principles to emerge from implementation

Policy, legislation and practice should be rooted in local needs

For CBNRM policy, legislation and practice to be successful, they need to be rooted in the needs of local people. Policy arose as a response to the needs identified by communities during the early socio-ecological surveys. Importantly, there was a considerable overlap between the agendas of government officials, NGOs and local communities. Further, policy and legislation were not developed as a reluctant response to a condition attached to a donor project, they were developed because government saw the need for reform and had initiated this reform before donor involvement.

The policy and legislation that resulted provides a flexible framework to which communities can respond (Jones, 1999). The element of choice provided for by the policy and legislation is an important aspect of empowerment and control over a community's own affairs, allowing them choice over whether or not to gain rights over wildlife and tourism by forming a conservancy, about who should represent a community on the conservancy committee, and how communities can use their income.

CBNRM is dynamic and requires adaptation of responses by external agencies

In the years since the legislation was passed, there has been evolution and learning within conservancies as well as in support agencies. Conservancies initially required support in the formation and registration processes and to establish some form of natural resource management. Once registered, the focus of support remained with resource management as a foundation of conservancy activities then shifted to developing wildlife- and tourism-based enterprises in order to generate income. As conservancy institutions developed, the need for greater emphasis on promoting good governance and accountability became evident. This meant that support agencies needed to develop and adapt their own skills and services to meet the emerging and shifting needs of conservancies. It has also meant that such a programme is difficult to fit into the typically short project timeframes of donors (usually four to five years, with the possibility of an extension of similar length). There has been pressure on the programme from donors to reach a stage where external support is no longer necessary. However, newly emerging conservancies will continue to require support in meeting registration requirements. Also, as noted, the support needs of conservancies change over time, thus the 50 registered conservancies will also continue to require support for some time into the future.

The importance of scale in CBNRM

The development of the CBNRM programme in Namibia demonstrates some of the problems of dealing with scale in CPR management. In particular, there is a need to match appropriately sized social units with appropriately sized wildlife management units (Jones, 2001). Conservancies essentially provide social building blocks for natural resource management across a larger landscape and a demonstration of the benefits of such an approach in the Caprivi and Kunene regions are the joint elephant hunting quotas being issued by the MET to more than one conservancy (where it would be unsustainable to allocate quotas to individual conservancies). Cooperation and joint management by the conservancies are required for these quotas, and in Caprivi, for example, the conservancies that share a quota carry out joint anti-poaching patrols and wildlife monitoring.

Devolution to the lowest appropriate level

Although the devolution of rights over wildlife and tourism in Namibia directly empowers local communities, there is a strong argument for further devolution to sub-units within conservancies. Currently, conservancy committees tend to be accountable upwards to the organizations providing funding and technical support, rather than downwards to conservancy members. Devolution to appropriate sub-units within each conservancy would provide the opportunity to develop localized forms of participatory democracy. Such an approach could increase the likelihood of accountability downwards from the conservancy committee, place more responsibility on individual members and increase their involvement, enabling them to take a greater part in major decisions such as how to use the income generated by wildlife and tourism. It would also enable support agencies to provide capacity-building at the sub-unit level and to more individuals. The sub-units could take decisions over local resource management issues and could link very effectively to other resource management institutions being established, such as water-point committees and community forests. A number of conservancies in Kunene have begun to adopt such an approach and are revising their constitutions to reflect a more decentralized approach to their own governance.

The importance of intrinsic incentives for conservation

The experience of CBNRM in Namibia has helped to focus attention on intrinsic incentives for conservation, which are often overlooked or undervalued. Early conservation successes in Kunene Region were achieved without the prospects of large amounts of income being derived by residents from conservation. The exercise of responsibility, regaining of some control over wildlife and wanting wildlife for its existence value, appear to have provided sufficient incentive for many residents of the Kunene Region communal areas to conserve wildlife (Jones, 1999, 2001). Local leaders elsewhere have also been keen to see the reintroduction of wildlife that had disappeared.

The balance between process and product

An important principle that has emerged from the Namibian CBNRM programme is that implementation needs to be based on process rather than the achievement of predetermined products or outcomes. Experience has shown that, in the long run, a good process is more likely to lead to a good outcome than a quick-fix approach. A process-oriented approach implies participation in decision making by residents, giving them time to come to their own conclusions, enabling them to shape the outcome (rather than presenting them with a fixed package), and acknowledging and dealing with their concerns. Much of the flexibility of the Namibian policy and legislation described above enables a process-oriented approach to be followed in conservancy formation and operation.

The importance of light-touch facilitation

Experience has also shown that the best way to develop a process-oriented approach is through persistent and consistent 'light-touch' community empowerment and facilitation (Hitchcock and Murphree, 1995; Jones, 2001). This approach engages with communities directly and not only through local government institutions or traditional leaders. It requires staying in touch with community power shifts and internal dynamics, and ensuring that the various interest groups within a community have a voice in decision making. The approach also involves assisting communities to identify potential problems and other key issues, providing information and helping them to develop appropriate decisions, solutions and actions. A danger of providing regular support and facilitation over a period of time is that support organizations can become a 'gatekeeper' between the community and other outsiders. This needs to be managed by ensuring that support is demand-driven and that facilitators decrease their support over time and develop an appropriate exit strategy.

Role of government

The role of MET has been crucial to the success of Namibia's CBNRM programme. MET led the socio-ecological surveys and development of the conservancy policy and legislation in the early 1990s, has established a CBNRM support division and continues to support the programme through the ICEMA project. CBNRM is recognized as a major strategy of MET in its current strategic plan and is recognized in the third National Development Plan as well as in national poverty reduction action plans. An ongoing issue, however, is the tendency of government to hold on to power, to regulate against risk and to 'protect' communities, which contrasts with the devolutionary objectives of CBNRM, summed up in the statement that 'local management initiatives need state law, often more than their advocates like to think, though usually less than governments are willing to admit' (Lindsay, 1998, p4). Policy and implementation at any given time therefore reflect the outcomes of an ongoing conflict between these issues.

Lessons and constraints

Lessons

The evidence indicates that conservancies present an attractive institutional option for local residents to manage and gain income from the sustainable use wildlife. In some cases, competition over rights, revenues and resources has led to disputes and conflicts that have delayed conservancy formation for several years. However, local leaders and residents have persevered over time and have succeeded in resolving such conflicts and moving ahead with conservancy formation. Once formed, conservancies should be assisted to develop their own vision and plans for integrated rural development (not just wildlife management), through which service provision should be coordinated. There is a need to prevent the proliferation of sectoral organizations working in isolation and competition, so service providers should work jointly, responding to conservancy plans and visions, and conservancies need to be more proactive in developing formal and informal relationships with other resource management institutions (for example water-point committees, community forests, village development committees and so on).

The Namibian programme has benefited considerably from a strong partnership developed between various implementing agencies represented within NACSO. This body has proved to be a useful coordination mechanism for structuring relationships between different organizations and agencies, as well as preventing duplication of activities. Its members have developed a common vision for CBNRM in Namibia and one of the organization's major strengths is the sense of common purpose that has been developed among members.

Constraints

Rural communities have embraced the conservancy approach with enthusiasm so far, though this interest could wane if household-level benefits do not increase and if proprietorship over wildlife is not strengthened – the current financial benefits from wildlife and tourism to households remain low, while the costs of living with wildlife remain high and community proprietorship over wildlife remains relatively weak. This situation is the main constraint to implementing CBNRM in Namibia.

Another major constraint is the lack of secure and exclusive group land tenure. If communities cannot prevent other people using the land they wish to set aside for wildlife and tourism, then there remains little incentive to maintain wild habitats. A lack of secure land tenure also means that communities cannot easily raise finance using their land as security.

Traditional authorities are important in the Namibian context because they are 'owners' of the land in the minds of many residents, and most rural residents support the institution of traditional leadership (even though some individual leaders might lack honesty and integrity). Thus traditional authorities need to be included in conservancy development and management. In particular, conservancies need to work closely with traditional leaders in order to limit access to land, as these leaders allocate access to grazing.

Although CBNRM has been institutionalized within MET, it still lacks adequate resources to provide the necessary support to communities and NGOs, and despite the dedication of some individuals, much of the burden of field implementation has fallen to NGOs. Further, despite the number of organizations and the considerable financial and technical support from donors, NGOs and government, the capacity to provide adequate support to the growing number of conservancies is still lacking. It will be difficult to give new conservancies the same level of support that was enjoyed by the first to emerge, due to a lack of financial resources, but also because of the lack of qualified and experienced Namibians involved in the CBNRM sector.

Conclusions

CBNRM in Namibia has undergone considerable growth since legislation was passed in 1996 giving rights over wildlife and tourism to communities that formed a conservancy. Although not perfectly implemented, the approach to CBNRM has provided a strong foundation for delivering a range of conservation and development benefits. Although the rights of conservancies are conditional and limited, and government retains considerable control over some aspects of decision making, there have been very positive conservation results. Conservancies with wildlife and tourism assets are able to generate income that is increasingly being used for community benefit and improving livelihoods. Governance is improving within conservancies and as a consequence there is growing awareness at local and national levels in Namibia about the validity of wildlife-based enterprises as legitimate land uses, alongside subsistence agriculture and livestock production. However, there are further improvements to be made in governance within conservancies.

Considerable challenges remain, including increasing the impact of CBNRM activities at the household level and increasing the number of conservancy members who view CBNRM activities as contributing positively to their livelihoods. This is particularly important given the perceived increase in human–wildlife conflict and the costs of living with wildlife.

References

Barnard, P. (ed) (1998) *Biological Diversity in Namibia: A Country Study*, Namibian National Biodiversity Taskforce, Windhoek

Bromley, D. W. and Cernea, M. M. (1989) 'The management of common property natural resources: Some conceptual and operational fallacies', World Bank Discussion Paper No. 57, World Bank, Washington DC

Craig, G. C. (1999) 'Aerial census of wildlife in northern Namibia, August–November, 1998', unpublished data prepared for the Ministry of Environment and Tourism, Windhoek

Davis, A. and Kasaona. L. (2007) 'Supporting the management of conservancy generated income: Achievements and challenges for IRDNC Kunene', report prepared for IRDNC, Windhoek

Durbin, J., Jones, B. T. B. and Murphree, M. W. (1997) 'Namibian community based natural resource management programme (WWF NA 0004: Namibia): Project evaluation', report prepared for WWF, Gland

Hitchcock, R. K. and Murphree, M. W. (1995) 'Report of the field assessment team: Phase III of the mid-term assessment of the LIFE project', report prepared for USAID, Windhoek

Jones, B. T. B. (1999) 'Community management of natural resources in Namibia', IIED Drylands Programme, Issue Paper No. 90, IIED, London

Jones, B. (2001) 'The evolution of a community-based approach to wildlife management at Kunene, Namibia', in Hulme, D. and Murphree M. W. (eds) *African Wildlife and African Livelihoods: The Promise and Performance of Community Conservation*, James Currey, Oxford

Jones, B. (2005) 'Case study: Freehold and communal conservancies in Namibia', in Stolton, S. and Dudley, N. (eds) *Protected Areas in Eastern and Southern Africa: Reporting Protected Areas and Applying the IUCN Categories*, World Conservation Monitoring Centre, Cambridge

Jones, B. T. B. with Barnes, J. I. (2007) 'WWF human wildlife conflict study: Namibian case study', report prepared for WWF, Washington DC

Jones, B. T. B. and Mazambani, D. (2007) 'Managing growth and sustainability: Mid-term evaluation of IRDNC's community-based natural resource management programme in Kunene Region and Caprivi Region, Namibia', report prepared for

IRDNC/WWF UK, Windhoek

Jones, B. T. B. and Mosimane, A. (2007) 'Promoting integrated community-based natural resource management as a means to combat desertification: The Living in a Finite Environment (LIFE) project and the #Khoadi //Hoas and Nyae Nyae conservancies, Namibia', paper prepared for IRG, Washington DC

Jones, B. T. B, and Murphree, M. W. (2004) 'Community based natural resource management as a conservation mechanism: Lessons and directions', in Child, B. (ed) *Parks in Transition: Biodiversity, Rural Development and the Bottom Line*, Earthscan, London

LIFE (2005) 'Semi-annual report for the period: April 1 through September 30, 2005', report for WWF LIFE Project, Windhoek

LIFE (2006) 'Update information on conservancy development and income generation for the period 1 January to 31 December, 2006', report for WWF LIFE Project, Windhoek

LIFE (2007) 'Semi-annual progress report, April 2007', report for WWF LIFE Project, Windhoek

Lindsay, J. (1998) 'Designing legal space: Law as an enabling tool in community-based management', paper prepared for the International CBNRM Workshop, May 1998, Washington DC

Long, S. A. and Jones, B. T. B. (2004) 'Conservancy institutions and governance: Implications for livelihoods', in Long, S. A. (ed) *Livelihoods and CBNRM in Namibia: The findings of the WILD Project. Final Technical Report of the Wildlife Integration for Livelihood Diversification (WILD) Project*, WILD Project/Ministry of Environment and Tourism/Department for International Development, Windhoek

Murphree, M. W. (1997) 'Congruent objectives, competing interests and strategic compromise: Concept and process in the evolution of Zimbabwe's CAMPFIRE programme', paper presented at the conference Representing Communities: Histories and Politics of Community-Based Resource Management, 1–3 June 1997, Georgia

NACSO (2006) *Namibia's Communal Conservancies: A Review of Progress and Challenges in 2005*, NACSO, Windhoek

NACSO (2007) *Namibia's Communal Conservancies: A Review of Progress in 2006*, NACSO, Windhoek

Ostrom, E. (1990) *Governing the Commons: The Evolution of Institutions for Collective Action*, Cambridge University Press, Cambridge

SASUSG (1996) *Sustainable Use Issues and Principles*, Southern African Sustainable Use Specialist Group/IUCN Species Survival Commission, Gland

Stander, P. (1995) 'Aerial survey of the Khaudom Game Reserve, Eastern Tsumkwe District and surrounding areas', unpublished data prepared for the Ministry of Environment and Tourism, Windhoek

Stander, P. (2004) 'Aerial survey of wildlife in the Nyae Nyae conservancy, Namibia', unpublished data prepared for the Ministry of Environment and Tourism, Windhoek

Stander, P. (2006) 'Population ecology and demography of Kunene lions, 2006: Towards resolving human–lion conflicts with applied research and pro-active management', Research Paper 2006/1, Predator Conservation Trust, Windhoek

15

CBNRM in Botswana

Nico Rozemeijer

Botswana is very fortunate in terms of the diversity and range of natural resources in the country, boasting mineral resources as well as rich flora and fauna. Approximately 18 per cent of the country's land area is designated as protected areas and an additional 24 per cent is classified as wildlife management areas (WMAs), which are further subdivided into controlled hunting areas (CHAs) (Broekhuis, pers. comm.). Population density is low – 1.68 million people in 2001 – spread over a vast area (582,000km^2) and the economy is driven by a profitable diamond trade (CSO, 2003). Relatively high national wealth and low population density has limited competition over valuable (renewable) natural resources and allowed for the decentralization of community user rights and benefits to those resources. This chapter provides a brief description of the development of the CBNRM approach in Botswana and its status today, addressing the concerns of a growing chorus of local critics who are doubtful about the impact of CBNRM.

Given the relatively short implementation time of the CBNRM programme in Botswana, it is argued that, despite tangible outputs in terms of increased community incomes and improved rural employment numbers, it is not easy to measure the impact of the approach on rural development or the conservation of the country's natural resources. However, with the data available, and in the light of recent policy and strategy discussions, it is possible to reassess the assumptions that underlie this innovative natural resource management approach.

CBNRM – combining conservation with rural development objectives

CBNRM began in Botswana with the USAID-funded Natural Resources Management Project (NRMP), which ran from 1989 to 1999. In the late 1980s in most districts, land use and settlement plans were in place and WMA and CHA boundaries designated, and subsistence hunting and gathering made an important contribution to the livelihood security of a sizeable proportion of the rural population in the remote areas of the country. These factors, combined with the stable economy and a decentralized governance tradition proved to be conducive to CBNRM implementation in Botswana.

The Department of Wildlife and National Parks (DWNP), advised by the NRMP, realized that conservation policies could not be effectively implemented in the huge expanse of western Botswana without the commitment of resident communities to actively contribute to this goal, thus the CBNRM concept was adopted by DWNP to promote natural resource conservation. The adoption was based on the assumption that by allocating natural resource user rights to the community living with those resources, the economic incentive provided (i.e. the right to sell natural resources, which was previously not allowed) would encourage the conservation of the very natural resources the user communities depended upon.

The (draft) CBNRM policy and other policy documents and implementation guidelines aimed at more complex and sophisticated activities than simply selling hunting quotas, assuming that CBNRM opportunities would encourage the growth of natural resource-based community enterprises and the reinvestment of returns from subleases in productive development.

It could be argued that DWNP was so generous in decentralizing management responsibility because its prime conservation efforts were directed at the management of the national parks and game reserves that constitute 18 per cent of the land area. The WMAs, which cover approximately 24 per cent of the land are all located adjacent to protected areas and have a clear 'buffer zone' function (see Figure 4.1, page 57). By allowing communities to manage buffer zones, the DWNP would thus be relieved from costly involvement in their management and protected areas would be relieved of community pressure, making their management much easier. The introduction of CBNRM in WMAs explains, to a degree, the current absence of participation of local communities in the management of national parks and game reserves and (with the exception of the Central Kalahari Game Reserve that is subject to claims by resident San Bushmen) the lack of demand for such involvement.

The introduction of CBNRM in Botswana had a second objective: to enhance opportunities for communities to earn benefits from natural resources in order to initiate rural development, and reduce poverty and dependence on handouts in a sustainable and equitable manner (GoB, 2001). During the implementation of the programme, this objective quickly took prominence over the first, conservation-related, objective. The most popular and hitherto most successful activity to achieve this objective has been subleasing user rights to the private sector. In 16 community areas, part of the hunting quota and concession rights are sold to commercial safari companies and the revenues and employment generated so far are significant contributors to local livelihoods in the participating rural economies (Arntzen, 2003).

The NRMP was instrumental to the development of CBNRM policy including the Wildlife Conservation and National Parks Act, 1992, the joint venture guidelines of 1996 and 1999, resource user leases, CBNRM implementation guidelines and the draft CBNRM Policy. The project piloted community-based natural resource-related enterprise development and was closely involved in the establishment and monitoring of the first CBNRM project in Botswana in the Chobe Enclave. (The Chobe Enclave community was the first community that assumed natural resource management responsibility in 1993.) The NRMP also facilitated the development of management plans for community-managed WMAs.

With U$S25 million (National CBNRM Forum, 2000) and an army of consultants working with government agencies such as the DWNP, it has taken more than a decade for CBNRM to evolve into the approach we know today.

Natural resource user rights defined

The current rights and responsibilities of communities over natural resources in Botswana are mainly defined in the 15-year 'Community natural resources management lease' between a community-based organization (CBO) and the land authority (in most cases the Land Board). In order to obtain a 'head lease', the communities are required to establish a legally registered CBO, in most cases a trust, whose constitution and by-laws have to show proof of fair representation and accountability.

The 'head lease' gives the community exclusive rights to use an area (normally a CHA that the community lives in or adjacent to) for tourism-related activities, hunting, game capture and the commercial consumptive use of veld products. The lease does not confer ownership over the land or natural resources but only gives the leaseholder (temporary) rights to use them, in accordance with laws and policies of Botswana. Hunting and game capture, for example, must be done according to the Wildlife Conservation and National Parks Act, and the DWNP and not the community sets the annual hunting quota. Nevertheless, the lease gives important rights to communities, and as such offers opportunities for involvement in resource management, income generation and employment.

In return for giving communities the opportunity to earn money from natural resources, the government expects communities to actively manage the land and resources. To do this, communities have to prepare and implement a land use and management plan, which must be acceptable to government.

It should be recognized that not all communities who potentially have access to 'an area' (i.e. a CHA zoned for community management) have secured formal access to that area. Currently, 18 of the 37 zoned community-managed CHAs are covered by leases; in some cases one community manages more than one CHA and in some cases several communities jointly manage one CHA.

Management of WMAs and changes in resource values

The management of WMAs has changed dramatically since 1989, when a WWF mission visited Botswana to advise upon the implementation of a sustainable wildlife utilization model in WMAs. The mission found huge wildlife areas in seven districts covering 129,000km^2 where very little management took place – the primary use of these areas was largely uncontrolled citizen hunting and state-controlled commercial hunting, with no community management involvement whatsoever. (Permits for citizen hunting were accessible to all citizens of Botswana, with licences allocated on the basis of a raffle system.)

By 2006, 91 CBOs had registered constitutions (up from 46 in 2001) and were, in different ways, using natural resources in a controlled manner. These 91 CBOs covered 150 villages involving more than 135,000 Batswana (almost 10 per cent of the population). However, only 35 of the CBOs were actively implementing CBNRM activities and generating revenues (National CBNRM Forum, 2007).

The nature of (legal) hunting in community areas has also changed dramatically. Based on an analysis of the 2003 hunting quota (excluding wild cats, hares, vervet monkeys and porcupines), 43 per cent of species were hunted in community concessions where all of the revenue in terms of concession and quota fees accrues to the community and 33 per cent of the species were hunted in private concessions on state land, where resource fees accrue to the state. Citizen hunting has also decreased substantially over time, to just 24 per cent of the 7168 animals on the 2003 annual quota.

By 2006, 16 CBOs had signed joint venture agreements with the private sector, mostly for commercial hunting activities (see Table 15.1), which generated BWP15 million (US$2.3 million). In addition to the income benefits to communities, employment is also important, and direct employment rose from 750 in 2001 to approximately 980 in 2006 (National CBNRM Forum, 2007).

Associated with this change in WMA management, the actual and perceived value of wildlife has increased tremendously over the years. In 1989, a kudu could be bought through a raffle system for BWP5 (US$0.76). Special Game Licences that allowed hunting of a 'bag of animals' were freely available to a large part of the population in western Botswana, so everybody hunted everybody's resources. At this time, communities were not aware of the value of a 'concession area' or 'photographic attractions', income and employment generated from wildlife utilization at community level was very low and, as a result, the perceived value of wildlife was low, especially in rural communities. Today the situation is dramatically different. A citizen licence for a kudu costs BWP300 (US$46) and BWP1000 (US$154) for a non-citizen. The 16 communities that are partners in joint venture agreements receive fees for the quota, for the hunting concession and in some cases for photographic safari rights.

Table 15.1 *The growth in community revenue from joint venture agreements*

	1993	1997	1999	2001	2002	2006
Number of CBOs in a joint venture agreement	1	3	5	9	14	16
Joint venture income generated by CBOs (BWP)	24,000	1.41 million	2.27 million	6.42 million	9 million	15 million

Source: National CBNRM Forum (2007)

The increase in the financial value of wildlife resources can be attributed to both the government, which played a role in raising fees for hunting licences and for state-controlled concessions, and to the NRMP, which was instrumental in demonstrating how undervalued quotas were. One can assume, however, that the increased bargaining power of the communities holding the resources has also resulted in increased quota and concession fees. For example, the annual quota (excluding elephants) for the Chobe Enclave prior to 1993 was much larger than it is today, and was leased to a safari company for about BWP8000 (US$3150) per year, with the revenue generated going to the Land Board. The NRMP, jointly with the Chobe Enclave Community Trust, tendered the (much smaller) quota annually from 1996, and its value had increased to approximately BWP200,000 (US$70,050), which went to the community. By initially offering tenders for only one year, communities were not locked into a long-term lease, forgoing income while prices were still rising. The demonstration of this principle in Chobe had positive impacts for all subsequent tenders throughout Botswana.

One of the most positive impacts of the increased financial value of wildlife has been the changed perception of community members towards these resources. Wildlife in Botswana today is perceived to be very valuable – for example, an elephant is not longer 'just a problem animal'; it is also the target of a hunting safari with a minimum community income of BWP75,000 (US$11,500).

Questions remain regarding the extent to which the increased positive attitude towards wildlife at community level has translated into an increased awareness of conserving these natural resources. There is some evidence that poaching in community-managed areas has diminished (Mbaiwa, 2004), though there is little evidence that land use patterns in resource-rich areas have changed to the benefit of wildlife (for example, by applying less conflicting arable agriculture and livestock grazing models).

Impact on rural development

It is difficult to assess the exact impact of CBNRM on the development of the participating communities, particularly as there is no systematic collection of data on existing CBNRM projects. Information about revenue and employment generated through joint ventures is available, but data on veld product management projects, small-scale tourism ventures and craft production are largely lacking. It is even more difficult to estimate the value of subsistence use of resources such as thatching grass, building material, firewood, food and medicinal plants, fish and meat.

Table 15.2 attempts to give an indication of the value added to natural resources through CBNRM on the basis of a 'National CBNRM Survey' conducted in 2006. Box 15.1 presents information about the Nqwaa Khobee Xeya Trust and the Okavango Community Trust illustrating some of the opportunities and difficulties of measuring the impact of the CBNRM programme on remote rural communities.

Table 15.2 *Estimated natural resource value added*

Natural resource use	Estimated income	Employment at CBO level
Commercial hunting	BWP12 million (US$1.85 million) in 16 CBOs	560 people
Commercial and community-based tourism	BWP3 million (US$460,000)	420 people
Production of crafts using natural resources	BWP600,000 (US$92,300)	> 4000 people
Marketing of veld products	BWP710,000 (US$109,000)	> 3100 people
Subsistence hunting and fishing	Unknown income, subsistence use	
Firewood and timber collection	Unknown income, subsistence use	

Source: National CBNRM Forum (2007)

Though the 2006 survey provides a good indication of the potentially important contribution of CBNRM to the rural economies of (especially western) Botswana, it also demonstrates the dire need for an established CBNRM monitoring system.

Box 15.1 *Nqwaa Khobee Xeya Trust and the Okavango Community Trust*

Nqwaa Khobee Xeya Trust

The Nqwaa Khobee Xeya Trust (NKXT) represents the communities of Ukhwi, Ncaang and Ngwatle (835 people) in the management of a CHA (called KD1, 12,180km^2) in the Kgalagadi District of south-western Botswana. The population consists of 80 per cent San Bushmen and the more socially and economically dominant Bakgalagadi (20 per cent). The development of the NKXT's constitution and management plan took five years, arriving at a decision-making structure based on family groupings to ensure fair representation of the clans (tribes) between the three villages and in the overall board. The facilitation of this process involved local and foreign NGOs and was expensive.

In 2001, the NKXT was in the second year of a joint venture with a safari company. The trust sold 30 per cent of their quota (desert species without lion) dividing the remainder between the family groups for subsistence hunting. Income from quota and concession fees was BWP185,000 (US$28,500), and the safari hunting company employed 25 people during the hunting season. Roughly 75 per cent of the income was used to hire trust staff (financial management, natural resource monitors and guides), to run a trust vehicle and convene meetings. The remainder was invested in the development of a community campsite.

Problems of mismanagement of project resources arose in 2002 when the support organizations left the area and a new board was elected lacking exposure to trust plans and by-laws. These problems were exacerbated by the reduction in the quota that reduced income below its operating costs.

Okavango Community Trust north of the Okavango Delta

The Okavango Community Trust (OCT) covers five villages and 6431 people around Seronga in north-west Botswana and was conceived by the resident safari company in the area. The OCT's constitution was drawn up in 1995 by the company lawyers, which gave far-reaching decision-making power to a (small) board. A small village elite has dominated the board ever since, and the company has been able to renew contracts without going to tender. The OCT area is rich in wildlife and sells their whole quota, including 12 elephant, to the company. The income of OCT in 2001 was an estimated BWP1.4 million (US$215,000). In six consecutive years virtually no benefits accrued to the five participating villages, with the exception of an aborted effort to establish five community shops and financial contributions to funerals. Nearly all of the funds went to the 'administration' of the project: offices, vehicles, training of board members and meeting costs (transport and allowances).

Source: National CBNRM Forum (2002); unpublished data

Stakeholder participation

Communities have gained legal access to natural resources and generally consider that an important step in their empowerment process (Taylor, 2000), and CBOs have become important players in Botswana's development process. Policy development over the last decade has also showed the commitment of government to the concept.

There has been participation in the implementation of the CBNRM programme by a variety of other stakeholders, including government, the NRMP and NGOs. In addition to the roles played by government and the NRMP outlined above, at least ten NGOs in Botswana offer CBNRM-related services to communities. These services range from providing technical advice regarding natural resource management to organizational development and fund raising. In most cases, NGOs help CBOs to prepare and implement a CBNRM project in return for a 15–20 per cent management fee charged to cover the expenses of the NGO (for example staff time, transport, overheads and other expenses). The importance of the roles NGOs can play at community level as honest broker and long-term CBNRM process facilitator cannot be underestimated. In addition to CBO support roles, other international organizations offering financial and technical assistance, as well as local NGOs facilitating projects at local level, have been instrumental in further developing and implementing CBNRM in Botswana as one of the most decentralized natural resources management models in southern Africa. The well-funded NRMP kick-started the process of CBNRM development in Botswana, but the question of whether there is enough fuel to keep the engine running remains.

Community-based management – community who?

The communities participating in CBNRM in Botswana are in most cases clearly defined in a geographical sense and conflicts over natural resources between communities are rare due to the low population density, but intra-community conflicts are a significant problem.

The decision-making model applied in CBNRM in Botswana is a participatory democracy at village level. The rules of the democracy are described in the CBO constitution that is, theoretically, designed and approved by all community members. Most constitutions prescribe that important decisions are made by the general membership and all constitutions institute executive boards to be responsible for daily management and implementing community decisions. Financial management is also the responsibility of the board, which is accountable to the general membership and the use of income generated by CBNRM activities is generally decided upon by the membership. Income is usually set aside for administration and reinvestment in projects. In very few cases, cash is distributed at household level. The generic reasons against cash distributions that have been put forward are that: reinvestment in productive development generates more revenue and employment in the long run; existing food-for-work programmes and food rationing by the government makes cash handouts less pertinent; and past experiences with cash handouts resulted in exorbitant consumptive behaviour.

The board has the difficult task of ensuring representative decision making, which in theory goes further than merely being elected by the general membership every second year. Communities in Botswana are not homogeneous entities. They consist of men and women, rich and poor, young and old, healthy and sick, and people of different ethnic origin, all of whom have different interests in natural resource utilization. These interests are not fixed and may vary by issue and over time. As natural resources are vital inputs in securing a livelihood in rural Botswana, it would be romantic to think that community decisions can be made by consensus. There will always be compromises, which is acceptable if all community members have an equal opportunity to participate in decision making and if leadership has a sense of responsibility to protect the interests of the weak. While both moral principles have been infused in the (draft) CBNRM Policy (GoB, 2001) their application is left to community leadership. More recent drafts of the CBNRM Policy and accompanying guidelines assume more government control, the implications of which cannot yet be ascertained.

Waning support for the further development of CBNRM

During and immediately after the ten years of the NRMP (1989 to 1999) there was considerable support for CBNRM in terms of financial and technical assistance for the establishment and management of CBOs, for training programmes and for the NGOs facilitating the growth of the programme.

However, since Botswana has become a medium-income country, international donors and support organizations have phased out their interventions or are in the process of doing so. This has impacted negatively on CBOs and has had rather disastrous consequences for the fledgling NGO sector in the country, which does not have a local financial resource base and depends entirely on international financial support. The government has not stepped in to fill the financing gap and virtually all environmental NGOs currently face severe financial problems. As a result, these NGOs can no longer afford to play the role of broker in the further development and consolidation the of CBNRM programme.

Governments tend to think in terms of projects and preferably implement them in a blanket coverage fashion. There is a tendency among extension officers to assume that by using a few necessary instruments (developing and agreeing a constitution and management plan, acquiring a lease and formalizing a joint venture agreement) and providing a few training sessions to CBO staff, they have developed a community organization that is able to give sustained direction to the community members 'as managers of their natural environment'. Government does not seem eager to make resources available for the long-term capacity-building efforts that are necessary to make these institutions effective and sustainable.

The assumptions revisited

More than a decade of design, refinement and implementation of the CBNRM approach justifies some reflection. While the objectives of CBNRM in Botswana are relatively clear, the approach was developed on the basis of assumptions that were not always clearly spelled out. As mentioned above, it is very difficult to measure the extent to which CBNRM is achieving its natural resource conservation and rural development objectives. The main constraining factors to being able to make this determination are a limited implementation period in most participating communities, the lack of baseline data and the absence of adequate monitoring systems (Jansen and Molokomme, 2002). However, on the basis of experience to date and anecdotal evidence, it is argued that the assumptions on which the CBNRM approach in Botswana is built are rather unrealistic, thus making it unlikely to achieve an enduring positive impact.

Assumption 1: following its introduction by a foreign donor as an appropriate approach to conservation and rural development, CBNRM would, over time, be politically accepted and institutionally embedded, which would guarantee continued support.

CBNRM in Botswana is neither fully politically accepted nor institutionally embedded. The Ministry of Environment, Wildlife and Tourism has long been reluctant to send the draft CBNRM Policy to parliament for fear of political opposition, and the instrumental department (DWNP) does not seem to be able to push the policy through. Because it was DWNP that was the focus of capacity-building efforts by donors such as NRMP, there is no call from other ministries or local authorities such as district councils or land boards for CBNRM support measures. In fact, the potentially powerful local authorities are largely oblivious to CBNRM, as they were bypassed when management authority was devolved from state to community. This is not to say that councils and land boards were not involved in design or did not give their consent to the national government-driven initiative. However, in the implementation and further development of CBNRM projects, these institutions did not play a role and there was no institutional embedding (for example in district planning and extension services). Thus ownership by these institutions is limited and there is a danger that local political support to sustain CBNRM will not be there when it is most needed. Non-governmental stakeholders are disorganized and their voice is getting weaker as donor support is reduced.

Assumption 2: community members would be eager to accept devolved management authority since 'user rights' mean 'direct benefits'.

In most communities there might be the will, but there is rarely the capacity to exercise management authority, which came from above and was not a total devolution of power. Traditional rights were not restored; instead the user rights came along in the form of a complicated recipe – to organize communities, design by-laws, ensure representative decision making, account for decisions, make plans, write them down and stick to them. It is argued that most communities did not realize what deal they signed up to when accepting management responsibility.

Assumption 3: accrued benefits would make community members aware of the (commercial) value of the natural resources, which in turn would be a strong incentive for their conservation.

The assumed link between a community benefiting from natural resources and having an incentive to conserve them was very optimistic in the circumstances. All CBNRM projects (with government steering) opted for commercialization of the natural resources. The benefits in these cases accrue to the community (for example revenue from quota sales), while the costs of conserving (for example not hunting) are borne by the individual. In all cases, this has been to the advantage of the village elite and at the cost of those who are most dependent on natural resource use for survival (the poor, traditional hunters, women and the elderly). There are virtually no CBNRM projects in Botswana where cash dividends have been paid out, and it is very unlikely that in these projects benefits to the average community member will exceed the costs to the same individual. It is therefore equally unlikely that this will prompt the conservation of natural resources, especially by those who have reduced access to them for subsistence purposes (Taylor, 2000).

Assumption 4: remote areas have investment potential and community structures are appropriate entities to manage these businesses.

CBNRM concerns the management of communal resources that are expected to benefit the entire community. As cash distribution to households was not really considered an option – the distribution of small amounts were thought to have less impact than the investment of larger amounts – there was a drive towards reinvestment of income from quota and concessions into productive community enterprises.

Community structures do not seem to be the right entities to manage a profitable business. During the design of the programme, it was realized that direct revenues from natural resource use would not be enough to stimulate rural development, but it was thought that productive reinvestment of these revenues and employment creation would be more likely to set this process in motion. That this would happen through community-managed enterprises has proved so far unrealistic (Jones, 2002). The remote areas in question have very little development potential and business skills are lacking, while community entities with collective decision-making structures do not have the characteristics necessary to run a successful commercial venture. This is not to say that no enterprise development can take place at community level, but to expect that community revenue would automatically result in productive reinvestment has proved to be somewhat naïve.

Assumption 5: participatory democracy (where the general membership, guided by an elected executive CBO board, would take important decisions in line with the agreed constitution) would guarantee that decisions would be representative of all the interests in the community.

Communities are complex social structures and equitable representation does not come easily. For representative decision making, skilled leadership and an empowered membership is necessary. In the absence of the latter in most remote communities, it is very difficult to avoid the development of village elites (mainly literate men from dominant village factions) monopolizing the decision-making process. To avoid the rise of a local CBNRM elite requires enabling policies and regulations (control mechanisms) as well as capacity-building of the constituency. In line with the rural development agenda of the CBNRM approach, emphasis should be placed on the latter in order to avoid government stepping in with sweeping nationwide control measures that may negatively affect decentralized decision making, empowerment and commitment.

Assumption 6: the process of gradually building community capacity to exercise management responsibility would be necessary only for a finite period and the stage of maturity would be reached 'in a few years'.

Building the capacity of communities to become managers of their natural resources for the benefit of all is a long-term process. Ten years of CBNRM implementation has shown that this capacity-building process takes much longer than was originally anticipated. This has to do with the complexity of the approach (multiple sectors and multiple actors), the complexity of making management decisions over natural resources that are not fully controlled by the CBO, and the complexity of the community social fabric. The assumption that communities will be ready to take on management responsibilities at short notice is unrealistic and holds the risk of early failure.

Conclusions

The CBNRM approach in Botswana has the potential to contribute meaningfully to the conservation of natural resources and to trigger sustainable rural development in remote areas. The data on current community participation and benefits to date support this. While most building blocks are in place, further refinement and adaptation remains necessary. The extensive review of the Botswana CBNRM programme that took place in 2003 confirmed this picture (Arntzen et al, 2003).

With hindsight, the assumptions upon which the approach was built were too optimistic and underplayed the complexity of the socio-economic context within which CBNRM was applied. Stakeholders might have also been somewhat naïve in embracing CBNRM as the panacea for their problems. The time is ripe to address the criticism openly and learn from past experiences, to adapt the approach and improve it.

Processes of multiple stakeholder engagement and empowerment of weaker stakeholders towards the wise use of natural resources underpin CBNRM. This basis needs to be enforced by involving more sectors and resources (beyond the current focus on wildlife), local government structures and traditional authorities (and knowledge).

While the sustainable utilization of natural resources has to move beyond selling wildlife quotas, there is potential for increased involvement in tourism (including co-management arrangements in and around protected areas) and veld product marketing. Forms of utilization need to focus on employment creation and small-scale business development to maximize the spread of benefits among involved communities. Thus community trusts will have to change their focus from management into providing a range of opportunities related to natural resource-based activities.

The benefits from CBNRM are meant primarily to compensate natural resources management costs. As most opportunity costs are borne at a household level, benefits need to be felt at that level, for example through direct employment, social security schemes, contributions to improved food security and so on.

Methodologies for community capacity-building need to be long term and more appropriate to suit specific local conditions, and the roles of facilitating organizations playing the honest broker in complicated change processes at community level need to be secured on a long-term basis.

Acknowledgements

Valuable contributions to this paper were made by Jan Broekhuis, who has worked in Botswana since 1991 and was until recently the Assistant Director National Parks and Game Reserves at the Department of Wildlife and National Parks in Gaborone; by Masego Madzwamuse, formerly the country representative of IUCN in Botswana and in that capacity managing a range of natural resources-related programmes; and by Ruud Jansen, independent consultant, who has worked in the field of natural resources management in Botswana since 1983.

References

Arntzen, J. W. (2003) 'An economic view on Wildlife Management Areas in Botswana', CBNRM Support Programme Occasional Paper No. 10, IUCN, Gaborone

Arntzen, J. W., Molokomme, D. L., Terry, E. M., Moleele, N., Tshosa, O. and Mazambani, D. (2003) 'Review of CBNRM in Botswana', DFID/WILD/CBNRM Support Programme, Gaborone

CSO (2003) *2001 Population and Housing Census*, Central Statistics Organisation, Gaborone

GoB (2001) 'Draft CBNRM Policy', Government of Botswana, unpublished policy draft

Jansen, R. and Molokomme, K. (2002) 'Mid-term review of the German Development Service Natural Resources Management Programme in Botswana', report prepared for the DED, Gaborone

Jones, B. T. B. (2002) 'Chobe Enclave, Botswana: Lessons learnt from a CBNRM project 1993–2003', CBNRM Support Programme Occasional Paper No. 7, IUCN, Gaborone

Mbaiwa J. E. (2004) 'The success and sustainability of community based natural resources management in the Okavango Delta, Botswana', *South African Geographic Journal*, vol 86, no 1, pp44–53

National CBNRM Forum (2000) 'Minutes of the fifth District CBNRM Forum, Maun, March 29, 2000', unpublished meeting minutes

National CBNRM Forum (2002) 'Proceedings of the second National CBNRM conference in Botswana, 14–16 November 2001 and CBNRM Status Report 2001', unpublished report for the National CBNRM Forum

National CBNRM Forum (2007) 'Proceedings of the fourth National CBNRM conference in Botswana 20–23 November 2006 and the CBNRM Status Report 2006', unpublished report for the National CBNRM Forum

Taylor, M. (2000) 'Life, land and power: Contesting development in northern Botswana', PhD thesis, University of Edinburgh, Edinburgh

16

CBNRM in Mozambique:
The Challenges of Sustainability

Isilda Nhantumbo and Simon Anstey

Approximately 51 per cent (40.1 million hectares) of land in Mozambique is covered by forests, with another 19 per cent covered by wooded grassland and 12 per cent by mostly grasslands. There are three forest land categories in the country: 26.9 million hectares of productive or high commercial value timber (mainly forest concessions); 14.7 million hectares of multiple use (mostly for subsistence fuel and construction materials, and also where competition between many uses and many users is noticeable); and 13.2 million hectares of conservation reserves (DINATEF, 2006). The last of these categories covers approximately 15 per cent of the surface area of the country and is managed by the National Directorate of Conservation Areas in the Ministry of Tourism.

Over 60 per cent of the population in Mozambique is rural and dependent on natural resources for their livelihoods (PROAGRI, 1997). However, much of this use of resources can be considered illegal as it is conducted without the formal requirements of permits or licences. As a result, the relationship between the natural resources agencies and communities has been that of police and law breaker, an approach that was clearly a failure due to the continued high – and unsustainable – use of many resources.

Lessons from the region suggested that giving communities ownership and control over resources (as the government and private sector already had) would be a critical factor in changing attitudes and practices, thus community participation in natural resource management has been embraced in Mozambique as a strategy for empowering communities to make decisions about the use and management of their resources. Policies and legislation were enacted and support systems established in order to create an environment conducive to people's participation.

For more than a decade, Mozambique has been striving to implement resource devolution, concentrated on forest resources. The way the forest categories were established in Mozambique and the requirements for gaining access to forest resources implicitly provide the productive forests for the private sector, while the biodiversity-rich forest areas are earmarked for the state (forest reserves). Multiple use forest areas remain, in which CBNRM can be implemented. This presents a dilemma for the communities and practitioners of CBNRM: how do communities derive sufficient income from relatively low-value resources and continue to be motivated to engage in sustainable management and use of the resources? This chapter looks at the policy and legal framework as well as case studies to illustrate some of the challenges of implementing CBNRM in the country.

The Mozambican context

The experience of CBNRM in Mozambique is quite different to that of other countries in the southern African region, primarily due to the different starting points of each country. One of the major differences is that during colonial times, Mozambique did not have a dualistic land tenure structure that included 'native reserves' or 'Tribal Trust Lands', nor after Independence was communal land tenure a feature of the rural landscape, as was the case in much of southern Africa. The post-Independence government nationalized all land (which remains the property of the state), and grants user rights to non-state entities. In the 1990s, a key element of land tenure reform was policy and law (GoM, 1997a; 1997b) that recognized the rights of occupancy, provided the subject(s) had resided in the area for 10 years or more, and which provided options for granting private land use rights for up to 100 years and provided self-identified 'local communities' with land use rights in perpetuity for which a Certificate of Use and Improvement Rights (*Direito de Uso e Aproveitamento de Terra*, DUAT) could be issued. While there is no obligation to register land, the option for doing so is provided. Critically, in Mozambique, both private and 'local community' land use rights are equal rights in law – unlike the differential land rights and legal status of communal and private tenure elsewhere in the region. (The legal framework and process of reform is discussed in more detail in the next section.)

The resource base of Mozambique has influenced the focus of CBNRM approaches – the country has richer forest resources (particularly of commercially valuable timber) than most of its neighbours, significant marine and freshwater fisheries, and a coast with high tourism potential, but it has relatively poor wildlife resources compared to elsewhere in the region. These factors (in conjunction with institutional aspects noted below) have resulted in a high number of CBNRM initiatives focusing on 'community forestry' or based around the management of multiple resources – in contrast to the predominantly wildlife- and tourism-based CBNRM experiences regionally.

Additionally, CBNRM in Mozambique has not benefited from an existing wildlife and tourism industry, where demand, capacity and marketing of resources were already available.

The implementation of CBNRM activities inside and outside of protected areas is another difference of the Mozambican CBNRM experience. The challenge of managing protected areas with resident human populations was recognized during the expansion of both terrestrial and marine protected areas in the early 2000s. In response, Mozambique has been faced with developing CBNRM options, policy and practice not only in 'community areas' but also within expanding state- or multiple partner-protected areas (see also Chapter 21).

Outside of protected areas, weakly implemented resource laws meant that relatively uncontrolled or illegal timber harvesting, trophy hunting and informal bush meat and charcoal harvesting were the most persistent and developed uses of resources. In fact, CBNRM emerged as a strategy to curb this unsustainable use of natural resources by devolving the rights and responsibilities of control over the resources to the local communities.

CBNRM activities in Mozambique are highly project based compared to the programmatic approaches taken in other countries such as Namibia and Zimbabwe, which has resulted in a diversity of aims, objectives, implementation processes and results. This diversity has meant that, at times, ad hoc institutional arrangements have been necessary (for example the Ministerial Diploma setting the percentage of benefits to be shared among stakeholders in the Tchuma Tchato project, and the system approved by the Niassa Province Governor that allocated proportions of revenue to management of the programme, to local government and to the community). The variety of arrangements contributes to the variation in processes and results.

Mozambique underwent intensive development of the legal framework in the area of environment and natural resources in the 1990s. Most of the policy and legislative instruments are considered progressive as far as the granting of rights to communities is concerned, especially with respect to the recognition of customary rights and boundaries, rights of occupancy and the community participation in the process of allocation of land to private investment through a process of consultation (see Nhantumbo, 2000, 2002). During this period, policies regarding the definition of rights to natural resources and more access and control to all citizens were also enacted. The current debate, however, is whether the rights gained by the communities are indeed giving them the sense of ownership, that is an understanding of the rights, ways to use them and to exploit the economic opportunities thereafter. There seems to be a missing link for the realization of the expected social, political and livelihood benefits, such as strengthening of local institutions, participation in decision making over the use of resources, creation of employment and income generation, which should combine to help reduce poverty among communities. When these expectations are not fulfilled, the natural resource management institutions disintegrate and communities resort to unsustainable harvesting of natural resources to realize short-term economic benefits. The Madjadjane and Goba CBNRM initiatives are illustrative of this problem.

Implementing CBNRM

CBNRM initiatives differ in terms of the types of resources managed, the threats to be addressed by the initiatives, the strategies for engaging communities and the means of providing benefits as incentives to those communities. The productivity of resources, the success of enterprise development designed to add value (where relevant), the scale of production and processing are key challenges to the success of these initiatives.

Some experiences started from a drive by facilitators to secure land rights for communities, later introducing means of maximizing the use of other resources. Other experiences were initiated with community members involved in resource assessments, production of management plans, application for licences or concessions and detailed studies on the potential of resources for alternative income-generating activities. Where both forest and wildlife resources exist, an integrated approach to the use of these resources is often taken. Still others were driven by the need to address conflict over the use of resources. Most CBNRM projects now being implemented address both security of resource tenure and the generation of benefits from the sustainable use of resources. In fact, the nature of activities is also expanding into tourism. Other income-generating activities being pursued include harvesting and selling firewood, poles and to a lesser extent valuable timber, woodcarving and other crafts and beekeeping.

Following the establishment in 1995 of the first initiative, the Tchuma Tchato project, there was an increase in the number of CBNRM initiatives to 32 in 1998 and 42 in 2001. In that year, a study of the nature and focus of the different initiatives concluded that only the minority (12 per cent) had specific objectives relating to local empowerment (resource or land transfers to communities) while the majority were concerned with providing some level of secondary benefit sharing, environmental education or achieving conservation ends through instrumental involvement of the local community (Matakala and Mushove, 2001). Most initiatives were concerned with indigenous forest management (55 per cent), wildlife (26 per cent) and soil, water and fisheries management. Most projects were being implemented in de facto community areas (35 initiatives), 11 in or around protected areas and 6 in peri-urban zones. The most recent statistics indicate the existence of 68 initiatives countrywide (Foloma, 2006). Concomitantly, there has been a process of community land delimitation pursuing secure tenure for communities, which provides the means for negotiation with the private sector and for undertaking various rural development activities, including CBNRM. Although the forest and wildlife legislation provides equal opportunities for the private sector and communities as far as access to forests is concerned, there are, as yet, no community forest concessionaires fully established in productive forests; CBNRM is mainly implemented in the relatively poor resource areas, which poses challenges in realizing the promised and/or expected economic benefits.

Threats to CBNRM initiatives remain – particularly with regard to the security of tenure over resources. For example, Chipanje Chetu, a promising CBNRM initiative due to its endowment of both wildlife and forest resources, has been 'leased away' to a private investor in a recent process driven by government technocrats rather than community initiative. That community rights are overlaid with private sector rights undermines the former and sets back the sense of resource ownership that communities were gaining. This conforms to an earlier observation by Matakala and Mushove (2001) that the devolution of power and authority from state to community was not genuine and did not sufficiently empower the communities to make their own decisions.

Another concern is that resources available for investment in CBNRM initiatives have been declining, with waning interest and support from donors and government agencies. This is, ironically, at just the phase when the fundamental legal and practical experience from past initiatives provides the highest chance for successful CBNRM implementation in terms of full land and resource right options.

Tangible economic benefits from most initiatives remain relatively limited to date due to the combination of low investment in enterprise development and establishment, limited capacity to add value to natural products at community level and underdeveloped markets. These problems have been compounded by the increasing privatization or inclusion of the highest-value resources (commercial timber, trophy wildlife, prime coastal tourism zones) into state-protected areas. The scale of areas covered by CBNRM initiatives addressing land and resource tenure and control remains relatively small – less than 1 per cent of the country – and growing at a rate far less than the private sector or state-protected area growth (Anstey, 2004). The high transaction costs of securing resource use rights for local communities in particular remain far beyond the investment possibilities of most initiatives.

The policy and legal framework:
Focusing on devolution of land and resource rights

Box 16.1 *Key policy provisions enabling implementation*
of CBNRM in Mozambique

The 1990 Constitution establishes that land and all other resources belong to the state on behalf of the people, which holds the rights of adjudication to the different users. The forestry, wildlife and land policies reinforce this notion.

Under the 1995 Land Policy, the 1997 Law of Land Use and Management Rights, the 1999 Regulation of the Law on Land Use and Management Rights and the 1999 Technical Appendix on procedures for community consultation and delimitation and demarcation of land:

- The land law reinforces that the land belongs to the state, and it cannot be sold, alienated, mortgaged or seized.
- The concept of 'community' adopted by the land law also applies to the forestry law.
- Secure tenure arises from the recognition of customary law and rights of occupancy (for those resident in an area for more than ten years) and provides men and women with equal rights.
- A DUAT provides rights that can be privately held, that are inheritable and transferable. The DUAT also enables collective land rights with the possibility of land delimitation and demarcation and the election of local representatives to be signatories of the DUAT.
- Community participation in forest resource management is fostered by consultation with communities before land allocation to third parties – particularly in concessions over productive areas, and in the development of partnerships.

Under the 1997 Policy for the Development of the Forestry and Wildlife Sector, the 1999 Forestry and Wildlife Law, the 2002 Regulation for the Forestry and Wildlife Law, and the 2005 Joint Ministerial Diploma for revenue distribution to local communities:

- The social objective of the forestry and wildlife legislation and policy is to enable community participation in managing natural resources through improved use and management practices.
- Communities are eligible for an annual licence or long-term concession for the exploitation of forest products. The requirements include a forest inventory (a simplified version for the annual licenses), management plan and the establishment of a processing plant.
- Communities can establish the boundaries of their area and can be granted formal rights to use and manage the resources, including high-value (timber) forests.
- The law establishes the need for benefit sharing between the government and the local communities as an incentive for community participation, and also for addressing poverty in rural areas that are rich in natural resources.

- 20 per cent of government revenues from royalties paid (mostly by the private sector) for extracting resources constitutes the government contribution to communities residing in areas where forests are harvested.
- Communities should be consulted before allocating resources under all the management regimes (concession, annual harvesting licences or gazettement of protected areas), and the consultation process should be used as an opportunity to foster resource management partnerships between the parties.
- The procedure applicable both for the communities and private investors for securing access to forest resources include among other requirements, the undertaking of a resource assessment, design of a management plan and proof of capacity for adding value to the resources.

Despite the development of progressive legislation and policy (see Box 16.1) with regard to land and other natural resources in Mozambique, a number of key challenges relating to their design and implementation remain, hampering the ability of CBNRM projects to maximize their positive outcomes.

In Mozambique, both private and 'local community' land use rights are equal rights in law, but the challenge of realizing community benefits – even a decade after being legislated – is that these rights remain poorly understood at most local community and local administration levels and poorly defended within a generally weak justice system (Tanner et al, 2006).

The principle of participation in decision making is entrenched in the legal framework governing forestry, wildlife and land (for example through creation of land, water, natural resources committees or councils and the recognition of customary institutions and their role in exercising power at local level).[1] However, there is a gap between the legislation and its implementation, raising questions about whether government is prepared to fully devolve administrative, political and management powers over natural resources to the local level and about whether there are strong and capable recipients at local level (Mansur and Nhantumbo, 2000; Matakala and Mushove, 2001).

The high financial and technical costs associated with applying for a DUAT cause difficulties for communities that do not have external support. The apparent streamlining of the bureaucratic process can have the effect of undermining community rights. For instance, where the private sector seeks resource rights, it has to consult for 90 days, which is often not long enough for communities to become aware of plans and gain sufficient knowledge and understanding to make informed decisions. The consultation is often carried out in a manipulative way so that communities sign the consultation document with different expectations of what the private investment will bring them, contrasting with what they are likely to receive in reality. This is also a problem with respect to the legislation relating to forestry and wildlife. Internal conflicts within participating communities can also cause problems.

The implementation of this legislation occurs in a context of poor and often undemocratic local governance structures, where decision-making power is skewed away from local communities, the people that the legislation is meant to empower. Reforms to promote elected local government have been reversed and under Decree 15/2000, traditional chiefs have been essentially reappointed through administrative mechanisms as the legally recognized community representatives. In addition, the multi-sectoral institution proposed under the forestry and wildlife law, the Participatory Natural Resource Management Council (*Conselho de Gestão Participativo*, COGEP) is not likely to strengthen local community decision-making power over resource management, land use and revenue allocation. It is a co-management institution, made up of community representatives, local government, the private sector, NGOs and local associations, rather than a 'producer/community' devolved institution that can take up the opportunities of land and resource reforms. Because of the presence of other stakeholders (particularly government and the private sector) the COGEP is likely to skew the power dynamics away from local communities.

Another challenge facing community institutions is that most are not formally constituted, and they are required to register as an association in order to be able to conduct any formal business initiatives (for example opening a bank account).

Outcomes of CBNRM programmes across Mozambique

In addition to the direct involvement of communities in natural resource management for the generation of benefits, the government introduced other forms of benefit sharing in an effort to instil a sense of resource ownership by local communities, to promote sustainable use practices and eventually contribute to local development. As mentioned above, the forestry and wildlife law regulations stipulate that 20 per cent of government revenue generated from forest and wildlife harvest royalties should revert to the communities living in the vicinity of the resource. Table 16.1 outlines those communities currently eligible to receive these funds and the income allocated to communities in the different provinces for 2005–2006.

Table 16.1 highlights the difficulties that provincial and national governments, as well as communities face in ensuring this revenue distribution occurs fairly. Both provincial and national government are involved, with provincial government needing to retain the revenue to be distributed, rather than

Table 16.1 *Beneficiary communities*

Province	Eligible communities	Organized communities*	Communities that received the 20%	Value of the 20% in 2005/2006 (MT)	Amount allocated to communities (MT)	Amount allocated to communities (%)	Amount allocated per community (MT)
Maputo	87	15	3	15,409	1562	10	521
Gaza	97	27	27	17,346	7111	41	263
Inhambane	132	20	6	87,242	12,696	15	2116
Sofala	52	15	8	386,900	16,728	4	2091
Manica	108	24	12	179,078	36,968	21	3081
Tete**	55	9	0	159,537	0	0	0
Zambézia	118	12	6	406,641	13,121	3	2187
Nampula	81	43	16	75,581	19,538	26	1221
C. Delgado	155	16	10	270,362	52,752	20	5275
Niassa**	71	0	0	21,577	0	0	0
Total	956	181	88	1,619,675	160,476		

Note: * With a constituted committee and a bank account; ** Both Tete and Niassa retained the revenue (20 per cent) at source (i.e. at provincial level) and any revenue distributions made are not reflected here

Source: Adapted from Nube et al (2006)

sending it to national government and requesting its return – an often difficult and lengthy process.

For communities, the difficulty is often in meeting the prerequisites necessary to qualify for receiving the benefits. Organizing the communities – although some of them already have CBNRM initiatives and related institutions – is still a big challenge for the government and NGOs. As a result, only 10 per cent of the available funds were allocated to only 9 per cent of the eligible communities. In addition, only Gaza Province has allocated more than 40 per cent of the available funds, which incidentally is one of the provinces with the lowest average amounts to be allocated per community. A high proportion of potential income per community reflects the high volume of activities that the private sector undertakes in some provinces, particularly in forest concessions in Cabo Delgado, Manica, Zambézia and Sofala, and tourism activities in Inhambane.

Another challenge associated with the distribution of this 20 per cent is that the private sector claims that the funds are a direct contribution to local communities, whereas in reality the government is forgoing these funds in order to directly benefit the communities.

A pioneering CBNRM initiative in Mozambique

The government-led approach was developed through a five-year 'Support to the implementation of community-based forestry and wildlife management' project, implemented jointly by National Directorate for Forestry and Wildlife (DNFFB) and the FAO between 1998 and 2002. The approach was implemented in five communities in Maputo, Nampula and Manica provinces.

One of the aims of the initiative was to use pilot areas to test CBNRM implementation methodology and analyse the constraints and potential for replication throughout the country. Significant investment was made in undertaking relevant socio-economic and biophysical studies, analyses of resource potential, determining opportunities for income-generating activities, research on different technologies for sustainable use and so on. Each of the chosen sites was endowed with different resource values and different resource uses – one supplied (unsustainably harvested) charcoal to Maputo, others used timber for carpentry and carving – and in most cases the intervention aimed at regulating rather than preventing ongoing activities.

Key features of project implementation were gender equity (so men, women, the young and elderly were represented on the management committee and in different interest groups), securing the participation of the communities in all stages of the project, and the devolution of land and forest resources control (by securing a DUAT and an approved management plan). The establishment of self-governing structures (for example natural resource management councils, interest groups, etc.) to enforce the rules of sustainable use and management of the resources was also important.

The efficiency of traditional activities (for example carpentry and charcoal production) was increased by improving resource management and introducing more efficient production and use technologies (for example improved kilns and stoves). Where possible, non-traditional activities were also introduced, including beekeeping and production of charcoal from waste paper. The project also provided support to enterprises that lay outside the realm of natural resource management (for example cloth making).

The steps depicted in Figure 16.1 illustrate a process based on knowledge transfers in an attempt to construct an appropriate approach and methodology to suit the Mozambican context. Unfortunately, the poor forest resources base was overlooked in the evaluation of potential community benefits from the investment in conservation, and as a result the affected communities were left disillusioned and sceptical about the benefits of CBNRM.

Foloma and Zacarias (2003) conducted an analysis of the sustainability of the project-supported enterprises and found that 31 out of 59 business groups were still active and generating income. The authors observed that the surviving enterprises were almost invariably those that were traditionally undertaken by the communities, such as carpentry, woodcarving and charcoal production. This finding raises the question of what failed in the introduction of new and innovative income-generating activities, and whether it was associated with the lack of continuity of the support services (technical, financial, marketing).

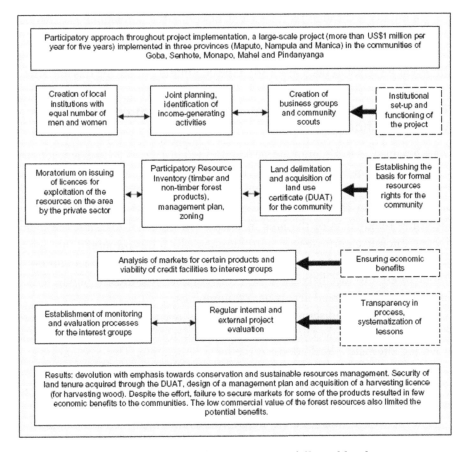

Figure 16.1 *Some of the main steps followed by the DNFFB/FAO/Netherlands project*

Institutional sustainability is key to determining the fate of CBNRM initiatives when direct external assistance ends. A 2007 visit to the Mahel initiative (in Maputo Province) by one of the authors showed that though not all business groups are functional, the community scouts are still active almost five years after the project ended, the community council is active with most of the original members still leading/facilitating the decision making, and the community remains a pioneer in receiving the 20 per cent share of government royalty revenue. As the area has potential for game farming, the community is using part of this revenue to prepare the area for game farming and is seeking partnerships with potential investors.

National-level activity implementation, complemented by technical and political support from district and local government was only possible because the community – through the management committee – maintained a strong commitment to the initiative. Furthermore, the fact that earlier studies and zoning had identified an area for game farming with potential for generating significant income has been motivating the community to pursue this initiative.

The Chipanje Chetu Programme: An example of potential success cut short

The Chipanje Chetu Programme was established in 1998 in an area adjacent to the buffer zone of the Niassa Game Reserve (in Niassa Province in the north of Mozambique). The area covers approximately 650,000ha of Miombo wood-lands and is also endowed with significant wildlife and fisheries resources. Given the remoteness of the area in relation to potential markets, the main viable economic enterprise was safari hunting, which was promoted through a partnership with a private operator. Other activities included timber and non-timber forest product harvesting. The benefits are distributed among fewer than 3000 people, spread across five communities.

Poaching and indiscriminate bush fires threatened wildlife and forest resources, and the project hoped to control these unsustainable practices while providing alternative livelihoods to the local population. The core aims were explicitly concerned with piloting the linking of local democracy and rural development through new governance rights over an envisaged ten-year process, rather than achieving conservation via community participation.

The approach was one of long-term commitment, consolidating the knowledge and experience of the community and the facilitators – IUCN, the provincial/district government, local NGOs and community members (see Figure 16.2). The initial arrangement of a multi-stakeholder collaborative group is probably the only example that may have come close to establishing the participatory natural resource management council envisaged by the Forestry and Wildlife Law.

Since the programme was established, the challenge has been to maintain momentum where funding has been relatively small and erratic. Nhantumbo (2004) conducted an analysis of the potential net gains of the Chipanje Chetu Programme. The analysis showed that the project was not an economically viable investment as the net present value (NPV) of the revenue level at the time (nearly US$12,000) was negative.[2] However, it was predicted that if revenues were to increase by at least 25 per cent, the project could be self-sustaining and generate some surplus, and could reach an NPV of US$134,654 if revenue were to increase by 35 per cent (Nhantumbo, 2004).

An important result of the study was the fact that the positive net returns could only be realized from year seven, and if cash dividends were paid to each individual in the community, then each would only receive just over US$50 or

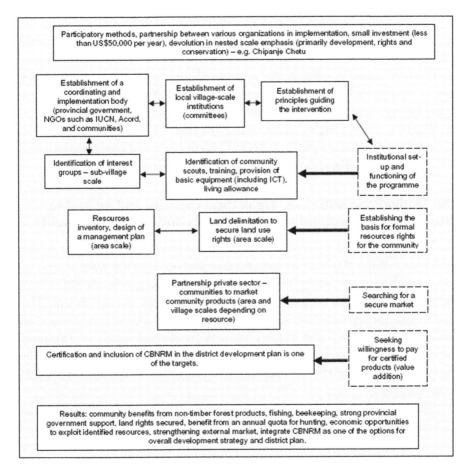

Figure 16.2 *Implementation steps of the Chipanje Chetu Programme*

around US$250/household (in an area where mean household income prior to the initiative was around US$45/household) (Nhantumbo, 2004). The critical question is whether the community can wait for seven years to obtain the gains of sustainable practices.

A paradox of this project is that the apparent security of land and other resource tenure was taken away from the community. From 2005, this initiative ceased to function with the community as the major decision maker as regards the management of natural resources and types of partnerships it would like to make. A process of experimentation based on the rights and semi-autonomous local institutional development resulting from community land certification and innovative distribution of pilot sport hunting revenues has been curtailed by government, which replaced it with a more formal and technocratic concession process to a private investor. The community had invested in sustainable management practices, with visible increases in wildlife populations. Increasing benefits were being generated and distributed among the community (57 per cent of the trophy fee revenue, amounting to US$24,000 in 2004) and the rest to the management of the initiative and the Sanga District authorities.

The fact that the 'rights' given to communities were cancelled/changed at will sends an unsettling signal to CBNRM practitioners and communities alike, as it demonstrates that tenure security is not guaranteed in reality.

The Madjadjane initiative: Weak local capacity and an end to project funding jeopardizes investments and potential benefits

Madjadjane is located in Maputo Province in the south of Mozambique, in the buffer zone of the Maputo Elephant Reserve and occupies a surface of around 7000ha with a population of 500 people. The area is close to South Africa, and the youth and most men tend to migrate there in search of employment. The area is biologically very important, being part of the Maputaland Centre of Endemism, and is thus rich in flora but very fragile and unable to withstand intensive harvesting of timber or non-timber forest products. Despite being in the buffer zone of the Maputo Elephant Reserve, the area does not have significant wildlife.

The Madjadjane community acquired their DUAT in 2000. A project was then initiated to help build on these rights, seek economic development opportunities and address conservation priorities in response to threats to the area. The main threats to the area were the intensive harvesting of wood for charcoal production, and poor park/community relations due to incursions of elephants from the reserve, destroying crops and community property. Figure 16.3 highlights the main steps undertaken during the implementation of the project.

An important aspect of this project – and one that is common to many initiatives in which NGOs facilitate the acquisition of a DUAT – was that the signatories to the certificate had little understanding of their role as de facto guardians of the land rights on behalf of the rest of the community. As a result, even the identification of the signatories to the DUAT took months to be completed. It would seem that security of rights is not sufficient to change people's livelihoods if, subsequent to achieving tenure security, they do not have capacity to add value to their natural resources.

Unlike Chipanje Chetu, Madjadjane has accessible markets for its products – in this case tourism activities (a community lodge associated with attractions such as bird watching, bush walking, and catch and release fishing), which were at the centre of the development and conservation efforts. Additional activities such as woodcraft, batik, dance, theatre and beekeeping were all aimed at

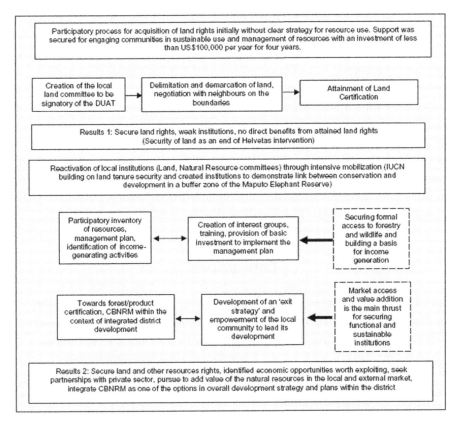

Figure 16.3 *The process followed in the Madjadjane CBNRM initiative*

reducing the involvement of the community in charcoal production, considered to be incompatible with tourism. Diversification of household activities was considered to be key in filling in the gap resulting from charcoal production.

The viability of the various enterprises for a period of 20 years at a 10 per cent discount rate was estimated by Nhantumbo (2004). The estimated NPV based on expenses and revenues at the time was US$32,465. The analysis determined that positive net returns would be realized only after five years of investment in capacity and enterprise development. Despite the investment and the focus on capacity-building, community empowerment continues to be a challenge. In fact, one of the key aspects of the apparent failure of the initiative to inculcate community ownership of resources was a lack of education, impairing a basic understanding of business management – particularly basic bookkeeping and financial management. The options of a partnership with the private sector and employing a manager were considered, but the community chose not to take up such a partnership, despite their own very limited business and financial management skills.

The results of this project raise questions about whether the devolution of resources to communities will ever empower them sufficiently to contribute to conservation and to reduce poverty through job creation, whether devolution perpetuates the current status quo of communities as passive recipients of benefits, and if this is not desirable, then what are realistic means of achieving the goals of CBNRM? If a decade of CBNRM in Mozambique had been built on the basic foundation of the education of rural communities, then there would certainly be local champions and managers who could have brought the business management skills desperately needed to ensure that CBNRM impacted positively in the livelihoods of the communities.

Conclusions

CBNRM has great potential to improve the living conditions of the majority of the poor in rural areas and to reduce the intense harvesting of natural resources in Mozambique. The approved legal framework provides an opportunity for this to become a reality. However, a number of key challenges remain.

Mozambique is rich in forest resources and less so in wildlife. The revenues from the former are relatively small as communities tend not to have the necessary investment in institutions, access to capital or capacity to exploit so-called 'productive forest' areas, which are typically retained by the state for private sector concessions. The forest managed under CBNRM initiatives is more often medium to low productivity, an enormous burden as far as the profitability of resource exploitation is concerned. In most CBNRM initiatives, communities exploit less valuable products (for example, the non-timber forest products such as honey and materials for crafts) and though some are branching out into tourism activities, managerial capacity still limits the realization of benefits. Therefore access to finance and diversification remain important to the success of initiatives. As mentioned above, education is also a vital component. It is unclear whether development partners supporting conservation and natural resource management are prepared to include this component, or whether CBNRM practitioners are prepared to be more innovative in building partnerships with sectors that have little to do with conservation but have other development experience.

The diversity of CBNRM approaches illustrates the limited guidance of government and the lack of a standard implementation framework. Between 1995 and 2002 the main legal instruments were produced and now sufficient time has passed so that lessons from implementation can be learned. This should allow both the development of guidelines to promote more effective implementation and also promote a re-examination of the mechanisms of access to natural resources. There are clear lessons to be learned about the practicality and social justice of the process of acquisition of land use rights and about high transaction costs associated with current resource use regulations. As highlighted above, local communities are still some way from reaching full resource control.

Nevertheless, CBNRM represents considerable potential for community development – poverty reduction is one of the major goals of government policies – and can promote resource conservation across large areas of the country. More attention should therefore be paid to the development of a functional and long-term strategy for the implementation and continued experimentation of CBNRM in order to address and overcome the remaining weaknesses.

Notes

1   1995 Land Policy; 1997 Land Law; 1999 Regulation of the Land Law; 1997 Forestry and Wildlife Policy; 1999 Forestry and Wildlife Law; 2002 Forestry and Wildlife Regulation.
2   Net present value was based on actual costs and revenues up to 2003 for a 20-year period, discounted at 10 per cent.

References

Anstey, S. G. (2004) 'Governance, natural resources and complex adaptive systems: A CBNRM study of communities and resources in northern Mozambique', in Dzingirai, V. and Breen C. (eds) *Confronting the Crisis in Community Conservation: Case Studies from Southern Africa*, Centre for Environment, Agriculture and Development, University of KwaZulu-Natal, Durban
DINATEF (2006) *Avaliação Integrada das Florestas de Moçambique. Inventario Florestal Nacional*, Ministério da Agricultura, Maputo
Foloma, M. (2006) *Directório das Iniciativas de Maneio Comunitário*, DNFFB, Maputo
Foloma, M. and Zacarias, A. (2003) 'Relatório do levantamento da situação actual dos grupos de interesse nas áreas piloto apoiadas pelo projecto GCP/MOZ/056/NET. DT No.13', UMC/FAO, Maputo
GoM (1997a) *Política de Desenvolvimento do Sector de Florestas e Fauna Bravia*, Government of Mozambique, Maputo
GoM (1997b) *Lei sobre o Uso e Aproveitamento da Terra*, Government of Mozambique, Maputo
Mansur, E. and Nhantumbo, I. (2000) 'A experiência do Projecto Moçambique FAO/Holanda em Áreas Piloto de Maneio Comunitário de Florestas e Fauna Bravia em Maputo e Nampula', paper presented at the First Conference on Common Property Resources Theory, Zongoene, Mozambique
Matakala, P. W. and Mushove, P. T. (2001) *Arranjos Institucionais para o Maneio Comunitário dos Recursos Naturais (MCRN): Perfis e Análise de 42 Iniciativas de MCRN em Moçambique*, FAO, Maputo

Nhantumbo, I. (2000) 'The new resources tenure framework in Mozambique: Does it really give the tenancy to the rural communities?', paper prepared for the 8th Biennial Conference of the International Association for the Study of Common Property, 31 May–3 June, Bloomington, Indiana

Nhantumbo, I. (2002) 'Environment and development implication of the implementation of the new land law in Mozambique', paper presented at Round Table Discussions, Ford Foundation Office for Southern Africa, 25 April, Maputo

Nhantumbo, I. (2004) 'The apparent fallacy of transforming common property resources into a poverty reduction strategy', paper prepared for the 10th Biennial Conference of the International Association for the Study of Common Property, 9–13 August, Oaxaca, Mexico

Nube, T., Sande, L. and Moises, N. (2006) *Memórias do Seminário de Reflexão sobre a Canalização dos 20% da Taxas de Exploração dos Recursos Florestais e Faunísticos*, DINATEF/MINAGRI, Maputo

PROAGRI (1997) *The Forestry and Wildlife Sector*, MINADER, Maputo

Tanner, C., Baleira, S., Norfolk, S., Cau, B. and Assulai, J. (2006) 'Making rights a reality: Participation in practice and lessons learned in Mozambique', LSP Working Paper No. 27, FAO, Rome

Part V

Integrating Wildlife and Parks into the Social Landscape

17

Recent Innovations in Conservation[1]

Brian Child

Part V presents examples of innovative actions designed to solve specific conservation problems using unconventional approaches. The chapters tend to focus on more well-known and better-documented examples from South Africa, so an attempt has been made in this introduction to list and summarize some of the lesser-known examples from elsewhere in the region. The South African examples reflect a shift in state policy and action towards more inclusive and economically viable modalities of conservation. Many of the other cases are linked less directly to the state, although it is true to say that they would seldom have happened without an opportunity being offered by one or two motivated and enlightened officials. The common feature of what may appear as a disparate set of examples is a new way of thinking about conservation that gives credence to the rights and benefits of society, and especially local people, and to the financial and economic viability of conservation land.

North West parks in South Africa: Parks as socio-economic engines

Pilanesberg National Park in the North West Province of South Africa was the first state-protected area established and developed primarily as a strategy for rural economic development. Several events conspired to make this experiment possible (Davies, 2000). Pilanesberg is the first example where park managers deliberately set out to manage a park for financial and economic goals as well as ecological ones. Proving that wildlife conservation is an effective economic use of land resulted in a demand for more and larger state-protected areas.

A window of opportunity for experimentation was opened up by the 'Independence' granted to the Bophuthatswana homeland in late 1977. Full advantage was taken of this by a series of innovative park managers who moved to Bophuthatswana, often to test new strategies that were not possible in the more conventional and restrictive park agencies where they came from (Davies, pers. comm.). Showing how ideas are threaded through history, Johnson et al's chapter traces the underlying philosophy for Pilanesberg back to the Fulbright Scholars who were so influential in Zimbabwe in the early 1960s, as well as to Ken Tinley who represented Mozambique in the SARC-CUS forums discussed in this book's Introduction. Tinley and Jeremy Anderson, thinking out of the normal conservation box, formulated some of the early ideas that drove the Pilanesberg experiment, many of which had been discussed by the Standing Committee for Nature Conservation and the Management and Utilization of Wildlife (MUNC) of SARCCUS (see Chapter 1). Pilanesberg, by virtue of being a 'new' conservation jurisdiction provided a location to test these ideas without incurring opposition from within entrenched park agencies.

Pilanesberg set out unashamedly to optimize the economic returns from wildlife, believing that wildlife's future lay in its economic potential. Breaking convention, high-value safari hunting was introduced within the park, live animals were sold and tourism was outsourced to the private sector when most parks still had only government rest camps. Pilanesberg's managers also redefined the core competencies of state and private actors in park management, a mind-shift reflected in the performance metrics and language used in this chapter and that, as we shall see from subsequent chapters, has begun to infuse conservation thinking more widely. Johnson et al score the success of Pilanesberg (and Madikwe, which pushed these ideas even further) in terms of wildlife conservation, but they introduce metrics for investment, financial viability, economic growth and employment into the park manager's scorecard. These metrics may seem out of place from the perspective of conventional conservation biology but, as we will see throughout the following chapters, they lie at the heart of a highly successful political and economic strategy for promoting conservation. Thus, in the 1990s, the Bophuthatswana government invested in building a new park, Madikwe, specifically as an engine for economic growth. Then private land began to be incorporated into these parks on a contractual basis, and in 2006 the Bakgatla baKgafela community voted to establish 33,000ha of communal land as a game reserve. The goal to link Pilanesberg and Madikwe into a 250,000ha park through a number of partnership arrangements will double the land for wildlife conservation in the province, while providing it with the economic strength of a mega-park.

Pilanesberg provides a powerful lesson in the sequencing of the conservation and development argument. It shows how a carefully crafted approach, selling parks in terms of their economic and employment impacts, was used to turn unproductive agricultural land into an economically valuable conservation area, and to then generate momentum for more. Scarce land was obtained for conservation in a manner that was financially and economically efficient, as it had to be in the face of strong demographic and economic pressure. It harnessed forces that in many other countries have seriously damaged the integrity of state-protected areas, as we see in the litany of under-managed or 'paper parks' (Cumming, 2004).

SANParks: Making parks relevant to society

National parks in South Africa are managed by South African National Parks, a statutory authority that retains its own revenues and employs its own staff. SANParks has jurisdiction over many (but not all) of the jewels in South Africa's protected area estate including, importantly, Kruger National Park. However, South Africa's nine provinces run a parallel system of provincial protected areas, and even municipalities set aside conservation areas. The balkanization of protected area management in South Africa has been an important structural factor allowing for innovation that would not have been possible under a single, all-powerful protected area agency.

Prior to multiracial democracy in South African in 1994, SANParks' precursor, the South African Parks Board, and several provincial conservation agencies were very conservative in their approach to conservation, including towards commercial opportunities. Most innovation took place on private land, but some provinces or homelands were imaginative, including the Natal Parks Board and Bophuthatswana National Parks Board, as they were then known. Natal encouraged private landholders to become responsible for their wildlife and provided some restocking and extension services; 'Bop-Parks', as we have seen, experimented with new models for state-protected areas. This background is contextualized in the chapter by Castley, Patton and Magome, which builds on these experiences.

Their chapter demonstrates the considerable technical competence in biodiversity conservation in South Africa, and reminds us that the people who have written the chapters in this book are conservationists who have retooled themselves by extending their competences to fields such as economics and politics. This emanates from the recognition that conservation, on an ever more pressured planet, is an outcome of the way incentives are aligned, must respond to the requirements of society at large, and is not the exclusive domain of a small group of self-appointed champions.

Having identified gaps in the protection of South Africa's biodiversity using good science, Castley et al outline South Africa's intention to invest significantly in extending the network of protected areas, with a vision of doubling them. In a developing country facing the typical demands of population growth, poverty, disease and unemployment, this is a remarkable goal. Some would question why expanding parks so extensively is even imaginable in the democratic environment of post-1994 South Africa, when the general narrative in the academic literature is about the conflicts between parks and people and the trade-offs between conservation and development. The political impetus for expanding parks is a result of the fundamental change in conservation philosophy articulated by, for example, the Pilanesberg narrative, from the massive growth in private conservation, and from a clear demonstration that wildlife provides more jobs and greater economic growth than alternative land uses. Indeed, we see metrics such as revenue and employment, and narratives such as community empowerment, beginning to enter the lexicon of South African park managers. We can also perceive some level of discomfort in this chapter at being forced to expand beyond the comfortable frame of conservation biology into the unfamiliar and complex arena of people, economics and politics.

The juxtaposition of new and old narratives in Castley et al's chapter suggests a conservation agency in the throws of rapid change. For instance Kruger National Park was famously self-reliant to the extent that it even employed its own gunsmiths (Child, 2001), so talk of 'outsourcing elements of non-core competency' is indicative of radical change, not only in the conservationists' language but in their intentions. The need to make parks more efficient recognizes the dangers of relying on state financing in the face of greater national priorities such as health and education. The search for policies and practices to make parks relevant to society recognizes the importance of political sustainability, and the chapter reflects an agency aspiring to move away from high-handed bureaucratic precedent towards greater accountability to society and commercial opportunity: Castely et al discuss 'product diversification', the 'unbundling of rigid and restrictive rules', and the formation of local park boards and management information systems. Thus, SANParks is beginning to carve a new social contract between parks and the society it is supposed to serve. SANParks is also developing innovative partnerships for expanding protected areas and, in so doing, is providing new models for conservation landscapes that are much bigger than the parks that lie at their core.

Building a mega-park in an agricultural environment: Greater Addo Elephant National Park

Castley, Knight and Gordon describe how the greater Addo Elephant National Park is being sewn together from disparate pieces of conservation area, farms and coastline, to form a mega-park covering five biomes and branding itself around the theme of the 'Big 7' – the 'Big 5' terrestrial land mammals as well as whales and sharks. Underlying this initiative are important changes in conservation philosophy – the definition of conservation is being broadened beyond charismatic fauna to emphasize representational ecosystems and ecological processes. Similarly, we see the political strategy of linking biodiversity goals to the economic platform provided by protected areas. Thus, data on employment from the private sector is beginning to emerge as a political bargaining tool for the expansion of parks, with the chapter noting that conservation provides one job per 100ha compared to one per 367ha for small-stock farming, and that one new job is created for every ten foreign visitors.

In an emerging trend in South Africa, private land is being voluntarily incorporated into parks through various incentives, most of which do not involve cash transfers. We note (Knight, pers. comm.) that parks can raise the value of adjacent land very considerably. As we see from Addo, and the following example in the Cape, the pioneering ideas of the Pilanesberg experiment are gaining traction.

Building an urban park: Table Mountain National Park

Daitz and Myrdal chart the complicated political processes required to consolidate wild land and biodiversity within a single protected area, this time amid urban Cape Town. While implementation invariably involves complicated detail, the lesson of their chapter is the power of the broad economic vision that parks can provide. Building Table Mountain National Park hinged on leaving behind an entitled and bureaucratic approach to park management and demonstrating that the park was in the communities' best interests, economically, socially, financially and aesthetically. The early goal of making this park financially viable from entry fees and concession fees proved impossible. However, park officials recognized the broader economic values of the park and persuaded local, national and international agencies to pay for the environmental and social services that the park could provide. Their case was strengthened by effective publicity and by using politically astute pro-poor solutions to rehabilitate the park's biodiversity. For example, public funding was used to employ a lot of people to clear huge amounts of flammable invasive (exotic) plant species from the park's rich fynbos communities, which also reduced the risk of fire to nearby high-value real estate.

This case study demonstrates that Table Mountain National Park, like most parks, has a strong and positive economic value (using 'economic' in its proper sense as the total value of tangible and non-tangible benefits provided at the level of society). However, park managers need to deliberately and thoughtfully articulate the full range of environmental, social and economic values to the public.

Giving parks back to people: Contract parks in South Africa

Prior to South Africa's transition to multiracial democracy, nature conservation was associated with racial injustice. Conservationists were worried that the new government would use parks for non-conservation purposes, such as agriculture and livestock production. Grossman and Holden's chapter describes three examples of communities that agreed for some or all of their land to be used for conservation purposes, under contractual national parks, including the northern section of Kruger. Using the same economic logic that justified building and expanding parks – i.e. that parks can be managed as an engine for economic growth and employment – negotiations were struck whereby ownership of parts of parks was retained by, or returned to, communities but the land was contractually managed for nature conservation. The communities were intended to benefit from business opportunities in the parks, and to have a say over the management of the parks, while continuing to live where they were. Where this has been done thoughtfully, as with the Makuleke land claim in Kruger, it has shown signs of working. New tourism and hunting operations in the park have increased revenues considerably.[2] However, Grossman and Holden also demonstrate the considerable challenge of working with cobbled-together 'communities', including problems of internal differentiation and conflict, especially when the park agency is relatively inexperienced in these matters and its commitment to environmental justice may not reach down through the agency to field staff.

Parks as economic engines: A new political strategy

The chapters in Part V are focused mainly on South Africa, which has strong property rights and a sound legal system that reduce economic distortions on prices. Additionally, the government is remarkably cognizant of upstream and downstream economic multipliers and employment, and is prepared to pay for these environmental services. Consequently, South Africa is an excellent place to evaluate the economy of protected areas, at least in broad terms. Private landholders have clearly demonstrated that wildlife conservation pays in certain agro-ecological zones (see Part III). Led by the example of Pilanesberg, park managers in South Africa have also gradually adopted an economically instrumental approach to conservation and are demonstrating how valuable many protected areas are.

Taking a proactive stance that recognizes the economic as well as the conservation values of protected areas, has injected considerable financial and political energy into conservation in South Africa. Parks are not only resisting expropriation to other forms of land use but new parks are being constructed and many parks are being expanded. While savanna parks are inherently able to pay for themselves, most parks provide a suite of environmental services over and above tourism that are commonly not traded in the marketplace. The South African government, with its sophisticated economic outlook, recognizes many of these values and provides substantial grants to parks on this basis. We see this most clearly in the case of Table Mountain National Park. In the Maloti-Drakensberg complex of parks, which hosted the 2008 SASUSG conference, parks are beginning processes to enable the generation of considerable revenues by selling environmental services, in this case improved water quality to the Johannesburg conurbation (Mander, pers. comm.).

These opportunities have arisen because of a mind-shift by park managers: while biological knowledge is still vital, park managers have now placed biodiversity within an economic framework. They are capitalizing on the fact that parks can generate more income with sound commercial practices, and that parks supply a range of economic and environmental services for which society is now prepared to pay. While parks are still wise to seek the safety of financial viability, park agencies are getting better at understanding and explaining economic effects such as environmental services and the multiplier effects of tourism dollars (a subject that certainly needs more research and quantification). As we will see, they have learned to argue the economic case for vistas, alien vegetation clearance, fire prevention and job creation in political forums, and have obtained government grant financing on this basis. In other words, in a sophisticated economy such as South Africa's, where institutions are better able to internalize the true costs and benefits of land use, the case for park and wildlife conservation is becoming ever stronger. The turning point was a fundamental change in conservation philosophy: if parks are managed as a form of common property to provide value to society in a multitude of appropriate forms, the likelihood that society will value parks increases. Creating this range of values, and ensuring that society is aware of them, becomes the means by which biodiversity and natural values are achieved.

Innovation north of the Limpopo

Innovation is not limited to South Africa, despite the bias in Part V. Nevertheless, in the absence of strong legal systems and with a general mistrust of private sector motivations, innovation north of the Limpopo River has often taken a different form. It has usually involved the restructuring of park agencies or outsourcing management functions to NGOs. These issues were discussed at some length in *Parks in Transition* (Child, 2004), but there has been progress over the last four years so brief comments about innovation in other southern African countries are in order.

Two examples from Zambia suggest that park agencies can rapidly improve performance with simple but fundamental changes in organizational management. Funded by the Norwegians, South Luangwa National Park was a politically ring-fenced project that was managed as a cost centre that controlled its own income and expenditure, and was staffed by government employees under the nominal authority of the wildlife department. Initially, this model took on the characteristics of a donor-funded 'spending organization' and was expensive and ineffective, although for a time it was good at law enforcement and stopped the massive elephant poaching. In the late 1990s, as donor funding ran out, staff were taken through a four-year participatory process to agree the goals of the park and ways to measure them, within realistic budgetary guidelines. By devolving management responsibilities to nine sub-sections on the basis of clear work plans and budgets, and regularly subjecting performance indicators to internal peer-review, output was doubled and costs halved. Clarifying responsibility and authority, and locating it at low levels in the administrative hierarchy (a common business practice), radically improved performance. The real lesson was that staff were not 'useless' as many commentators observed, but were unable to fulfil their potential in the centralized and bureaucratic systems that typify park agencies (Dalal-Clayton and Child, 2003). In a similar example, this time in Kafue National Park, 'useless' staff became effective once performance management was introduced. Instead of sitting at the office while poachers ravaged wildlife, scouts were paid US$10 for each day on patrol plus bonuses for anti-poaching successes. These anti-poaching operations were carefully monitored (as the basis of payments and strategic deployment) and poaching was controlled in three months with scouts patrolling for 20 days each month (Child, 2002).

Elsewhere in Zambia, the supervision and financing of parks and park staff have been outsourced to NGOs. In this way, Frankfurt Zoological Society controlled poaching in North Luangwa National Park (Borner and Longwe, 2002), and the Kasanka Trust went a long way towards rehabilitating the Kasanka National Park, a small miombo park famous for its bat populations. The Achilles heel to these models, however, is their dependence on donor funding. South Luangwa, by contrast, designed tourism zones and tourism concessions to earn at least US$1 million, while making organizational changes (see above) to cut costs from US$3 million to US$1.2 million (Dalal-Clayton and Child, 2003).

In Mozambique, too, we see similar models emerging, based on outsourcing tourism and even management services, and on commercially viable management plans. Not-for-profit partners are now the primary managers of Niassa Reserve and Gorongoza National Park (see Chapter 6). Donor funding is playing an important role in rehabilitating the resource base (Booth, pers. comm.), and the search is now on for models of economic viability by outsourcing tourism and hunting concessions to the private sector.

Helena Motta's chapter describes the rehabilitation and development of marine parks in Mozambique. Although Mozambique's formal aspiration is to develop national parks in the mode of IUCN's Protected Area Category II, the truth is that all parks in Mozambique support people and the solution has to be more complicated. Consequently, we see parks emphasizing traditional functions such as biodiversity conservation, reaching for financial sustainability and economic justification by the means that we have described for Pilanesberg, and simultaneously using community management principles with the people living inside them. Park management is clearly a complicated business and is getting more complicated (Cumming, 2004). We also see this complexity of stakeholders and management and scaling arrangements in Whande and Suich's chapter on transfrontier conservation areas, but note how such high-level initiatives tend to ignore community issues at their peril.

So far, the chapters we have introduced describe how protected area agencies are redefining their contract with society, and new practices related to this. In the final chapter, Fearnhead describes a model of a private enterprise that offers a full range of park management services to governments across Africa, which continue to own the parks and set conservation policy in them. The creation of this company was a response to the opportunities presented by the gap between park performance and potential, and the irony that many state-protected areas are declining at precisely the time that private conservation is proving to be so viable. Fearnhead, who has worked in both the public and private sector, discusses how 'legislated mediocrity' arises from the absence of competitive pressures, and from weak human resource management and performance accountability in the public agencies. He provides personal insight into the development of a private park management agency (African Parks), and describes the resistance to, and suspicion of, this model by several African wildlife agencies. There are echoes of the earlier resistance to new ideas such as game ranching and models such as Pilanesberg. The organizational and technical arguments for private management of state-protected areas are strong (see also Reed, 2002), and this model adds enough value that it is likely to work in the long term. However, the private sector is used to the clear-cut logic of business deals and economic efficiency but can be deaf to important factors such as representation, politics and social justice. Moreover, issues of 'ownership' and control exert powerful emotions in post-colonial states, especially when the cleavage between the private and state sector also follows racial lines. It is true, as Fearnhead argues, that the principles of private management would serve parks well, however, it is equally true that the private sector needs to become more comfortable, conversant and patient with issues of politics and stakeholder process. We just have to hope that these transaction costs will not prevent this new model developing further because, as Fearnhead points out, it offers significant conservation, economic and employment advantages.

Conclusions

Traditional park managers championed ecological perfection and sustainability but neglected the socio-economic parameters needed to justify the parks to society in economic and political terms. Nevertheless, many park managers felt these social and financial pressures acutely in the daily performance of their duties. However, it was only when they broke with conventional conservation dogma and grasped the nettles of social, economic, financial and political sustainability that they began to turn this complexity of issues to their advantage. This bind was broken in southern Africa when well-trained, innovative and experienced practical conservationists used the opportunities associated with political upheavals in the region to launch imaginative approaches to many old problems. This has involved openly seeking social and political acceptability and economic viability, alongside ecology, in the search for viable conservation strategies. Indeed, this is reflected in the logo of SASUSG, the learning network that commissioned this book, where the resilience and adaptability of social ecological systems is seen to lie at the nexus of people, ecology and economics.

Notes

1   I would like to acknowledge, with thanks, the technical insights and editorial improvements provided to the introductory chapters by my father, Graham Child.
2   Although recently the community has been deprived of promised high-value hunting in the park and has lost a lot of potential income, the ramifications of which are not yet clear.

References

Borner, M. and Longwe, I. (2002) 'North Luangwa Conservation Project, FZS Project 1027 / 86, assessment', Frankfurt Zoological Society, Lusaka

Child, B. (2001) 'Policy recommendations for the use of resources by South African National Parks', Consultant's Report for South African National Parks, Pretoria

Child, B. (2002) 'Performance based resource protection Kafue National Park', Zambia Wildlife Authority, Lusaka

Child, B. (ed) (2004) *Parks in Transition: Biodiversity, Rural Development and the Bottom Line*, Earthscan, London

Cumming, D. (2004) 'Performance of parks in a century of change', in Child, B. (ed) *Parks in Transition: Biodiversity, Rural Development and the Bottom Line*, Earthscan, London

Dalal-Clayton, B. and Child, B. (2003) *Lessons from Luangwa: The Story of the Luangwa Integrated Resource Development Project, Zambia*, International Institute for Environment and Development, London

Davies, R. (2000) 'Madikwe Game Reserve: A partnership in conservation', in Prins, H. H. T., Grootenhuis, J. G. and Dolan, T. T. (eds) *Wildlife Conservation by Sustainable Use*, Kluwer Academic Publishers, London

Reed, T. (2002) 'The function and structure of protected area authorities. Considerations for financial and organizational management', Summer Internship Program Report (1999), World Bank, Washington DC

18

Changing Institutions to Respond to Challenges: North West Parks, South Africa

S. R. Johnson, W. Boonzaaier, R. Collinson and R. Davies

Following decades of stringent conservation rules in the post-colonial period in southern Africa, which tended to marginalize and exclude indigenous rural inhabitants, in some countries new perspectives on conservation began to creep in during the 1970s. Specifically, three Fulbright Scholars from the Leopold school of thinking at the University of California – Thane Riney, Ray Dasmann and Archie Mossman – visited Zimbabwe and looked into aspects of the wildlife industry in the prevailing conservation environment, game ranching and adaptive management, among other issues. Riney in particular stressed the critical links between ecological, economic and socio-political forces in successful conservation. The formation of the Southern African Regional Commission for the Conservation and Utilization of the Soil encouraged free thinking, and greatly facilitated the spread of the ideas that had emanated from these three Fulbright scholars among the southern African conservation and wildlife fraternity. These ideas found fertile ground in the minds of the ecologists and managers employed to develop the fledgling Pilanesberg National Park in South Africa during the early 1980s.

In South Africa, this new thinking also took root in the private sector when a number of cattle ranchers began to experiment with 'mixed' farming – destocking cattle and increasing populations of wildlife. The Natal Parks Board encouraged this trend by offering excess populations of game to Zululand ranchers free of charge; other authorities encouraged and supported the creation and development of game ranches focusing on meat production and hunting. Similar initiatives sprang up on ranches in the Northern Cape and the former Transvaal, and by the time the Pilanesberg National Park was started, wildlife utilization on private land was exploding.

The Pilanesberg National Park

The idea of establishing a protected area in the Pilanesberg hills – the remnants of an enormous volcano approximately 60km north of Rustenburg – and which was home to communities from the Bakgatla baKgafela and Bakubung tribes, originated in a regional planning report commissioned by the transitional homeland government and produced by Potchefstroom University in 1969. The report and subsequent economic studies recommended that, given the low agricultural potential of the area, consideration should be given to the establishment of a recreation resort and nature reserve as a rural development strategy rather than an area designed to protect specific biodiversity attributes.

Following the 'Independence' of Bophuthatswana (one of South Africa's supposedly independent homeland states, which were in fact an expression of the segregationist and exclusionary policy of forcing black communities into marginal areas of the land) in December 1977,[1] control of Pilanesberg National Park and its development was given to Agricor, an agricultural development corporation, and funds were raised for the planning of the new reserve. Landscape architects were appointed to draw up a feasibility study and later a development plan for the reserve, which was submitted in August 1978. The planning report made a number of suggestions regarding the purpose and management of the proposed reserve, including that:

> *Natural areas require internal management to ensure survival of the features for which they are being protected but also need to be seen in the regional context. The success of conservation measures and the survival of wildlife and natural areas depend largely on the attitude of the rural inhabitants surrounding those areas and who would, therefore, derive immediate tangible benefits from their protection and utilization. They should therefore be regarded as an integral part of the regional economy.* (Tinley, 1978, p2)

The report also stated that the necessity of cropping ungulate populations in most national parks 'in some sense exposes the inadequacy of the notion that national parks should be absolute sanctuaries' (Tinley, 1978, p4), and suggested the zoning of the park to enable various land use practices and activities to be managed, including hunting, intensive use tourism, wilderness, buffer and development areas.

The landscape architects oversaw the initial dialogue with the communities resident in the area, negotiating a deal in which marginal tribal land within the identified area would be swapped for an equivalent amount of more fertile agricultural land outside of the proposed park. An agreement was eventually reached and families in the area were moved to other land, adjacent to the proposed park. Fencing of the land began in January 1979 and the 48,000ha Pilanesberg Game Reserve was officially opened in December, 1979.

The initial management personnel were drawn from South African conservation agencies and were thus relatively conservative and bureaucratic. The limitations facing the project arising from its operation within a government bureaucracy, led the head of Agricor to motivate for the formation of an independent park steering committee, under the auspices of Agricor. This move was successful, and the fact that the Bophuthatswana National Parks Board (BNPB) started off under the control of Agricor gave the steering committee the institutional capacity to develop the park along the lines of a development, rather than conservation, project. The personal interest, participation and strong leadership of Agricor management and other individuals serving on the steering committee of the BNPB laid the cornerstone for the business approach that was taken by the project.

The emergent years

Following expressions of concern regarding the slow pace of park development, new staff were recruited in October 1980. Jeremy Anderson was appointed as the new park director, tasked with overseeing the overall development process, while Willie Boonzaaier, an accountant was tasked with streamlining the finance, planning and administration and preparing the park for becoming a parastatal. Anderson immediately initiated trophy hunting in the reserve in an attempt to increase financial returns to the park, given that very few wildlife-viewing tourists were being attracted to the area in the early stages of park development.

In 1981, the new team brought in ecologists Peter Goodman and Roger Collinson to develop a full park management plan and Collinson was subsequently recruited to oversee the implementation of the management plan, with the special task of obtaining optimal economic potential from the park's wildlife populations. Collinson and Boonzaaier immediately started developing new game production models in an attempt to improve the returns from wildlife through hunting, live sales and culling, based upon economic successes achieved in the private sector game-ranching industry. The pair became excited by the idea of manipulating game stocks to optimize returns and studied the wildlife utilization philosophies and wildlife production models developed by scientists such as Stuart Eltringham. They became convinced that wildlife and other resources, as well as tourism and business opportunities had to be identified, developed and utilized to their full potential with benefits flowing to all legitimate stakeholders – especially neighbouring communities.

A fundamental change of philosophy introduced – under major objection from conservative South African conservationists – was the idea of placing an economic value on stocks of wildlife and to incorporate this value in annual financial statements to allow a true reflection of the financial success of the primary business of conservation. At the time, it was regarded as sacrilege to place a price tag on the head of wild animals, as they were a non-tradable national asset. In fact, the conservative establishment labelled BNPB staff 'the cowboys of conservation' in response to the perceived radical approach.

A major stimulus for further change came from actions taken by the new Bophuthatswana Minister of Finance who expressed his dissatisfaction with the slow progress towards Pilanesberg's financial self-sufficiency. The minister unilaterally entered into a contract with the private sector on behalf of government for the development of a commercial lodge operation inside the park, the KwaMaritane Game Lodge. The first five-year Development Plan for Pilanesberg was subsequently prepared and became the first formal blueprint for overall development processes in the park, embodying the principles of sustainability, community beneficiation and private sector participation. The plan included a detailed zonation of the park to optimize utilization, prevent conflicts of interests between different users and to prevent unplanned and unauthorized development taking place.

The development plan made provision in the short term for hunting to be a primary income generator until (non-hunting) tourist densities and higher tourism yields could overtake the returns from hunting. This eventually became standard procedure in other parks, including the Madikwe Game Reserve, as discussed below.

The next step was to contemplate the role of the private sector. Although the board initially developed and ran tourism enterprises, the domestic market for game reserve accommodation was distorted by state-run subsidized enterprises such as South African National Parks and Aventura resorts, which adversely impacted on the pricing of tourism products and services in protected areas. The foreign tourism market was not, at that time, a lucrative one due to South Africa's apartheid policies. The BNPB recognized that its primary objective was conservation and that facilitating tourism development and socio-economic benefits was secondary, and it was felt that the only way in which the park could compete against subsidization was to form 'smart partnerships' with the private sector and rely upon market forces to level the playing field. The board, accepting that the private sector was best equipped to run business operations, wanted to avoid running tourism enterprises in the long run, contributing to market distortions. Thus, while government initially had a stake in newly developed tourism enterprises, it was with the clear intent of selling off its shares as soon as the destination was established and operations proved to be viable.

In 1984, the BNPB was officially designated as a parastatal organization, and a new, commercially oriented executive management team was recruited to manage the entity using commercial principles. BNPB saw its mission as contributing towards the improvement of the quality of life in Bophuthatswana by conserving wild plants, animals and landscapes for the satisfaction of people's present and future needs. The basis of the 'conservation with development' approach was a utilization strategy that sought to optimize the economic potential of tourism through non-consumptive activities such as ecotourism, as well as the commercial harvesting of wildlife resources. Increasing community participation in park management and conservation decision-making processes and increasing the emphasis on environmental education and conservation extension were also important innovations.

A key aspect of this approach, from a learning perspective, was the interaction with the privately run KwaMaritane Game Lodge in the park. As a result of the close relationship with the management and staff of KwaMaritane, park management realized that the private sector could operate its activities effectively and responsibly within a state context. Such operations were not necessarily a threat to the integrity of the conservation role of government if managed within clearly agreed parameters, which were fairly monitored, with a minimum of bureaucratic interference.

Over time, intensive financial and economic analysis of the tourism and conservation activities in the park demonstrated that all state-run tourism and economically based conservation activities could potentially be run by the private sector, which would allow park staff to concentrate on core conservation activities. The BNPB then started exploring commercialization models applied in other industries and developed the first true public–private partnership (PPP) agreements within a southern African protected area agency.

Achievements

By 1990, Pilanesberg was a very successful tourism destination with tourist numbers increasing each year. Visitor numbers grew from 70,000 in 1992 to around 400,000 in 2002, an average of 24 per cent growth per annum over that period. Revenue from tourism activities increased alongside this growth in visitor numbers (see Figure 18.1 and 18.2). The hunting enterprise developed and matured and a sustainable live harvest was sold annually into the thriving local game market (see Figure 18.3). These factors indicated that 'sustainable use' of resources could generate significant revenues and could provide other socioeconomic benefits. The early contracts were not without criticisms from the private sector game-ranching industry relating to 'unfair competition'. Some of these early deals with the private sector, in both Pilanesberg (for example KwaMaritane) and Madikwe (for example River Lodge), were weighted in favour of the developers ('sweetheart' deals), which was justified by the need to kick-start investment with a small number of risk-tolerant pioneer investors

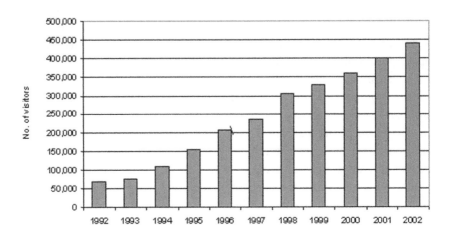

Source: Boonzaaier (pers. comm.)

Figure 18.1 *Visitor numbers, Pilanesberg National Park, 1992–2002*

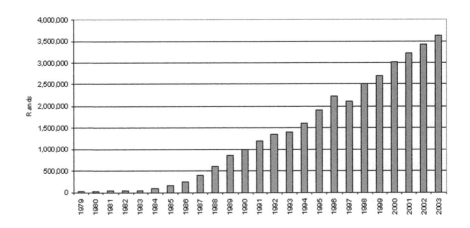

Source: Boonzaaier (pers. comm.)

Figure 18.2 *Tourism revenue, Pilanesberg National Park, 1979–2003*

who would pave the way and demonstrate to other investors that the levels of risk were acceptable.

According to the new management plan, Pilanesberg National Park (PNP) intends to expand its boundaries. This expansion initially hinged on a proposed community–public–private partnership for the incorporation of

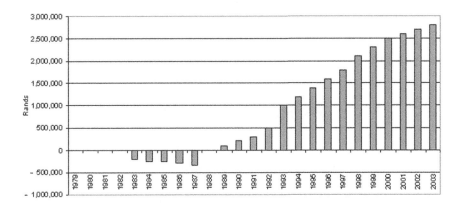

Source: Boonzaaier (pers. comm.)

Figure 18.3 *Revenue earned from hunting, culling and capture,*
Pilanesberg National Park, 1979–2003

private and community land to the north of the existing PNP. With renewed interest in mining immediately to the north of the PNP, this proposal was derailed. However, in 2004, the first private property (the 1700ha Zandspruit) was incorporated when the fence with PNP was dropped and traversing rights over PNP were acquired by the owners of Zandspruit. The agreement was made in accordance with the Land Incorporation Policy of the North West Parks and Tourism Board (NWPTB, formerly the BNPB), which is based on South African National Parks' agreements for contractual national parks.

Unfortunately, mining exploration is taking place on the community land intended for incorporation, though the NWPTB has negotiated with the mining companies to minimize the impact of mining on the land and maximize the potential for converting the land to conservation and possible incorporation into PNP at a later date. It has now been agreed with the mines that this 'mining area' will be game fenced and stocked with non-dangerous game. The community and the mines will jointly control the land use, which will cater for a combination of mining, controlled cattle grazing and game production until mining ceases and the area can be fully incorporated as part of the park.

During a public meeting in 2006, the Bakgatla baKgafela community voted in favour of setting aside an additional area of 33,000ha for the development of a 'Big Five' game reserve immediately to the north of the mining area as described above. The Bakgatla baKgafela, in partnership with a private development company, completed a feasibility study during early 2007 and drafted a development and business plan that is now being discussed with affected neighbouring villages in preparation for acquiring a favourable community resolution towards implementation of the plan. If this plan is implemented, it is expected that 1000 temporary and 1600 permanent jobs will be created, and an investment of R700 million (US$99.1 million) will be made in the area.

Madikwe Game Reserve

The success of the Pilanesberg model allowed additional parks to be developed in the North West Province, using the same philosophical approach. In 1991, a parcel of land, of approximately 700km^2, was identified on the South African side of the Botswana border, just south of Gaborone. It had been used for cattle grazing with some arable agriculture and further agricultural production emphasizing cattle grazing was planned. However the potential of the area for wildlife and tourism was recognized and the NWPTB requested an independent study be undertaken looking at all the natural resources use options. This was commissioned and the study found in favour of 'upmarket' wildlife tourism and hunting (Setplan, 1991). The report found that a game reserve could deliver significantly more benefits in terms of jobs, investment and returns to government than alternative agricultural uses and so, on the basis that it was the most suitable land use, Madikwe Game Reserve was proclaimed. Ensuring socio-economic returns has underpinned all the decision making of the reserve.

Once the decision was made to establish the game reserve, a budget for the project was allocated to the NWPTB, and the lessons learned in Pilanesberg were refined and improved. A decision was taken to introduce the 'Big Five' into Madikwe, as they added significant value through game viewing tourism, rather than hunting – a value that was not directly dependent on the ability of the natural resources to provide production, which translated into value through consumptive use. It diversified and increased income, increased employment opportunities and added value on-site to the product.

The game reserve targeted the upmarket niche because it created more jobs and was more cost effective – the required infrastructure was less expensive to develop and maintain, especially roads that only had to be suitable for 4x4 vehicles, as only guided tourism was going to be allowed in Madikwe. The mid-market in South Africa at the time was heavily dominated by the state, making competition by the private sector extremely difficult, and the upmarket segment at the time appeared to be strong and undeveloped.

Several elements were key to achieving the positive financial returns that were required by the NWPTB. The project built partnerships with the private sector, communities and the state (through the NWPTB), and only private investment was used in the tourism and service sector. It was recognized that extensive community support would be required to ensure the communities participated as equal partners in the project and that benefits would have to be optimized at the local level with due regard to quality and market forces. This was essential as, if the communities did not receive some financial benefit, their long-term support may be jeopardized, the implications of which could result in increased poaching as people felt alienated from the park. The necessity of effective communication, equitable access and sharing of benefits also required that community structures were built through collaboration with community leaders.

Biological and socio-economic benefits

The reserve is fully enclosed by a 2.3m high electrified game fence and as it is relatively small, virtually all of the game populations need to be actively managed. To this end there is a periodic (usually annual) game census of the more abundant ungulate species. Valuable species (both in financial and ecological terms) and potentially dangerous species are monitored more intensively to ensure their productive capacity is optimized and/or that inbreeding does not occur. Lion, wild dogs, black rhino and white rhino are managed as meta-populations of a wider gene pool of animals in other reserves (particularly Pilanesberg). The cost of game management, conservation infrastructure maintenance and security is approximately R7.5 million (US$1 million) per annum (Davies, 2003).

Table 18.1 outlines the game count figures for Madikwe between 1995 and 2005 (excluding the figures for black and white rhino). The positive biodiversity benefits of the establishment of Madikwe are demonstrated by the increased populations of most game species, with few exceptions (mostly marginal species for the area that did not adapt to the conditions).

Park income is derived from two major sources – levies or rentals from privately developed lodges and the sale of game. In the early business feasibility models, tourism income was predicted to be the major source of income, though game sales were recognized as important, especially in the early years. Revisiting these business models demonstrated some significant deviations from the initial projections (see Figure 18.4) (Davies, 2003).

It is apparent from Figure 18.4 that concession fees have been significantly lower than projected and game income significantly higher. Lower than expected concession fee income was the result of significant delays in the roll-out of the concession process. A land restitution initiative was launched by the government following the 1994 elections, which allowed communities who had been dispossessed of their land during the apartheid era to lodge applications to have it restored, and concession agreements could not be concluded while the restitution process was under way. It was not until 2000 that the long

Table 18.1 *Madikwe Game Reserve game count results (1995–2001) and annualized population growth rates, 1995–2005*

Species	1995 (estimated)	1996 (estimated)	2001 (estimated)	Total introduced	Estimated annual increase (%)
Blesbuck	250?	<100	2?	199	−50
Buffalo	76–85	90–100	236	66	17
Eland	850	1060	700	519	4
Elephant	246	255	320	227	10
Gemsbok	520	650	500	330	13
Giraffe	135	150–160	200	163	5
Red hartebeest	800	900	500	490	1
Impala	2750	3000	3200	1865	7
Ostrich	70–80	90–100	50	97	−7
Kudu	1700	2000	1700	801	12
Sable	n/a	35	15	50	−25
Springbok	230	250–350	50	505	−16
Tsessebe	n/a	27–31	30–50	50	−16
Waterbuck	550	650	600	555	1
Blue wildebeest	1650	2100	3500	807	19
Zebra	1600	1900	2500	1199	12
Lion	11	21	50	12	n/a
Wild dog	6	12	19	19	n/a
Cheetah	20–25	20–25	25	19	n/a
Spotted hyaena	8	26	35	19	n/a
Brown hyaena	20–40	20–40	20–40	2	n/a
Leopard	15–30	15–30	25	2	n/a

Source: Adapted from Hofmeyr et al (2003); unpublished data

process of public tender of opportunities, assessing bids and concluding agreements began again. In addition to the length of time taken to conclude agreements, operations had to finalize construction and start trading before they could pay any rentals.

The higher than expected game income was due to the increase in value of a few species, especially buffalo and white rhino. Buffalo showed significant increases in value, as well as better than projected population growth rates, having an annual compounded growth rate of 16 per cent. While the buffalo were purchased for approximately R25,000 (US$3350) per head, they were subsequently conservatively valued at R100,000 (US$13,350) per head (Davies, 2003).

Approximately R160 million (US$23.7 million) of investment has been made by the private sector to date to develop more than 28 lodges, and income from concession fees is approximately R5.2 million (US$769,000). Over 400 permanent jobs have been created in local communities and numerous small

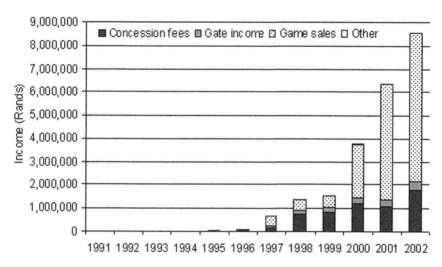

(a) Actual income

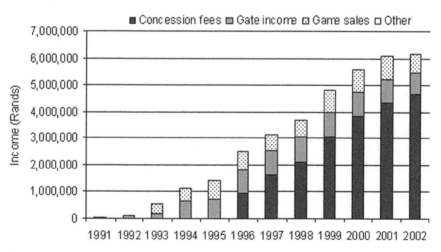

(b) Projected income

Note: * Actual and predicted income from game sales, concession fees, entry fees and 'other'. Predicted income estimations corrected for inflation (percentage monthly change in consumer price index for all commodities), based on estimates made in 1993.

Source: Davies (pers. comm.); Leitner (pers. comm.)

Figure 18.4 *Predicted and actual income, Madikwe Game Reserve, 1991–2002**

businesses are operational. Income from hunting, live sales and concession fees exceeds annual operational costs, and has been predicted to increase by more than 50 per cent on projected committed income (Davies, 2003).

A framework for additional land to be incorporated through contractual agreements and managed under conservation has been finalized by the NWPTB with neighbouring landowners. Under this framework, over 5000ha of land has already been incorporated into the reserve and agreements have been reached regarding an additional 15,000ha. The realignment of the fences is currently under way.

A tender process was held, under which rights to build and develop lodges on 45-year lease terms were offered exclusively to local communities in the reserve. So far two have been finalized: one has been operational for some time, while the second has only recently opened for trading. The community own the development rights, a portion of which they cede to raise capital, which is used to leverage additional funds (grants or loans). The community thus hold both the lease rights and the assets under the lease. Operators have been contracted for 10 to 15 years to operate the lodges, which pay asset rental and a percentage of turnover from trading. This form of community–private joint venture offers an exciting new model for community-owned tourism. In this process, communities have ownership rights and play a role in the operational aspects of the company. There has also been significant training and development aimed mostly at local people. Although the initial intention was to encourage the development of local small businesses, there seems to be a preference for employment and earning a wage (Koch and Massyn, 2003).

While the development model within Madikwe Game Reserve may not be universally applicable, the partnerships built to date have unlocked significant development opportunities and have diversified and added value to an otherwise locally depressed economy. Such developments help to demonstrate that conservation can be economically beneficial if the conditions and operating environment can be structured appropriately. Thus, more land has been gained for conservation in a financially and economically efficient manner.

Madikwe has proven that wildlife can compete with agriculture as a land use and can in fact offer significantly higher economic returns. The capital costs are high and such development is not without risk, being heavily dependent on tourism income. However, this risk can be reduced through appropriately structured agreements where risks are shared, and a diversification of markets (i.e. not relying solely on foreign tourism, which often offers the highest reward but can be very fickle). Alternative local tourism models based more on 'owners' contracting into long leases have reduced the reliance of the operators and the NWPTB on typical seasonal tourism. Within this framework the 'owners' are responsible for paying a fixed rental that is not dependent on occupancy and therefore irrespective of the income earned by the facility; it is therefore a form of timeshare but on a very small scale. The policy framework has also allowed the park to expand through contractual arrangements over private land and at the cost of private investment, while increasing the range and extent of the habitat available for wildlife. This is particularly important for species that require large areas (for example wild dogs, lion, black rhino and elephant), reducing the need for intensive management of game populations.

The Pilanesberg Corridor

The lessons learned from PNP and Madikwe Game Reserve are now being taken a stage further, based upon the proven economic model of this form of land use in the North West Province (see Table 18.2 for the economic impacts of the two protected areas). The province commissioned a study to explore the possibility of linking up the two protected areas and including an additional area of some 167,500ha, forming a protected area of nearly 250,000ha, including Pilanesberg and Madikwe. The initiative, known as the Pilanesberg Corridor, would increase the area under conservation in the North West Province from 4 per cent to 8 per cent (Boonzaaier, 2002).

Table 18.2 *Economic impacts of the Madikwe Game Reserve and Pilanesberg National Park*

	Madikwe Game Reserve	Pilanesberg National Park	Total
Private sector investment	R50 million (leveraged on state outlay)	R121 million (excluding Sun City)	R171 million
Forthcoming private investment	R60 million (6 new lodges over 18 months)	R10 million (lodges of Pilanesberg)	R70 million
Turnover from concessionaires	R25 million per year (to double in 18 months)	R87 million per year	R112 million per year
Taxes paid	R2.5 million per year	R6 million per year	R8.5 million per year
Direct jobs (indirect)	200 (plus 60 NWPTB jobs)	530 (plus 100 NWPTB jobs)	730 (temporary jobs to be added)
Economic empowerment	Minimum community equity in bids 25%/outsourcing	60% community equity in new lodge/potential in Dirapeng	Empowerment through operations, secondary products.

Source: Boonzaaier (2002)

Local, regional, provincial and national authorities have recognized the area as a primary nature-based tourism growth node, and the area is already recognized as a tourism destination as a result of the success of Pilanesberg and Madikwe reserves. New development is required, tourism demand is proven, and existing parks are saturated and the land use is in line with local development objectives and integrated development plans of local government. Minimal financial contribution is required from government as the bulk of the tourism investment will be undertaken by the private sector. Relationships between NWPTB and communities are improving, though more needs to be done to create awareness of the initiative. One major advantage of the proposed corridor is that no settlements would have to be moved, although alternative grazing may be needed in some cases. However, land claims will need to be resolved before further action can be taken.

Estimates suggest that the corridor could make a direct contribution of R178 million (US$17.1 million) per annum to the local economy (59 per cent of tourism turnover in the district) and that expected knock-on effects could add 20 per cent to gross geographical product. Where agriculture in the study area only sustains 93 jobs, it is estimated that the corridor project can sustain 1657 permanent jobs, with an average annual remuneration of tourism-related jobs in the area being R51,376 (US$4930) versus R8822 (US$850) for agriculture. Improved career paths, training and small and medium-sized enterprise opportunities would also benefit local communities (Boonzaaier, 2002). In the medium term, the corridor has been estimated to be able to attract in the order of 750,000 visitors per annum, rising to approximately 840,000 visitors in the long term (approximately 40 per cent overnight visitors and 60 per cent day visitors) (Boonzaaier, 2002).

The annual operating costs of the intensively used and managed PNP are R178 (US$17) per hectare and that of the more exclusively used and extensively managed Madikwe Game Reserve are R66 (US$6) per hectare (Boonzaaier, 2002). (The main difference can be attributed to the higher density of self-drive visitors in PNP who require a higher level of bulk infrastructure and maintenance.) The operating costs of the Pilanesberg Corridor are not expected to grow proportionately with land size, as certain costs remain fixed. An average cost of R50 (US$7) per hectare for additional exclusive use areas and R150 (US$22) for intensively used recreation areas are thought to be realistic estimates of operating costs (Boonzaaier, pers. comm.). Figure 18.5 shows estimated income and expenditure projections for the Pilanesberg Corridor. The figure indicates that the income of the Pilanesberg Corridor has the potential of covering operational costs, provided the project is paced to match the growth in tourism demand.

Conclusions

An analysis of the strengths and weaknesses of various institutional frameworks identified four main criteria that enable a conservation organization to provide

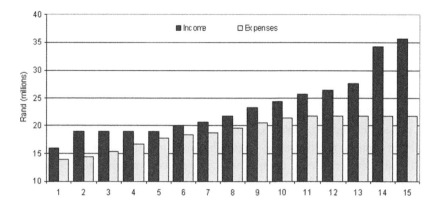

Source: Boonzaaier (pers. comm.)

Figure 18.5 *Projected income and expenditure, Pilanesberg Corridor*

a superior level of service to its clients, and to contribute to an improvement in the quality of life of communities (BNPB, 1991; CNC, 1994). First, financial autonomy and powers of attorney allow the organization to use and dispose of its own income, reducing its dependence on provincial budgets. Capital funding should also be accessible from external (i.e. non-governmental) sources to establish revenue-generating facilities and activities. Second, where decision-making powers are delegated to an organization, management is able to respond more quickly, efficiently and effectively to environmental and market opportunities, allowing conservation officials to be seen as more responsive to the needs of communities. Third, the ability to structure, grade and position staff independently of Public Service Commission processes allows an organization to obtain high-quality staff through competitive processes, while offering competitive remuneration and benefits, also allows management to keep staff motivated, productive and efficient. Fourth, the organization must be subject to public accountability and auditing through the legislature to ensure proper control over public funds.

A number of strengths and opportunities allowed the NWPTB/BNPB to develop and implement such innovative and successful projects. These strengths included the parastatal institutional framework of the board and its adoption of modern commercial business principles in park and organizational management. This institutional structure optimized performance in terms of its efficiency and ability to meet stated objectives. A number of visionary people were involved and the value of human resources was recognized, allowing freedom to experiment, which was key to optimizing staff performance.

The financial flexibility of the NWPTB enabled the organization to motivate, plan, promulgate, stock and develop the Madikwe Game Reserve in less than two years. Such development opportunities in other public service-based conservation organizations have been known to take between 10 and 20 years.

Further innovations of the NWPTB were its recognition of the value of available resources – land, wildlife, plants, landscape – and their encouragement of partnership processes to assist in achieving objectives. Communities have been seen as partners in conservation, not merely passive recipients of benefits, and the private sector has been viewed as a valuable ally rather than an 'external burden' to be tolerated.

This is not to say that there were no weaknesses and constraints in the process and institutions – working with multiple stakeholder processes can be very complex and time consuming, and can create tension between the dynamic nature of commercial approaches and the protracted nature of community management processes. Thus, dynamic and appropriately qualified and motivated staff are required to make it work. The exercise of caution is also required, as decisions can sometimes be made reacting to market pressures when the precautionary principle might be more appropriate.

The innovations of the NWPTB were based on the new protected area management paradigm that does not see biodiversity conservation as the single objective of protected area management; rather that sustainable development (and sustainable use of resources) is key. The institutional structure of the NWPTB meant that protected area management processes could be more efficient and cost effective in the long term, as fewer bureaucratic restrictions were imposed and dynamic development and implementation processes generated enthusiasm and commitment. That decision making could be based on the best available information – informed choice – allowed genuine financial, social and cultural benefits to accrue to stakeholders.

Notes

1   The homeland of Bophuthatswana was reincorporated into South Africa after the 1994 elections, which effectively ended apartheid rule. The Bophuthatswana National Parks Board was amalgamated into the North West Province's new conservation organization called the North West Parks and Tourism Board.

References

Boonzaaier, W. V. (2002) *Pilanesberg Corridor Concept Plan*, 2nd edn, North West Parks and Tourism Board, Rustenburg, South Africa

BNPB (1991) *Development of Mission and Strategies*, Bophuthatswana National Parks Board, Mmabatho, Bophuthatswana

CNC (1994) *The Northern Cape: A Conservation Opportunity*, Cape Nature Conservation, Cape Town

Davies, R. J. (2003) *Madikwe Game Reserve: A Decade of Progress*, Madikwe Development Series, North West Parks and Tourism Board, Rustenburg, South Africa

Hofmeyr, M., Davies, R. J., Nel, P. and Dell, S. (2003) 'Operation Phoenix: The introduction of larger mammals to Madikwe Game Reserve', in Davies, R. J. (ed) *Madikwe Game Reserve: A Decade of Progress, Madikwe Development Series*, North West Parks and Tourism Board, Rustenburg, South Africa

Koch, E. and Massyn, P. J. (2003) 'The Madikwe initiative: A programme designed to optimize local benefit by integrating the conservation of wildlife with local economic development', in Davies, R. J. (ed) *Madikwe Game Reserve: A Decade of Progress*, Madikwe Development Series, North West Parks and Tourism Board, Rustenburg, South Africa

Setplan (1991) 'Dwarsberg: A Development Assessment', unpublished report for the BNPB

Tinley, K. (1978) 'Pilanesberg National Park: Planning and management proposals for the Department of Agriculture, Republic of Bophuthatswana', unpublished report for the Bophuthatswana Department of Agriculture

19

Making Conservation Work: Innovative Approaches to Meeting Biodiversity Conservation and Socio-economic Objectives (an Example from the Addo Elephant National Park, South Africa)

Guy Castley, Michael Knight and Jill Gordon

Recent developments within South African National Parks (SANParks) have seen a shift in focus from its flagship parks, such as Kruger National Park, towards the consolidation and expansion of many of the smaller parks, primarily to meet the conservation objectives of the major biomes that have been under-represented (thicket, grassland, fynbos, Succulent Karoo, Nama Karoo and forest) (Driver et al, 2005). Consequently, many of these smaller parks have now taken their place in the spotlight, and considerable attention is being given to enhancing the biodiversity conservation potential within these parks and to the diverse range of socio-economic opportunities that should result from their further development. Conservation here is understood as the balance between the potentially contradictory concepts of 'biodiversity protection' and 'economic development' of protected areas within an integrated regional mosaic (Borgerhoff Mulder and Coppolillo, 2005), a balance that needs to be achieved, particularly if protected areas as an entity are to survive (McNeely, 1989).

In the changing management focus of SANParks – from conservation of charismatic fauna to representative ecosystems and ecological processes – the Addo Elephant National Park (AENP) was identified as a priority for expansion and development, and has therefore received significant support for expansion activities, with national government providing funds for land purchases, the first such allocation in decades. The greater AENP (gAENP) initiative, which specifically refers to the expansion process of the park, makes explicit linkages between ecological, social and economic viability – seen not only as part of a biodiversity hotspot (Cowling and Hilton-Taylor, 1997), but also as a potential economic engine for the region, as reflected in increased employment ratios and tourism numbers (CES, 2002).

The greater Addo Elephant National Park

The 2270ha AENP was proclaimed in 1931 for the prime purpose of preserving the last remnant population of elephants in the Eastern Cape, and since then small additions have been made to the park to accommodate the expanding elephant population. It was not until the early 1990s, when the park had increased to 11,000ha, that the focus of park management moved away from mega-herbivore conservation and associated threats (Johnson, K. et al, 1999; Johnson, C. F. et al, 1999; Lombard et al, 2001), to a broader biodiversity conservation focus (Hall-Martin and van der Merwe, 2003) (see Figure 19.1). This culminated in the motivation for the gAENP initiative, based upon conserving the areas' unique combination of biodiversity, with its representation of five of the country's nine terrestrial biomes (Driver et al, 2005; Mucina and Rutherford, 2006) and its links to an important marine area. The socio-economic opportunities that could be derived from further development were also crucial (Kerley and Boshoff, 1997). In addition, conservation as a land use with its associated nature-based tourism opportunities within the thicket biome (which makes up a large of the gAENP), was noted to be ecologically and economically more sustainable than goat-focused pastoralism, which is the predominant form of agricultural use in the area (Moolman and Cowling, 1994; Kerley et al, 1999). Goat pastoralism transforms the landscape significantly, directly contributing to an alteration in vegetation structure and the ability of thickets to assimilate carbon (Mills et al, 2005).

The AENP has taken a lead in fine-scale systematic conservation planning for protected areas in order to identify conservation priorities within a broader ecological planning domain (Castley and Knight, 2003). Systematic conservation planning itself is a relatively new science and has only recently been adopted by conservation planners and scientists as a strategic tool for directing conservation action (Erasmus et al, 1999; Pressey, 1999; Reyers et al, 2002). The bulk of the groundbreaking work in South Africa was undertaken for what is now known as the Cape Action Plan for People and the Environment (CAPE) project in the Cape Floral Kingdom (WWF, 2000; Younge and Fowkes, 2003). Other regional conservation plans, such as the Sub-tropical Thicket Ecosystem Planning (STEP) (see Pierce, 2003), Succulent Karoo Ecosystem Planning (SKEP, 2003) and gAENP programmes have further developed the tool. The gAENP project has differed from these other planning

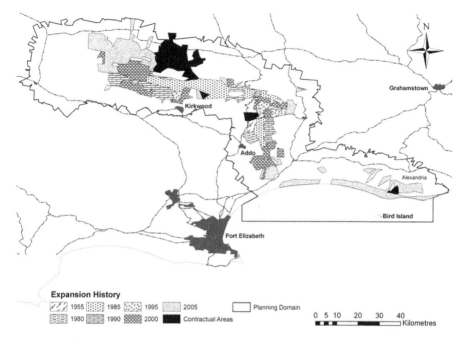

Source: Guy Castley, Michael Knight and Jill Gordon

Figure 19.1 *Expansion history and existing contractual arrangements around AENP in relation to the systematic conservation planning domain*

exercises in that it has seen implementation activities resulting in significant local change.

SANParks adopted a rigorous strategic approach, where an integrated systematic conservation plan was developed to provide the necessary guidance and motivational support for the AENP expansion programme (CES, 2002; CSIR, 2002). This was strengthened by SANParks' mandate to contribute to the expansion of the formal terrestrial protected area network from 6 to 8 per cent of South Africa's surface area by 2010 (Yawitch et al, 2005). The paradigm shift in the thinking behind setting aside protected areas from one that was purely preservationist to one that attempts to encapsulate ecological patterns and processes within a broader ecosystems conservation approach (Grossman and Holden, 2005) also influenced the SANParks approach. Linked to this shift was the recognition that the establishment and expansion of protected areas cannot take place in isolation from the neighbouring land uses and communities. So the challenge was set for the expansion of the AENP into a mega-biodiversity conservation area, with strong socio-economic links with the surrounding region.

Framework for the gAENP expansion

The gAENP project is directly linked to the efforts of the South African government to address national and global environmental priorities by reversing land degradation and enhancing biodiversity, conserving representative examples of varied landscapes, and improving local livelihoods by promoting sustainable development and tourism (DEAT, 2005). Key natural ecosystems and spheres of activity that are vulnerable to degradation have been identified as part of the governments' obligations under the Convention on Biological Diversity, and the AENP is one of these. Though the province's economic base is centred on agriculture, with some very localized manufacturing and industry, tourism is becoming more important in many rural areas because it can offer a productive land use alternative to unsustainable pastoralism (Kerley et al, 1995; Geach, 2002). The AENP has the capacity to stimulate and support a growing nature-based tourism industry as demonstrated by the rapidly increasing number of private wildlife reserves in the region (Sims-Castley et al, 2005; Langholz and Kerley, 2006). Thus, the gAENP is perfectly placed to address some of South Africa's conservation goals while providing a platform for socio-economic development, focused on the nature-based tourism industry. The park will be the only one in the world offering visitors the 'Big 7' experience, comprising the traditional African 'Big 5' terrestrial game species (lion, elephant, leopard, rhino, buffalo) plus whales and sharks (including the great white) (Castley, 2002) and thus is likely to continue to attract increasing numbers of tourists.

Tourist numbers for the AENP have been increasing at about 10 per cent per annum over the last decade and are expected to increase from 114,000 in 2001 to 180,000 by the end of 2010. Associated nature-based tourism expenditures and indirect economic impacts will also increase significantly. In 1995, a travel cost study of the AENP estimated the recreational value of the park at approximately US$60 million with only 100,000 visitors per annum (Geach, 1997). This is expected to be almost double with the current number of visitors to the park – in excess of 170,000 visitors in 2006 (SANParks, 2007).

Direct employment increases associated with conservation and tourism in the park represents about one job per 100ha, in comparison to the agricultural norm of one job per 367ha of small stock farming, the predominant land use around the park (CES, 2002). This suggests that conservation as a land use option may offer greater economic and social returns than small stock farming. Furthermore these figures could be considerably higher given the estimated 70 per cent knock-on effects of this tourism activity (Langholz and Kerley, 2006). Estimates suggest that the park has led to the creation of one third of businesses within a 30km radius of the park, including 430 jobs and the injection of US$10.3 million into the local economy (Saayman and Saayman, 2005). The relatively high ratio of foreign to local visitors (50:50) to the AENP compared to other national parks in South Africa also bodes well for regional job creation. As many as 14,000 jobs could be created in the region from current visitor numbers, with estimates suggesting that one new job is created for every ten foreign visitors (Hugo, 1992). Given that the park is already functional and there is to be greater diversification of the wildlife product, enhanced investment opportunities will undoubtedly attract further investment that will hopefully further increase job creation.

It is planned that up to 15 per cent of the expanded area of the gAENP will be privately owned land (see Table 19.1). Farms considered for inclusion into the gAENP will be evaluated according to a series of criteria and owners will be offered a suite of incentives (for example tax rebates, provision of management support for clearing alien vegetation, fencing, marketing opportunities, traversing rights and provision of wildlife) to convert from agricultural production to conservation (Knight, 2003). Importantly, national government supports the provision of these incentive measures.

The project will also implement SANParks' proposed sustainable natural resource use policy with local communities in specific zones. Commercial and recreational fishing in the proposed marine protected area (MPA) will be permitted only in specifically designated use zones, for which negotiated catch limits will be set at sustainable levels in line with park objectives (SANParks, 2006). Key to the success of these initiatives will be the need for continual stakeholder engagement.

Land incorporation

Ultimately, the gAENP will encompass habitats ranging from semi-arid Karoo plains, fynbos-covered mountain slopes, impenetrable thickets and montane forests, rolling coastal grasslands and coastal forests within its 372,000ha terrestrial planning footprint. The park will also encompass an MPA of 120,000ha that will include the Bird and St Croix Island groups in Algoa Bay.

To meet AENP's expansion objectives, SANParks developed a land incorporation framework that highlighted the importance of large areas in terms of their conservation value (as determined by the conservation plan) (CSIR, 2002) and their importance and potential as economic development nodes, able to stimulate nature-based tourism growth (Knight, 2003). The incorporation of land into the gAENP is dependent upon its relative value with respect to its ecological integrity, biological representativeness, its estimated socio-economic costs (such as resettlement and price) and potential to reduce threats to biodiversity.

The area required for incorporation into the terrestrial part of the park is based upon baseline targets for vegetative units and the habitat requirements of large herbivores and predators, some of which are yet to be reintroduced into the ecosystem. While the terrestrial zone has 44 per cent of the planned land area already included in the park (over the previous seven years, the park has almost doubled in size, to 164,233ha (SANParks, 2007)), the establishment of the proposed MPA has only just begun. In the meantime, the offshore islands with a buffer zone of 25 nautical miles were promulgated as an MPA in 2005, while the extensive public participation phase to review the establishment of the larger MPA was also initiated in the same year. These expansion plans will see AENP become the third largest conservation area in South Africa after the Kgalagadi Transfrontier Park and Kruger National Park.

Expansion challenges and new directions

Though the current AENP is a mosaic of landscapes that are not completely contiguous, the long-term vision is that of a park with a core area for wildlife that is significantly larger than it is at present. The project will achieve this vision by consolidating the existing park through land inclusions, focusing on priority parcels identified using a range of criteria (as discussed above) (CES, 2002). Given the need to address broader goals of greater public participation, expanded tourism development and the need for improved socio-economic development in the region, a cautious and phased approach is warranted. Such an approach has associated risks in that it potentially allows land speculators to take advantage of long-term expansion plans to achieve personal short-term financial gains.

Park expansion is planned in five regions that will see these sections developed initially in isolation and later be fully incorporated into the larger conservation area (see Table 19.1). Activities within each block will include boundary expansion, fencing, wildlife introductions, tourism and social development activities.

Public–private partnership (PPP) arrangements will be utilized to stimulate sustainable development and socio-economic growth. The contractual inclusion of land and development of PPPs will help SANParks to meet strategic conservation goals, as they can increase the area under conservation and use financial resources more efficiently. Thus, the implementation of the expansion programme will occur under three different management scenarios (and two different partnership arrangements):

1   SANParks land;
2   concession areas within SANParks land that are owned and managed by SANParks; and

Table 19.1 *gAENP consolidation regions*

Region	Nature of expansion
Coerney	Consolidation of areas and inclusion of new habitat to expand the (existing) primary tourist area where the majority of the elephant are currently located
Nyathi/ Zuurberg	Consolidation of linkages to the west of the (existing) Nyathi area and existing concession areas to the Zuurberg range. Further consolidation of areas to the south-west of the Kabouga section in the southern Zuurberg region
Darlington	Consolidation of the semi-arid region to the north of the Zuurberg range that includes the amalgamation of contractual areas as well as SANParks areas to the east and west of the Darlington Dam
Colchester	Consolidation of areas around the Sundays River mouth and creation of linkages between the Coerney section with the coastal thickets. Further development of contractual agreements
Coastal	Consolidation of the coastal zone between the Sundays River mouth in the west and the Alexandria forests in the east. Processes to establish the MPA that will establish the link between terrestrial and marine biomes

Source: Knight (2003)

3 contractual land adjoining the park, where SANParks may or may not manage the land, but where the prospective partners have contractually committed themselves to long-term partnerships in keeping with the SANParks conservation philosophy.

The concession process, geared towards sustainability (Fearnhead and Mabunda, 2003), is different to that of contractual land partnerships in that the concessionaires rely on SANParks-owned land for the development of tourism products, which are ultimately transferred to SANParks at the end of the concession lease, whereas contractual partners ordinarily bring a parcel of land to the agreement.

Contractual arrangements

There is an increasing trend to incorporate privately owned land on a contractual basis into national parks. In a shift from traditional park expansion programmes, the plans for gAENP make provision for contractual areas within the planning boundary, involving private landowners, local communities and the private business sector. Generally such arrangements are restricted to specific peripheral (non-core) areas. However, in some cases – where acquisition is unfeasible given very high-value or communal land – the contractual approach is used to incorporate land into core conservation areas.

Contractual arrangements allow SANParks to focus on their primary responsibility – conserving South Africa's biodiversity for current and future generations – while engendering an attitude of environmental custodianship among private partners that encourages a move towards compatible land use practices in areas adjacent to national parks. It also has the advantage of spreading economic benefits to a larger sector of society through the creation of new economic opportunities and investment. Where contractual arrangements are not feasible, other options for land inclusions may include agreements with weaker management conditions and/or buffer arrangements, which provide fewer incentives to private landowners (Knight, 2003). However, in cases where essential land parcels cannot be incorporated by other means, expropriation will be used as a last resort.

Allowing land to be incorporated into the park without necessarily requiring a transfer of title has a number of benefits: enabling the park to be managed collectively as an ecological whole; avoiding the overwhelming cost of large land purchases; and enhancing the opportunities for developments such as nature-based tourism, hunting and live game sales for landowners, under a SANParks banner.

The advantage of these arrangements to landowners is that a suite of nature-based tourism options open to them because of their (now unrestricted) size, which has built strong stakeholder participation and support for planning and monitoring, private sector involvement and community/social ecology components. However, the presence of large and dangerous species, such as elephant and lion, precludes the park from fully amalgamating with the surrounding lands, unless their land use is compatible with the presence of such animals.

Contractual parks demonstrate an opportunity for achieving conservation and social development objectives, but the models require further development, as facilitating community-based conservation is often time consuming and needs to be flexible to accommodate the variability within individual agreements (Reid et al, 2004). Also, in some cases, the current acquisition framework and protected area expansion models may not engender a desire within some local communities and neighbouring landowners to support biodiversity conservation philosophies, despite some communities receiving direct benefits from conservation efforts. This is an area where SANParks could refine their existing protected area management principles, for example through the zonation of parks (to incorporate multiple use areas) and allowing communities to utilize local resources (Naughton-Treves et al, 2005).

Another issue surrounding land acquisition relates to dispossessions that may result from land purchases. In order to address this, SANParks developed a Resettlement Policy Framework for AENP (Gordon, 2003) that is now in the process of being adopted as core SANParks policy. The Resettlement Policy Framework is a groundbreaking policy that ensures the implementation of the socially focused legislation (such as the Extension of Security and Tenure Act

(62 of 1997)), which protects the rights of affected farm workers – particularly in terms of accommodation and job security – in cases where land purchases result in a conversion from agricultural to conservation land uses.

Although the planning domain of the gAENP project covers some 372,000ha of land, the plans to consolidate and expand the park to capture the unique biodiversity of the region will focus on a core area of some 205,000ha of terrestrial (see Table 19.2) and 120,000ha of marine habitat until 2010. During this time, although some land will be acquired, increased emphasis will be placed upon contractual inclusions to shift the ratio of purchased to contractual land from 89:11 to 85:15. This objective emphasizes not only the important contribution that neighbouring private landowners can make towards meeting conservation objectives but also the commitment of SANParks to achieving their mandate by drawing on the strengths of PPPs.

Table 19.2 *Land acquisition planned under the gAENP with landholdings in 2007*

Timeframe	Purchased land (ha)	Contractual land (ha)	Total land (ha)	Purchased: contractual ratio
2007	146,625	17,608	164,233	89:11
End of project (2010)	175,000	30,000	205,000	85:15
Increment	28,375	12,392	41,767	70:30

Source: SANParks (2007)

Contractual partnerships remain a challenge given the expectations of what they can deliver (revenue, immediate tourism markets and so on), and the time-frames in which these objectives can be achieved. This highlights the need for continuous and constructive dialogue and partnership development to build positive relationships among stakeholders. Current contractual inclusions are primarily restricted to large land areas with absentee landlords. A further challenge is to shift this emphasis to include clusters of small on-site active farmers who do not warrant inclusion on the limited ecological value of their own land, but who do so collectively. Successful agreements would see associations of like-minded private landowners being included into the park under a single contract versus separate landowner agreements (Knight, 2003).

The gAENP development scenarios (Davies, 2002) demonstrate potential significant employment gains when land in the planning domain is converted from marginal agriculture (unsustainable small stock farming) to nature-based tourism, with direct benefits for employment of local labour. For example, the new Kuzuko tourism development (in one of the contractual areas to the north of the existing park) previously had only 20 workers on 14,400ha of pastoral farmland (Begby, pers. comm.). Estimates suggest that the development of this land into a tourism venture will see employment double in the medium term, in addition to the direct employment of 38–48 short-term workers. Furthermore, the Kuzuko development has also seen the opportunities for these employees (typically ex-farm labourers) to increase their skills through capacity-building programmes. Further evidence is observed in the expanding private game reserve industry in the region, whose economic and social impact have increased with wage bill and employment increases of 32 per cent and 4.5 fold per reserve, respectively, over the previous livestock-focused agricultural land uses (Sims-Castley et al, 2005; Langholz and Kerley, 2006).

The commercialization process

The gAENP project was the forerunner in the development of the commercialization process within SANParks. The Gorah Elephant Camp in the main tourist section of the park was the first concession awarded by SANParks, and this agreement formed the basis for negotiations in other national parks (SANParks, 2001). The concession is targeted towards the luxury tourism market and the enterprise has enjoyed considerable success since opening, indicative of the gAENPs ability to absorb such ventures. Since then, three tourism concessions have been awarded in the park and another two have been identified that will see the expansion of tourism into previously undeveloped areas of the park (SANParks, 2006). In addition, a new camp has specifically allocated a percentage of its turnover to support a local community trust fund, enhancing nature-based tourism opportunities and upliftment in previously disadvantaged communities.

As part of the commercialization process within SANParks, the management of restaurants, shops, games drives and other tourism adventure pursuits has also been transferred to private entrepreneurs. The AENP has been at the forefront in promoting participation from local communities to ensure that the benefits from tourism are fed back into these communities. An extremely popular initiative, spearheaded by the AENP People and Parks Department, saw the training of 'hop-on' guides to accompany self-drive tourists in the park. All of these guides are residents of the neighbouring communities and the programme draws extensively on their local knowledge of the area. The value of these guides is that they not only increase the exposure of tourists to the biodiversity assets of the park, but they also offer opportunities for community involvement and enhanced environmental education among both residents and tourists (Kerley et al, 2003).

Small, medium and micro enterprise (SMME) development opportunities within the gAENP project domain (including fencing, arts and crafts, charcoal production, tourism activities, alien vegetation clearing, rehabilitation, construction and sustainable harvesting) have remained an important focus of the project. Within three years, ten sustainable SMMEs have been trained, exceeding the target of two per year for the six-year project (World Bank, 2004).

The changing management environment

With new approaches to meeting conservation objectives come new approaches to protected area management. National parks are moving towards more consultative management processes in order to ensure community support, and stakeholder committees are being established across the country as part of the new national park governance regime, as directed by the new national protected areas legislation.

Active participation of project beneficiaries and other stakeholders from the initial planning process is important in identifying potential problems and solutions, generating support and fostering knowledge sharing. Public participation processes have been part of regular AENP management for many years and the gAENP project has strengthened these processes by the establishment of new and more effective institutions, such as the Addo Planning Forum (APF), that are focused on keeping stakeholders informed. The APF was established in 1998 and provides a broad platform for constructive and consultative dialogue between a wide spectrum of stakeholders in the project, including national, provincial and local government departments, NGOs, agricultural groups, local industry, labour groups and previously disadvantaged communities. As the newly selected official park forum, the APF will take on greater responsibility in advising the park of stakeholder's needs and perceptions. The recent public participation process associated with the development of the park management plan has provided another avenue to consult with the broader community (SANParks, 2006).

An outsourced Project Management Unit has been operational since 2004 and is responsible for project implementation including the establishment of a number of strategic working groups and stakeholder forums. One of these – the Project Steering Committee – will provide a high level of oversight of implementation for SANParks in collaboration with the APF.

A number of project working groups have also been established, which play a crucial role in integrating the project activities into the ongoing work of the park organization and staff. Additional groups will be formed as needed (and dissolved as assigned activities are completed) and their composition will remain flexible to allow members (individuals or from key organizations) to be co-opted as required.

Sustainability of the gAENP initiative

Several factors are linked to the sustainability of the gAENP in terms of strengthening protected area management and biodiversity conservation in the AENP area, as well as stimulating local business and improving local livelihoods.

Institutional sustainability

Institutional sustainability is important and has already come a long way. SANParks has built consensus among a wide range of stakeholders regarding the project design and scope. Appropriate institutions are being used to facilitate long-term local participation in park planning and management.

The gAENP is designed to ensure the continuous participation of the AENP staff in project execution through the frequent and regular Project Management Unit, Steering Committee and Resettlement Working Group meetings. This will ensure that the project is well integrated into park operations and that a seamless transition is made from the project deliverables to their incorporation into routine park operations and general stakeholder acceptance.

Financial sustainability

Financial sustainability is critical to ensure that the outputs of project investments can be maintained in the longer term. Although gAENP receives external funding from the Global Environment Facility, the project has been designed with the end of external funding in mind. The gAENP project was planned to generate a number of positive fiscal impacts (Davies, 2002), the primary impact being the higher net returns from the conversion of land from ecologically and economically unsustainable pastoral farms to nature-driven tourism based on the expanded park's valuable biodiversity (Antrobus, 2002). While average gross income for small stock pastoral land is about US$7 per hectare and game farming yields approximately US$8 per hectare, nature-based tourism can generate up to US$30, depending on the type of operation established (high-end client versus lower-value tourism) and generally enhances ecological services to the regional environment. The conversion of much of the highly unproductive pastoral land (of which a significant portion is abandoned) to higher-value nature-based tourism would represent a positive shift in economic efficiency at low opportunity costs (Langholz and Kerley, 2006).

Another important financial impact of the gAENP is the increase in revenues to the park from tourism and wildlife management operations. With tourist numbers expected to increase to 180,000 by 2010, revenues from entrance fees, concessions and game sales are projected to increase to US$1.4 million by 2010 (Davies, 2002). Initial financial simulations indicated that the project's investments would make the park financially independent by 2007, with an operating surplus of over US$200,000 per annum. In fact, in the 2005–2006 financial year, the park made a net profit of US$1.49 million (Moolman, pers. comm.). However, the challenge remains the conflict between conservation and agricultural land uses in a dynamic socio-political climate.

Social sustainability

Social sustainability will be addressed by establishing an economically viable park and by public participation in planning and management of the park. As noted above, it is anticipated that the establishment of an economically viable park will provide income and livelihood returns to local residents greater than those currently received from agriculture (Kerley et al, 1995; Langholz and Kerley, 2006). This will greatly improve support for the park and create more stable and sustainable livelihoods. The involvement of stakeholders in the park and project management through consultative forums should ensure that this support is maintained.

By 2007, the project saw the number of permanent jobs increase to 237, well above the six-year target (SANParks, 2007). In addition, over 900 contract jobs were created, paying wages significantly higher than those for farm labour. On pastoral and dairy farms, average monthly labour rates are between US$32 and US$60 per month. By contrast, average labour rates for SANParks general labour are between US$116 and US$250 per month (Antrobus, 2002). Private tourism operators are paying at least these rates, and workers displaced from unproductive pastoral farms who are subsequently employed in the park or private sector tourism tend to benefit from higher incomes and improved skill levels.

Consideration in this process must be paid to the need to employ affected farm labourers during the land use conversion. In this regard SANParks have had assistance from the national Extended Public Works, Working for Water and Coast Care programmes. These projects offered critical bridging financing and job creation for affected labourers, while the park further developed its extended tourism operations.

Environmental sustainability

Environmental sustainability has been addressed through the strategic environmental assessment undertaken during project preparation (CES, 2002) and the application of South African Environmental Impact Assessment requirements. The conservation plan developed by the gAENP project will ameliorate potential negative environmental impacts on the park through the application of appropriate management actions. Furthermore, the planned implementation of an integrated environmental management system for the park, rooted in adaptive management principles (Riley et al, 2003; Enck et al, 2006; McCarthy and Possingham, 2006), will see management being environmentally responsible, adaptive and accountable.

Conclusions

The gAENP project has learned from and built on experience gained during the CAPE and STEP processes. In turn, the lessons learned from the gAENP initiative will have applications for other protected areas in South Africa with respect to sound conservation planning, participatory planning and implementation, social assessment, resettlement, management information systems, monitoring, land incorporation strategies and PPPs. In particular, the gAENP project has demonstrated important and valuable lessons for other countries in the region and beyond in terms of the effective implementation of conservation plans being the result of a number of key elements such as the involvement of stakeholders, the adoption of conservation assessments by the implementing agency and the identification of clear and explicit goals.

References

Antrobus, G. (2002) 'Income from livestock farming', Specialist Report No. 10, Coastal and Environmental Services, Grahamstown

Borgerhoff Mulder, M. and Coppolillo, P. (2005) *Conservation: Linking Ecology, Economics and Culture*, Princeton University Press, Princeton, New Jersey

Castley, J. G. (2002) 'The greater Addo Elephant National Park: A vision in the making', *Timbila*, vol 3, no 5, pp87–91

Castley, G. and Knight, M. (2003) 'Conservation in action: Expanding the Addo Elephant National Park, South Africa', *Oryx*, vol 37, no 1, pp15–16

CES (2002) 'Strategic environmental assessment report for the greater Addo Elephant National Park', Coastal and Environmental Services, Grahamstown

Cowling, R. M. and Hilton-Taylor, C. (1997) 'Phytogeography, flora and endemism', in Cowling, R. M., Davison, D. M. and Pierce, S. M. (eds) *Vegetation of Southern Africa*, Cambridge University Press, Cambridge

CSIR (2002) 'Greater Addo Elephant National Park conservation project', CSIR Report No. ENV-S-C 2002-010, CSIR, Pretoria

Davies, R. (2002) 'Park forecasts: Strategic environmental assessments of the greater Addo Elephant National Park', Specialist Report No. 13, Coastal and Environmental Services, Grahamstown

DEAT (2005) *South Africa's National Biodiversity Strategy and Action Plan*, Department of Environmental Affairs and Tourism, Pretoria

Driver, A., Maze, K., Rouget, M., Lombard, A. T., Nel, J., Turpie, J. K., Cowling, R. M.,

Desmet, P., Goodman, P., Harris, J., Jonas, Z., Reyers, B., Sink, K. and Strauss, T. (2005) 'National spatial biodiversity assessment 2004: Priorities for biodiversity conservation in South Africa', *Strelitzia*, vol 17, pp1–46

Enck, J. W., Decker, D. J., Riley, S. J., Organ, J. F., Carpenter, L. H. and Siemer, W. F. (2006) 'Integrating ecological and human dimensions in adaptive management of wildlife-related impacts', *Wildlife Society Bulletin*, vol 34, no 3, pp698–705

Erasmus, B. F. N., Freitag, S., Gaston, K. J., Erasmus, B. H. and van Jaarsveld, A. S. (1999) 'Scale and conservation planning in the real world', *Proceedings of the Royal Society of London, Series B*, vol 266, pp315–319

Fearnhead, P. and Mabunda, D. (2003) 'Towards sustainability', in Hall-Martin, A. and Carruthers, J. (eds) *South African National Parks: Celebration*, Horst Klemm, Johannesburg

Geach, B. G. S. (1997) 'The Addo Elephant National Park as a model of sustainable land use through ecotourism', MSc thesis, University of Port Elizabeth, Port Elizabeth

Geach, B. G. S. (2002) 'Institutional arrangements and capacity needs for tourism management in the greater Addo area', Specialist Report No. 15, Coastal and Environmental Services, Grahamstown

Gordon, J. B. (2003) *Resettlement Policy Framework and Associated Resettlement Action Plans for Year 1 of Implementation of the Greater Addo Elephant National Park Project: Park Planning and Development*, South African National Parks, Port Elizabeth

Grossman, D. and Holden, P. (2005) 'Spreading the benefits of conservation: Indigenous communities, equity and parks', in Cowan, G. I., Yawitch. J. and Swift, M. (eds) *Strategic Innovations in Biodiversity Conservation: The South African Experience*, Department of Environmental Affairs and Tourism, Pretoria

Hall-Martin, A. J. and van der Merwe, J. (2003) 'Developing a national park system', in Hall-Martin, A. and Carruthers, J. (eds) *South African National Parks: Celebration*, Horst Klemm, Johannesburg

Hugo, M. (1992) 'A quantification of the role of foreign tourism in the South African economy', *Journal for Studies in Economics and Econometrics*, vol 16, pp41–51

Johnson, K., Cowling, R. and Lombard, M. (1999) 'Protecting plants from pachyderms: Conserving succulent thicket, a threatened veld type, at Addo Elephant National Park', *Veld and Flora*, vol 85, no 4, pp126–127

Johnson, C. F., Cowling, R. M. and Phillipson, P. B. (1999) 'The flora of the Addo Elephant National Park, South Africa: Are threatened species vulnerable to elephant damage?', *Biodiversity and Conservation*, vol 8, pp1447–1456

Kerley, G. I. H. and Boshoff, A. F. (1997) 'A proposal for a Greater Addo National Park: A regional and national conservation and development opportunity', Terrestrial Ecology Research and Unit Report No. 17, University of Port Elizabeth, Port Elizabeth

Kerley, G. I. H., de Kock, G. and Knight, M. H. (1995) 'Desertification of subtropical thicket in the Eastern Cape, South Africa. Are there alternatives', *Environmental Monitoring and Assessment*, vol 37, pp211–230

Kerley, G. I. H., Boshoff, A. F. and Knight, M. H. (1999) 'Ecosystem integrity and sustainable land-use in the thicket biome, South Africa', *Ecosystem Health*, vol 5, no 2, pp104–109

Kerley, G. I. H., Geach, B. G. S. and Vial, C. (2003) 'Jumbos or bust: Do tourists' perceptions lead to an under-appreciation of biodiversity?', *South African Journal of Wildlife Research*, vol 33, pp13–21

Knight, M. H. (2003) 'Land incorporation framework for South African National

Parks: Park planning and development', unpublished report for South African National Parks, Port Elizabeth

Langholz J. A. and Kerley, G. I. H. (2006) 'Combining conservation and development on private lands: An assessment of ecotourism-based private nature reserves in the Eastern Cape', Centre for African Conservation Ecology No. 56, Nelson Mandela Metropolitan University, Port Elizabeth

Lombard, A. T., Johnson, C. F., Cowling, R. M. and Pressey, R. L. (2001) 'Protecting plants from elephants: Botanical reserve scenarios within the Addo Elephant National Park, South Africa', *Biological Conservation*, vol 102, pp191–203

McCarthy, M. A. and Possingham, H. P. (2006) 'Active adaptive management for conservation', *Conservation Biology*, vol 21, no 4, pp956–963

McNeely, J. A. (1989) 'Protected areas and human ecology: How national parks can contribute to sustaining societies of the twenty-first century', in Western, D. and Pearl, M. (eds) *Conservation for the Twenty-First Century*, Oxford University Press, Oxford

Mills, A. J., Cowling, R. M., Fey, M. V., Kerley, G. I. H., Donaldson, J. S., Lechmere-Oertel, R. G., Sigwela, A. M., Skowno, A. L. and Rundel, P. (2005) 'Effects of goat pastoralism on ecosystem carbon storage in semiarid thicket, Eastern Cape, South Africa', *Austral Ecology*, vol 30, no 7, pp797–804

Moolman, H. J. and Cowling, R. M. (1994) 'The impact of elephant and goat grazing on the endemic flora of South African succulent thicket', *Biological Conservation*, vol 68, pp251–261

Mucina, L. and Rutherford, M. C. (eds) (2006) 'Vegetation map of South Africa, Lesotho and Swaziland', *Strelitzia*, vol 19, pp1–816

Naughton-Treves, L., Holland, M. B. and Brandon, K. (2005) 'The role of protected areas in conserving biodiversity and sustaining local livelihoods', *Annual Review of Environment and Resources*, vol 30, pp219–252

Pierce, S. M. (2003) 'The STEP Handbook. Integrating the natural environment into land use decisions at the municipal level: Towards sustainable development', Terrestrial Ecology Research Unit Report No. 47, University of Port Elizabeth, Port Elizabeth

Pressey, R. L. (1999) 'Applications of irreplaceability analysis to planning and management problems', *Parks*, vol 9, pp42–51

Reid, H., Fig, D., Magome, H. and Leader-Williams, N. (2004) Co-management of contractual national parks in South Africa: Lessons from Australia', *Conservation and Society*, vol 2, no 2, pp377–409

Reyers, B., Fairbanks, D. H. K., Wessels, K. J. and van Jaarsveld, A. S. (2002) 'A multi-criteria approach to reserve selection: Addressing long-term biodiversity maintenance', *Biodiversity and Conservation*, vol 11, pp769–793

Riley, S. J., Siemer, W. F., Decker, D. J., Carpenter, L. H., Organ, J. F. and Berchielli, L. T. (2003) 'Adaptive impact management: An integrative approach to wildlife management', *Human Dimensions of Wildlife*, vol 8, pp81–95

Saayman, M. and Saayman, A. (2005) 'Socio-economic impact of the Addo Elephant National Park', unpublished report for the North West University, Potchefstroom

SANParks (2001) *South African National Parks – Annual Report 2001*, South African National Parks, Pretoria

SANParks (2006) 'Addo Elephant National Park Project: Park management plan (draft)', unpublished report for the South African National Parks, Pretoria

SANParks (2007) 'The Greater Addo Elephant National Park Project: Mid-term annual report', unpublished report for the South African National Parks, Pretoria

Sims-Castley, R., Kerley, G. I. H., Geach, B. and Langholz, J. (2005) 'Socio-economic significance of ecotourism-based private game reserves in South Africa's Eastern Cape Province', *Parks*, vol 15, no 2, pp6–18

SKEP (2003) 'SKEP twenty year strategy. Biodiversity conservation and sustainable land-use in the Succulent Karoo: Succulent Karoo Ecosystem Planning report', www.skep.org/downloads/skep-2003-report.pdf (accessed 17 September 2006)

World Bank (2004) 'The Greater Addo Elephant National Park Project: Global Environment Trust Fund Grant Agreement', unpublished document of the International Bank for Reconstruction and Development, Washington DC

WWF (2000) *Cape Action Plan for the Environment: A Biodiversity Strategy and Action Plan for the Cape Floral Kingdom*, WWF South Africa, Stellenbosch

Yawitch, J., Mancotywa, S. and Naude, K. (2005) 'Governance: New ways of working together', in Cowan, G. I., Yawitch, J. and Swift, M. (eds) *Strategic Innovations in Biodiversity Conservation: the South African Experience*, Department of Environmental Affairs and Tourism, Pretoria

Younge, A. and Fowkes, S. (2003) 'The Cape Action Plan for the environment: Overview of an ecological planning process', *Biological Conservation*, vol 112, pp15–28

20

Table Mountain National Park

David Daitz and Brett Myrdal

Thixo, the god of the Sun, and Jobela, the Earth goddess, made love and conceived Qamata.

When Qamata created the world, he was attacked by the Great Dragon of the Sea, who was very jealous and wanted to stop the work of creation. In the battle with the Great Dragon, Qamata was crippled. In order to help Qamata, Jobela created a number of mighty giants to guard the world. She positioned one at each of the four corners of the world – in the east, the west, the north and the biggest giant in the south.

After many, many angry battles with the Great Dragon of the Sea, the giants were killed one by one, but they asked Jobela, the Great Earth Mother, to turn them into mountains so that even in death they could guard the world.

And so, the greatest giant of all – Umlindi weningizimu – became Table Mountain, the Watcher of the South. (Credo Mutwa, pers. comm.)

The Cape Peninsula consists of a chain of mountains that rise dramatically out of the sea, extending from the north, where Table Mountain forms the picture postcard backdrop to the city of Cape Town, to Cape Point in the south. This peninsula is an area of very high conservation value. Aesthetically, it has spectacular scenery with mountains in close proximity to the coastline, itself a succession of beaches and rocky shores of breathtaking beauty. In addition, the peninsula is endowed with an array of diversity almost unparalleled anywhere else on the planet. In the Cape Peninsula there are approximately 2200 species of plants; 25 per cent of the total number of species in the Cape Floral Kingdom, found in only 0.05 per cent of the surface area of the entire kingdom. There are over 100 peninsula plant endemics, a similar number (111) of endemic terrestrial invertebrates and even an endemic vertebrate, the Table Mountain ghost frog (Minter et al, 2004). The biodiversity in the offshore marine environment is just as remarkable for absolute numbers of species present as well as the level of endemism exhibited (WWF, 2000).

The establishment process of the Cape Peninsula National Park

In 1929, the Wildlife Society (now named the Wildlife and Environment Society of South Africa –WESSA) proposed that the Cape Peninsula should be declared a national park. With the 1929 Wall Street crash, followed by the Great Depression and the Second World War, it is easy to understand why nothing came of this suggestion. However by 1939, the Cape Point Nature Reserve had been proclaimed – an area of 8800ha at the southern tip of the peninsula. A form of weak legal protection was conferred in 1989 on the 29,120ha area that became known as the Cape Peninsula Protected Natural Environment (CPPNE) (SANParks, 2006). An institutional mechanism for cooperation between government authorities was established for the CPPNE, as it included land under the control of 14 different public authorities at national, provincial and local government level, as well as 174 private landowners. This loose coalition of interests never successfully coordinated their conservation actions, and, after a brief period of improvement the conservation status of the area continued to degrade.

Following a long process of investigation and significant interaction with civil society, the Western Cape Provincial Government decided in December 1995 to accept the creation of a national park that would include as much of the CPPNE as possible. This decision was hedged with some very problematic conditions. Among these were the requirements that if any one of the three local authorities[1] involved declined to make their land available for inclusion in the new park, or if SANParks failed to demonstrate 'significant progress' within six months, then the provincial government reserved the right to reject SANParks' involvement and implement their preferred solution – a provincial reserve. Their solution was at odds with the outcome of the most recent public consultations, and one can do no more than speculate as to the province's motivations. However, a convincing argument could be made for attributing it to a desire for parochial political advantage.

The decision of the provincial government to allow the creation of the new national park was made public in early December 1995. Dr Robbie Robinson, SANParks' chief executive officer at the time, reacted speedily by recruiting a project coordinator to take charge of creating the new park, who started work in January 1996 with a broad mandate to do anything and everything necessary to get the new park established. Within six months, all six members of the core project team were in place. There were several immediate problems confronting the project team: public concern surrounding the establishment of the park; funding of the park; and the fact that the province had declined to define 'significant progress'.

'Significant progress'

While there was no definition of 'significant progress', the politicians were serious about their ability to recognize it when they saw it. However, the elasticity of the measure was perhaps preferable in such a complex situation – it could be used to the advantage of both parties. Further, when SANParks was awarded the opportunity to establish the Cape Peninsula National Park (CPNP), there had been no thorough consideration within the organization about how to create or manage such an entity. It was therefore not difficult to fill such a vacuum with a strategy that was quite different from the norm and that could therefore suit circumstances that were clearly completely different. In achieving the desired outcomes of the park, a common-sense management strategy was used: in order to bring about fundamental change in complex situations characterized by extensive inertia, it is essential to develop the capacity to pursue multiple strategic actions in parallel, rather than adopting a sequential approach.

Local control of revenue had never before been permitted in a national park (though it was never explicitly conceded in this case), and it was foreseen that many issues ranging from municipal infrastructure inside the new park, to the right to walk dogs in the peninsula mountains would best be dealt with locally. The autonomy conceded by SANParks' head office to the Cape Peninsula project team during the park establishment was the price paid for the efficient creation and management of the park, even though it occasioned more than a little discomfort with many at the head office.[2] The issue of how much of the locally generated revenue stays within the park remains a bone of contention to this day.

The courting of public opinion

The peninsula is valued highly by visitors, with more than four million per year. Approximately 2.8 million of these visitors are residents of Cape Town visiting for outdoor recreational purposes, with the remaining 1.2 million being either foreign tourists or domestic visitors from beyond the boundaries of the Cape Metropolitan Area (CPNP, 2001). Thus, it is not surprising that an area of such value should have provoked serious expressions of public concern regarding its long-term conservation.

There were many damaging perceptions about the practical implications of creating a national park in an area that had historically been freely accessible. In the case of the Cape Peninsula, much public anxiety stemmed from the perception of a significant proportion of the community – despite solid evidence to the contrary – that SANParks would put a game fence around the park, stock it as quickly as possible with big and dangerous animals and make people pay to enter, as had been the case with most preceding parks.

The most important realization at the start of the process to create the park was that the 'real' success of the attempt to create the CPNP would not be judged by the elastic criterion of provincial government but would be determined by public opinion (with similarly inexact standards of measurement), as public opinion would undoubtedly lead the actions of provincial government.

Once SANParks' management recognized that this was likely, the primary strategy was built around winning the case for the park in the court of public opinion. Provincial government may have believed that it would decide, in private, whether significant progress had been achieved or not, but the project team bet on public opinion – as articulated in the media – carrying the day. Because public opinion would be based primarily on media coverage, letters to the editor, public meetings and so on, a deliberate process of engaging with NGOs and other interested groups and individuals was undertaken. This engagement emphasized the difference in the vision of the new national park from that of any existing national park in the country.

Between February and September 1996, the park project coordinator addressed 63 public meetings in and around Cape Town, and the project team worked hard to attract media attention to all of its activities. Given the context, these efforts were shrouded in a swirling mist of controversy – not always deliberately orchestrated by the project team – but that the media found virtually irresistible. As a result, there was conflict and controversy aplenty, and copious media attention.

Early on, a heaven-sent opportunity for gaining the support of the Cape Town public arose. On the Atlantic coastline between Llandudno and Camps Bay there was a landmark building known as the White House, originally a home of the van Breda family and subsequently used as an office. The house was the only visible structure on a pristine piece of coastline in an area where the vista – from the wild Atlantic crashing onto the granite boulders, rising dramatically to the cliffs of the Twelve Apostles – was otherwise undisturbed by intrusive developments. The White House site had been bought by a developer, at least two years prior to the launch of the park project, whose intention was to build a small hotel on the site. He had received rezoning and building plan approval from the Cape Metropolitan Council, notwithstanding the best efforts of 'green' NGOs to oppose it. The coalition of opponents wanted to take the planning authority to the High Court on review, but the costs of doing so were beyond their collective means. SANParks had solid reasons to oppose the development, even though it was outside the boundaries of the CPPNE. SANParks' long-term intention was to secure the remaining vistas from mountain peak to coast wherever they were still intact and undeveloped, and aspired to eventually include the White House site into the CPNP. The project team and Dr Robinson fully appreciated the public relations advantages of taking on this fight. Papers were filed and served on a number of respondents, including the provincial government.

Legal authority to approve rezoning and subdivision applications was held by the provincial government in terms of the Land Use Planning Ordinance (1985). This ordinance provided for the provincial government to delegate such authority to local authorities, but only in circumstances where a satisfactory master plan (known as a structure plan) existed for the area of jurisdiction of that authority. In order to delegate this authority – politically a much-prized possession because it allowed wealth to be created at the stroke of a pen – the provincial government produced a master plan for the entire former Cape of Good Hope Province. In SANParks' view, the basis of the province's delegation of authority to the Cape Metropolitan Council to approve the development was a blatant attempt to circumvent the provisions of the law. The structure plan used to justify the delegation of authority was devoid of any meaning because of its abstraction and vagueness – and therefore the delegation was incompetent. The championing of this cause turned the tide of support towards SANParks, and although the case was lost in the High Court, it was won in the court of public opinion.[3]

Funding and its leverage effects

From the outset, the challenge in establishing the CPNP went beyond the obvious need to consolidate the management of all of the conservation areas under a single authority. As such, an authority would be worth precious little if it could not fund the conservation action that had been lacking for so long. When Dr Robinson was asked how, in a bureaucratic organization like his, he expected the initiative to progress in the absence of any budget allocation, he responded: 'We'll find the money!' And he was as good as his word; the money was found.

The priorities of post-1994 South Africa were pressing very heavily on the three local authorities critical to the process, which failed to see that the CPNP had the potential to become a cash cow. The strategic approach of the project team was to negotiate with all three authorities in one forum, where SANParks' ability to leverage foreign funding and remove the financial burden from these authorities could be exploited to the full. The negotiations started in September 1996 and stretched over more than 18 months, moving progressively from the easiest issues to the most difficult.

In November 1997, the Global Environment Facility (GEF) approved a US$12.3 million (R56.7 million) grant for the Cape Peninsula Biodiversity Conservation Project, and tangible evidence of SANParks' ability to deliver funding was on the table. This development also presented the park negotiating team with an important opportunity for leverage. In negotiating the grant agreement with the World Bank (the implementing agent for the GEF), it was possible to include a clause setting a deadline for the grant to become effective, with a suspensive condition applying to the agreement in its entirety. This condition required that specific local authority land had to be proclaimed as national park before the deadline.

Notwithstanding careful and thorough negotiations, the local authorities found it incredibly difficult to let go of the control of their land. However, the threat of losing the US$12.3 million grant had the desired effect and the agreement was signed by the three authorities and SANParks in April 1998. The agreement provided for the land to be made available to SANParks under a 99-year lease and outlined a process that could eventually lead to freehold transfer. It catered for funding support for five years, the transfer of municipal staff, ongoing maintenance of municipal infrastructure within the park, boundary adjustments and the protection of the interests of local ratepayers, such as free access and the right to name the park. Also included was the transfer from municipal control of the only four entry points in the CPPNE at which entry fees were levied (out of four hundred entry points). By agreeing to arrangements relating to the change of pricing at these four pay points, a significant cash flow came directly to SANParks from the first day that it assumed management responsibility.

The local authorities at the time were hamstrung by a lack of imagination that failed to perceive the income-generating potential of the park, and by a view that gate tariffs should be affordable to the poorest residents of the city. This was inspired by a need to secure votes and took no account of the fact that the vast majority of the residents of Cape Town could not afford the transport costs to get to any of the four pay entry points.[4] Affordable access was provided by a cheap season ticket (which subsequently became the Green Card and then the WildCard), which enabled SANParks to increase tariffs dramatically at the two most popular pay points, further improving cash flow and therefore the conservation action that the park could undertake. However, this happy situation spawned a conflict within SANParks when head office tried to bring the CPNP's income under its control.

The agreement about the transfer of staff from municipal employ to SANParks was only struck at the eleventh hour, in accordance with the negotiations moving from simple to more complex issues. By the time this issue was on the table, in March 1998, there was a great deal of trust and goodwill between the parties, as well as serious momentum to complete the process. The park project team offered a solution that was provided for in the country's labour law, which was politically palatable and did not require negotiation with the powerful municipal trade union. A difficulty arose because the cash flow of the new park would not be able to bear the salary and associated costs of these staff for the first five years, so local authorities were asked to subsidize their salaries. Agreement was eventually made on a R10.3 million (US$1.72 million) subsidy, the deal was done and operational control was transferred from the local authorities to SANParks on 1 May 1998.

Activities following park creation

The CPNP was gazetted on 29 May 1998, with the deadline for effectiveness of the GEF grant being 1 June 1998. The gazetting did not complete the process of creating the new national park in the Cape Peninsula, though it did create a stable platform from which to proceed and a platform for growth, following a five-year strategic management plan.

After the intense activity that opened the way for the newborn CPNP, the team spent some time consolidating their activities. The operational management of the park was reorganized into an area-based management structure, and the social ecology (now called people and conservation) and visitor management functions were absorbed under the overall leadership of conservationists.

The five-year strategic management plan (2000–2004) outlined the vision for the park: 'A park, for all, forever'. More than a slogan, it has been translated into the vision for the way the three dimensions of environment, people and business are combined into a strategy for sustaining the park's core business of biodiversity conservation, while delivering benefits beyond the boundaries of the park.

The core business – biodiversity conservation – produces no direct financial return. It does, however, result in lowered operating costs as effective restoration of damaged ecosystems allows their maintenance in a steadier state. For example, the removal of alien plants reduces fuel loads and therefore reduces firefighting costs; the building of paths reduces erosion costs and enables the mountains to withstand increasing human use; and the enforcement of no-take marine zones replenishes fisheries and wins the support of fisher communities in opposing crime syndicates that poach marine resources.

The 'for all' in the slogan emphasizes the people-centred approach that underpins the park's way of undertaking its core business. Not only are biodiversity conservation activities done in a labour-intensive manner, but emphasis is placed on building skills and developing contractors in a learning and working environment that is worth more than the sum of the daily minimum wage. The reinstated and strengthened People and Conservation Department guides this process and undertakes vital environmental education work among the youth. The department also nurtures the public consultation and participation structures, the Park Forum and the City–Park Bilateral.

In the process of park development, benefits to communities were not the priority of park management, though not because they were viewed as unimportant (on the contrary, they are a requirement of success). It was believed that before a new park could engage with communities – particularly poor communities – about the delivery of real tangible benefits to them, it had to establish itself first. In the urban context of the Cape Peninsula, where there were no cases of displacing community land uses in favour of purist conservation, the engagement with communities in the early stages had a focus that had nothing to do with benefits. This engagement was extensive and with a wide spectrum of communities, but it concentrated on winning support for the process of creating the new conservation entity. The issue of benefits to surrounding communities became more important as the park began to consolidate its efforts, when it decided that benefits should be spread equitably among an appropriate mix of small and micro enterprises, along with investors in large-scale concessions through public–private partnerships (PPPs).

The slogan's 'forever' requires the growth of income from tourism revenues in order to finance core park activities. Consequently, tourism undertakings cannot compromise the core business of the park in any way, and are therefore guided by 'touch the Earth lightly' principles, by only developing sites that were already transformed in some way. A further aim of the tourism strategy is to provide and manage tourism infrastructure such as hiking trails, overnight accommodation and retailing with the objective of increasing the length of stay of tourists in Cape Town so as to increase bed night income to the city as a whole.

Biodiversity conservation

The completion of the GEF-funded alien clearing programme saw the achievement of 85 per cent of its goal of removing all seed-bearing woody alien invasive plants from the peninsula within five years (and produced some 400 temporary jobs over that time) (Standish and Boting, 2006). Other early achievements were the restoration of the Buffelsfontein homestead and the construction of the boardwalk at the Boulders penguin colony, necessary to manage the increasing interaction between people and penguins. The development at Boulders subsequently grew into an award-winning visitor centre.

The top priority of the park remains the consolidation of linkages between terrestrial and marine ecosystems, with the aim of proclaiming all of the 29,120ha that was part of the CPPNE as Table Mountain National Park (TMNP) (as the park was renamed in 2004) and 1000km^2 of inshore ocean as TMNP marine protected area.

The public land has been largely consolidated; the challenge lies with the remaining 164 private landowners whose landholdings fall within the original target area. A working group of the Park Forum, WWF South Africa (WWF-SA), Table Mountain Fund, the City of Cape Town (CoCT) and the then UkuVuka (a fire prevention and management campaign), set out a strategy for land incorporation and appointed a land negotiator. Options of donation, contract, acquisition or cooperative agreement and incentives known as the FAR-sighted approach (which offered support for *F*ire management, *A*lien clearing and *R*ates exemption) were offered. Using this approach, over a third of the privately owned land has been incorporated into the TMNP, with the most significant achievement being the acquisition of the 450ha Noordhoek–Kommetjie wetlands, linking the southern and northern islands of the park. To date, some 24,500ha of land has been proclaimed as part of the TMNP.

The proclamation of the 1000km^2 marine park was completed in 2004, in terms of the Marine and Living Resources Act, and this is now managed in terms of the Protected Areas Act as a contiguous and complete entity of the TMNP. The long-term plan is to include False Bay into the marine protected area. The TMNP is also exploring options of a False Bay coastal corridor linking the two 'horns' of False Bay, the Cape of Good Hope (SANParks) and the Kogelberg–Hangklip horn that is a proclaimed biosphere reserve (the core of which is under the management of CapeNature, the provincial nature conservation agency). With cooperation between these two authorities and the CoCT, the link will allow the terrestrial encirclement of Africa's largest bay, which holds a wealth of diversity and is an important breeding ground for the great white shark.

The restoration of fynbos veld and Afromontane forests through alien plant clearing is the next priority, with progress in forest restoration receiving growing attention. The aim of this restoration is to plant the same number of indigenous trees as are to be harvested and removed in the newly incorporated Tokai and Cecilia plantations (almost exclusively alien pine trees) by 2025. The Expanded Public Works Programme (a government infrastructure development and employment scheme run through the Department of Environment Affairs and Tourism) funds a team of 43 forest gardeners who have gathered seed from within the indigenous forests (avoiding genetic contamination) and have cultivated and planted out some 30,000 saplings in two years. Forest restoration is 100-year project and hence there is not a day to waste.

The reintroduction of indigenous fauna is the third priority, which entails the control and removal of all alien fauna. Aside from the game fence enclosed 8000ha of the Good Hope/Cape Point area, the Groote Schuur Estate set aside by Cecil John Rhodes will be expanded and will receive eland, red hartebeest, steenbok, rheebok and grysbok. This has been accompanied by the controversial but successful removal of the Himalayan tahr and the reintroduction of some 74 locally extinct klipspringers.

More than half of the anticipated 500km of footpaths have been completed and ongoing maintenance prevents erosion and allows for the use and enjoyment of the trails as a vital component of what is, after all, a walking park. Fire management through prescribed (initiated) burns to ensure the regeneration of fynbos, and a now completed circum-peninsula firebreak to control wildfires and prevent harm to life and property, is in place. The marine team is equipped and has an improving anti-poaching record, with increasing emphasis on building community support as the way forward.

Natural and cultural heritage is protected through the declaration of the Cape Floral Kingdom as a World Heritage Site in 2004 (one of the eight representative sites is the TMNP) and the declaration of the park as a grade one heritage site in 2005. The way forward is to create heritage tourism projects such as the completed Platteklip washhouse and the to-be-restored Peers Cave precinct, which is the site of the only San paintings in the peninsula.

Participation in management processes

The result of no longer having a dedicated social ecology manager following the reorganization of the management team in 2000, arguably led to the neglect of the advisory Park Committee and the City–Park Bilateral structures, which became obstacles rather than partners in action, a situation made worse by the turmoil in local government at the time. It was important to rebuild a new Park Forum as the organ of community perspectives on park matters, following the dissolution of the old Park Committee by ministerial decree due to repeated lack of quorum. The desire for a new forum reflected the people-centred approach of park management, which placed great store on building community ownership and channelling this custodianship into active working groups covering issues such as visitor safety, community employment, land consolidation and biodiversity advice. Currently, instead of the divided and vested interests of organizational representatives, each of the 17 Park Forum steering committee portfolios is based on a particular interest group, and must convene active working groups and seek a balance of perspective within each portfolio interest. In return for this tough balancing act, the advisory role of the Park Forum is respected by park management and its advice is sought before management embarks on controversial projects (such as the removal of alien fallow deer or the restoration of the recently incorporated 1000ha Tokai and Cecilia pine plantations). In the future, the Park Forum may well become a legal entity able to raise funds for its own volunteer-staffed implementation programmes. The model for the forum was the San Francisco Golden Gate conservancy, which has a volunteer staff of more than 200 people and runs museums, restaurants and tourism projects in support of conservation.

TMNP management has also taken the initiative to rebuild the City–Park

Bilateral. The status and standing of the City–Park Bilateral was lifted to the level of executive councillors who meet with park management three times a year. The Bilateral monitors progress on CoCT-funded programmes and has matured into a working group system, so that municipal and park officials jointly manage common issues and report on progress, building delivery momentum.

The continued nurturing of both structures is vital lest their neglect lead to them becoming obstacles to progress once again. The way forward may well be for the City–Park Bilateral and Park Forum working groups to combine, to strengthen civil society and local government partnerships in support of the park and to increase efficiency of delivery.

In the context of popular participation, park management decided to poll opinion regarding the proposed change of name of the park to the Table Mountain National Park. This had been the originally intended name, but the park was initially called the Cape Peninsula National Park due to the dominance of mostly white middle class residents of the Peninsula on the old Park Committee and the sidelining (through distance) of the mostly black residents who lived on the far-flung Cape Flats. The poll voted in favour of the change and popularized the park citywide. Not a single citizen was in doubt as to where Table Mountain was and what it meant to them, yet only a select few knew where or what the Cape Peninsula was. The name of the park was formally changed to Table Mountain National Park in 2004.

Some 34,000 children visit the park each year, through the park outreach programme and its CoCT-funded Table Mountain bus, *iBhasi Yentaba Yetafile*, and the Pick'n'Pay-funded Kids in Parks programme, through which 500 children have spent a weekend in the park, developing a taste for nature and desire to return for more.

The role of communication in building awareness of conservation debates cannot be overemphasized given that the park belongs to those over whom the mountain watches, and the debates over every management action are passionate indeed. Park management therefore takes care to communicate how and why management activities are undertaken – in a city of four million people, the impression that citizens have of park management influences many peoples' attitudes to conservation and their appreciation of the national parks network as a whole.

Financial sustainability

Early park management often expressed the goal of sustaining a positive cash flow, independent of donor funding. The test came in mid-2003, five years after the creation of the park, when all three donor-funded programmes (the GEF funding, the *Fonds Français pour l'Environnement Mondial* funding and the CoCT funding for staff transferred to the park) came to an end.

Both the CoCT and national government were persuaded to invest in the park in order to promote affordable access for citizens, produce tourism infrastructure to grow the local and national economies and contribute to poverty alleviation in communities surrounding the park. The CoCT agreed on a four-year programme of investment and the customized Cape Town WildCard was reintroduced to provide citizens with affordable park entry. Day tourism by locals was encouraged by opening previously closed park picnic and barbeque sites, with improved facilities and an alcohol ban intended to help manage hooliganism. Minor access points and signage were upgraded to welcome visitors and volunteers were mobilized by a dedicated volunteer manager.

The national Department of Environment and Tourism (DEAT) agreed on a three-year programme of funding from the Expanded Public Works Programme, employing 430 people to produce some 250km of stone and timber paths and to construct overnight tented and indoor accommodation for the Hoerikwaggo Trails. Person-day and training-day targets have been met and the department is keen to sustain the momentum of a successful programme with a further phase of funding. SANParks also secured ongoing funds from the Working for Water (WfW) programme, which has helped to follow up on the GEF-funded alien clearing programme. Through both the DEAT and WfW investments, TMNP now sustains the employment of 630 contract workers over and above the 135 permanent staff of the park.

In order to encourage continued investment in the park, park management ensures that it has the internal capacity to manage such projects, and therefore gains a reputation for excellent outcome delivery. Regular reporting and audits of delivery are emphasized so that investors can measure returns on their investments and consider further investment. For instance, the Graduate School of Business of the University of Cape Town was commissioned to survey the impact of the first five years of the park. The study found that the park secured R178 million (US$29.7 million) of direct investment into the city economy in the first five years, as well as a measured economic impact of a further R377 million (US$62.8 million) into the macro-economy (Standish and Boting, 2006). Progress against this baseline is measured bi-annually in order to demonstrate the return on local and national investment. These rates of return have increased by a minimum of 30 per cent in the last three years (Standish and Boting, 2006). Returns from the restoration of natural capital and from building social capital are more difficult to measure, but equally important. These are now being measured over time using a balanced scorecard, which measures achievement of targets such that internal processes are geared to meeting external customer requirements and the achievement of the biodiversity mandate and required financial sustainability.

The Table Mountain Aerial Cableway Company and the Cape Point funicular and restaurant operations have provided an important foundation on which park management is successfully building to ensure the receipt of its due percentage of turnover. Some distance behind these in terms of income is the Boulders penguin colony, which provides a steady and growing income source. In addition to growing merchandizing and retail opportunities, there is slow but progressive development of the Koeelbaai boutique hotel and Roundhouse restaurant concessions that are due to come on line in 2008.

Future tourism income is expected from the suite of Hoerikwaggo Trails, which enable tourists to hike the 'wilderness line' from Cape Point to Table Mountain, providing opportunities for all market levels – from budget to upmarket. The trails have brought increased security to users of the mountain through the presence of guides and the greater number of hikers on the mountain, and have become magnets for investment by the CoCT and DEAT, which have both seen their requirements for improved tourism infrastructure realized. In fact, the trails were the main factor that leveraged the DEAT Expanded Public Works Programme funding of the rebuilding of 250km of footpaths, as described above.

Conclusions

The TMNP is now recognized in virtually all households in Cape Town, and the process of its creation has increased debate and understanding among city residents of the conservation challenges of their heritage of mountains and sea. Opinions are increasingly informed on subjects as diverse as tahr removal, baboon management, pine trees and marine resource management. TMNP is opening access for citizens to their mountain and the pride in and awareness of it are at an all-time high.

The solution to the quandary of how much surplus to reinvest in the park and how much to contribute to SANParks' national biodiversity project may well lie in the TMNP being the economic engine of the envisaged Cape Floral Kingdom conservation business unit. This unit may result from a possible amalgamation of the mandates of the provincial conservation agency with SANParks' interests in the Western Cape, including TMNP, or may be established solely for the conservation of the entire Cape Floral Kingdom as a World Heritage Site. In the interim, the continued and growing biodiversity business viability of the TMNP allows it to contribute sustainable benefits to the citizens of Cape Town and the users of the national park system as a whole.

TMNP's goal of financial sustainability is no longer to be independent of donor funding. The goal of the park is to deliver benefits to the City of Cape Town, the surrounding region and nation in terms of positive triple bottom line returns on investment. This demonstration of the value of conservation to the surrounding communities is what will secure the future of conservation. It could be argued that, for as long as national and local governments achieve real returns and benefits in terms of poverty alleviation and improved tourism infrastructure, then public sector investment opportunities can be justified and should be realized. The key risk to this strategy is crime against visitors to the park and the perceptions of the tourist market regarding the safety of visitors. Additional staff and partnerships with CoCT, provincial and national government look set to strengthen security.

Solid conservation management systems are steadily reducing the running costs of the park and increasing the value of biodiversity. While the park has a comprehensive system of measurement of the state of biodiversity, this area of measurement is one that will need ongoing research and development. The improvement in the social capital of surrounding communities is in part measured by employment, with some 830 households currently earning an income from the park, and by the provision of 200 learnerships to students who could not otherwise afford to continue their education. The intangible gains due to the spiritual and recreational upliftment associated with walking, swimming and enjoying activities in the park by young and old is impossible to compute. The value of a hike that links the city to the mountain by foot and traverses a guided exploration of cultural heritage and history is hard to quantify, and this is what makes the national parks system so special and inspiring. This, ultimately, is what will make the TMNP realize its goal of being recognized as the world's premier urban national park.

Notes

1   The three local authorities, Cape Town City Council, South Peninsula Municipality and the Cape Metropolitan Council, no longer exist, having been subsumed into the City of Cape Town metropolitan-wide authority in a subsequent round of local government reorganization.
2   There was a strongly held belief within SANParks at that time that they were 'second only to God' within national parks and should be free to operate in whatever way they saw fit, legal agreements notwithstanding. In Cape Town, the Table Mountain National Park (TMNP) team rejected this and actively cooperated with partner organizations to ensure that the deals made were honoured in both letter and spirit.
3   A deal was done between SANParks and the provincial government – SANParks agreed to withdraw its point relating to incompetent delegation in return for the provincial government's wholehearted and public support for the initiative to create the park. The provincial government was desperate for this concession because, had the point been won in court, the potential for chaos in hundreds of thousands of property transactions over many years was alarming.
4   It is worth noting that affordable access can become a serious red herring. The vast majority of visitors at the popular pay points – Cape Point and Boulders – are either overseas visitors or domestic tourists from beyond the Cape metropolitan

area, who are quite price insensitive. To bow to the pressure for gate charges that locals can afford would be to forgo an important income-generating opportunity for the park – a recommendation to overcome this is to have a tiered pricing structure. Price-insensitive domestic and foreign tourists can be charged what they are prepared pay, while affordable access for locals can be provided by other means, for example a season ticket allowing unlimited entry for a set period at a very low price.

References

CPNP (2001) 'Cape Peninsula National Park: Visitor and user survey, 2000. Phases 1 and 2', unpublished report for SANParks

Minter, L., Channing, A. and Harrison, J. (2004) *'Heleophryne rosei'*, in IUCN, *2006 IUCN Red List of Threatened Species*, IUCN, Gland

SANParks (2006) 'Table Mountain National Park: Park management plan', SANParks, Pretoria

Standish, B. and Boting, A. (2006) 'Toward triple bottom line reporting: The environmental, social and economic contribution of Table Mountain National Park. Financial Year 2005/2006', report prepared for SANParks

WWF (2000) 'Cape Action Plan for the Environment: A biodiversity strategy and action plan for the Cape Floral Kingdom. Summary report', WWF-SA, Stellenbosch

21

A Network of Marine Protected Areas in Mozambique

Helena Motta

Considering the diversity of habitats along the Mozambican coastline and the economic importance of littoral waters, remarkably few coastal areas are under protection – only 4.4 per cent of the continental shelf (East African Marine Ecoregion, 2004). Although Inhaca and Portuguese Island Reserves (IPR) were established in 1965 and Bazaruto National Park in 1971, it was only after 2001 that a substantial area of marine environment came under effective protection. The little that had been under protection prior to that had not been administered and regulations were not enforced. For example, in Bazaruto, there was no permanent park warden before 1989 and the management plan was drafted only in 1990 (Dutton and Zolho, 1990). Very little marine environment is managed in the older coastal reserves, such as Marromeu, Pomene and Maputo that protect terrestrial endangered species, as the protected area usually ends at the high tide mark. One exception is the Maputo Special Reserve, which protects marine turtle nesting sites (see Figure 21.1) (Louro et al, 2006). However, with the recent enlargement of the entirely marine Bazaruto Archipelago National Park (BANP) and the creation of Quirimbas National Park (QNP) – one third of which is marine – the total marine protected area increased from 620km^2 to almost 3000km^2 (MITUR, 2002, 2004).[1]

Legal and institutional framework of marine protected areas

The country has a number of pieces of legislation and policy for the management of natural resources, which have a direct impact on conservation areas: the environmental law; land, forest and wildlife legislation; the strategy and action plan for the conservation of biodiversity; fisheries, mineral and oil regulations, among others.

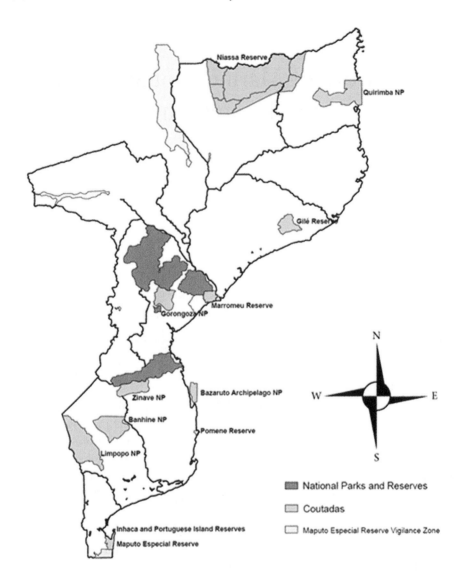

Source: WWF (2008)

Figure 21.1 *Mozambique's protected areas*

The responsibility for protected areas (national parks, reserves and *coutadas* or hunting concessions) was moved from the Ministry of Agriculture and Rural Development in 2002 to the newly formed Ministry of Tourism, under the National Directorate of Conservation Areas with the Purpose of Tourism (*Direcção Nacional das Áreas de Conservação*, DNAC). Forest reserves and wildlife outside protected areas remained under the jurisdiction of the Ministry of Agriculture.

Two major issues of credibility remain unresolved: the first is that responsibility for protected areas and wildlife is now split under two different ministries; the second relates to the perceptions that tourism is the sole purpose of conservation, thus excluding other economic sectors such as fisheries and forestry from participating in conservation. Meanwhile, legislation on conservation areas has been mainstreamed into the broader tourism legislation – the tourism policy and its implementation strategy, tourism law and strategic plan for the development of tourism (2004–2013) and the 'Principles for the Administration of Protected Areas' are legal instruments approved so far.

Marine protected areas (MPAs) in Mozambique can be created under the forest and wildlife law (Law 11/99) or under the general maritime fishery regulation (Decree 43/2003, Article 112), though to date, all parks have been created under the former. According to the forest and wildlife law, national parks are:

> *zones of total protection, clearly defined, with the aim of dissemination, conservation, protection and management of wildlife and vegetation, as well as for the protection of landscapes or geological formation of particular scientific, cultural and aesthetic value in the interest of public recreation, and representative of the national patrimony.* (Article 10)

The same article states that strictly no hunting, logging, agriculture, mining or livestock are permitted in national parks, nor is any modification of landscape or vegetation, pollution or introduction of exotic species. This means that the residents of marine parks cannot undertake subsistence fishing legally (or any other activities requiring consumptive use of resources), which takes no account of the needs of residents of protected areas.

According to the general land law (19/97), when a national park is being developed, communities have to be consulted and accept the park proposal, as well as the proposed zoning plan (there is no distinction between terrestrial and marine parks). However, the law is somewhat contradictory and creates confusion as Article 9 of the same law states that 'in total or partially protected areas ... no rights of land use and benefit can be acquired, except if a special licence is issued for that effect'. This means that as soon as a protected area is officially declared, community rights over land and other natural resources change significantly (Salomão, 2003). In practice, the impact of the changes to resource use rights are still being determined and the country is now in the process of redefining its conservation policy under the leadership of the Ministry for the Coordination of Environmental Affairs (*Ministério para a Coordenação da Acção Ambiental*, MICOA), with the participation of several other ministries and civil society.

According to the categories of protected areas of the IUCN, Mozambique's protected areas should meet the criteria of national parks (category II) – protected areas managed mainly for ecosystem protection and recreation. However, in reality, marine protected areas in Mozambique would more accurately correspond to a habitat or species management area, which is managed mainly for conservation through management intervention (category IV) (IUCN, 1990). This is because MPAs in Mozambique have been created with communities established within their boundaries and using their natural resources – BANP has approximately 3500 people living within its borders; QNP has around 22,000 people living on the coast and islands (MITUR, 2002, 2004); and though the IPR are only 20km^2, they are part of a larger system of islands with around 5000 residents (CNP, 1990; Kalk, 1995; Muacanhia, 2000). Part of the reason for the relatively high population levels within protected areas was that following Independence and during the civil war that hit the country, people moved into protected areas as safe havens.

Aims behind the creation of MPAs

In general, marine parks in Mozambique have been created in order to conserve the diversity, abundance and ecological integrity of biological resources, in order to be enjoyed and used productively by present and future generations (MITUR, 2004). In the case of IPR, the MPA serves to promote research and education and to improve public awareness, understanding and enjoyment of natural resources (Muacanhia, 2000). An additional purpose of some MPAs (for example BANP) is the improvement of social well-being through sustainable management (MITUR, 2002).

However, there is very little information available regarding the proclamation of some MPAs, for example the so-called Vilankulo Sanctuary, an area of around 25,000ha, which was declared by the Council of Ministers as a 'Total Protected Zone of Cabo São Sebastião', and there is no indication of the protected area category, there seems to be no management plan approved to date and the purpose of a 'total protected zone' is unclear.

Representative coverage of ecosystems

Mozambique possesses a coastline that extends 2770km and is characterized by a variety of habitats including delta estuaries, sandy beaches, mangroves, islands and coral reef systems. It can be divided into three regions – north, central and south. The northern coast extends 770km from the Rovuma River on the north (10°20'S) to Pebane in the south (17°20'S), and is essentially a coral coast with an almost continuous fringing reef. The central coast from Pebane to Bazaruto Island (21°10'S) spans approximately 950km and is classified as a swamp coast, where coral formation is limited. The southern coast stretches 850km from Bazaruto Island southward to Ponta do Ouro (26°50'S). This region is characterized by high parabolic dunes, north-trending capes, barrier lakes and patchy and deep rocky reefs with scattered corals (Rodrigues et al, 2000).

Not all systems are represented in the current network of MPAs. The northern coral fringing reef system is still very poorly represented, with only around 100km of coast inside the new QNP and there is little protection of the central coast of swamps. Of the southern coast of barrier lakes, coastal dunes and deep rocky reef communities, the BANP and IPR are the only protection so far. Discussions are under way to extend the Maputo Special Reserve to three miles offshore and a further one mile on the bay side. The Primeiras and Segundas Archipelagos are also being considered for protection, with support from local fisheries associations. Further proposals for new protected areas under consideration include an area close to the Rovuma River on the border with Tanzania, a reserve in the Mussoril area and the marine area between Inhaca and Ponta do Ouro. If these sites are eventually proclaimed, the area under protection will reach almost 10 per cent of territorial waters, and Mozambique will then be much closer to having a representative system of marine protected areas.

Are they legally defined?

Of the MPAs, the legal status and boundaries are clear for BANP and QNP. The management plans for both parks establish fully protected areas within their boundaries, according to zoning plans. Community consultations have been undertaken in both parks (MITUR, 2002, 2004), during which agreements were reached on how the parks, tour operators and investors would interact with local communities. Article 24 of the land law also mandates community management of their own resources. For these parks, communities must be involved in park management programmes as full partners.

However, the situation is rather different for the IPR, as they are not part of the national system of protected areas. The IPR are mostly used for research purposes and are managed by the Research Biology Station of Eduardo Mondlane University (Pereira, pers. comm.). The legal framework of the IPR is inadequate for the management of Inhaca Archipelago protected sites; there is no clear framework regarding management, law enforcement (for example fines) or research on the impact or the sustainability of the MPA (Muacanhia, 2000). However, a management plan is presently being drafted in collaboration with local communities and other stakeholders.

MPA performance

Park management policies

Management plans are not sufficient to regulate all the management issues of MPAs in Mozambique, and currently protected areas lack approved internal regulations, statutes and clear mandates for park wardens. Management has been made more difficult for park managers as their mandate to control activities within the park is incomplete, for example, a park warden is not entitled to interfere with the tourism developments inside the area under his responsibility. This caused difficulties in BANP in 2003 when relationships between key stakeholders started to deteriorate, primarily the result of development activities by existing hotel operators that were in apparent contravention of the limitations and procedures as stipulated within the management plan (Cunliffe et al, 2005). Things improved in BANP and QNP in 2006 with the appointment of chief law enforcement officers.

Meeting conservation objectives

There are mixed results concerning the status of marine habitats and species within MPAs in Mozambique. Results are available from several different surveys and though there is a lack of systematic and quality data collected on a regular basis, some of these systems have recently been improved. Given the lack of systematic data analysis, it is possible to present data relating to biodiversity indicators, but it is difficult to determine the implications of these results.

BANP had a comprehensive monitoring system, under which rangers regularly collected information on 12 different biological and socio-economic items. A recent analysis of the data collected between 1993 and 2004 concluded that some of the data are of questionable quality and have various gaps (Videira and Louro, 2003). An improved monitoring system, the management-oriented monitoring system (MOMS) has been in place since 2005. This system is an adaptation of the event book system developed and successfully implemented by the CBNRM programme in Namibia. The system is very simple and appropriate for both the existing conditions in the parks and the education level of the rangers (illustrations are used where possible). It has not yet been possible to determine whether the system is a success due to the relatively short period of MOMS implementation.

Data collected on nesting sites for marine turtles in BANP is thought to be reliable and shows that between 1994 and 2004 (excluding 1995–1996), a total of 152 nests were counted; an average of 17 nests per year. Around 20 per cent of all nests are destroyed every year by erosion and human interference, although the latter has been reduced in recent years and in the 2003–2004 season, 7 of 19 registered and protected nests were destroyed by natural beach erosion in Bazaruto Island. However, aerial surveys in 1995, 1999 and 2001 estimated the number of turtles in the park area at 565, 357 and 321 respectively, which represents a decrease in turtle sightings (Videira, 2004).

Aerial survey counts of dugong are thought to be less reliable. While the surveys of 1995, 1999 and 2001 estimated the number of dugongs in the BANP as 25, 72 and 104 respectively (Cumming et al, 1995; Mackie, 1999, 2001), other authors have reported a decrease in the population of dugongs from 80 individuals in 1990 to 23 in 2003 (Dutton, 2004), while yet another estimated the population of dugong in the Bazaruto area at around 100 individuals (Guissamulo, 2004). Recent aerial surveys showed larger estimates at around 153 individuals in the BANP (Findlay et al, 2006). The differences seem to come from the use of different methods of analysis. Clearly, more frequent and larger sampling at different seasons and tides would give a much better picture. In a positive sign, the killing of dugongs and marine turtles in BANP has recently resulted in court convictions, fines and confiscation of gill nets, which are illegal inside the park (park warden, pers. comm.).

Coral reefs are monitored by a national programme initiated by MICOA in 1999. The programme monitors some sites inside marine protected areas (Inhaca, Bazaruto and Quirimbas) and some sites outside protected areas, using indicators such as coral cover, fish abundance, diversity and size structure. Results already indicate significant differences between protected and unprotected sites (Rodrigues et al, 1999; Motta et al, 2000; Pereira et al, 2003). Coral cover, mostly hard coral, and fish abundance, diversity and size at the monitoring stations in Inhaca Island's reefs and Bazaruto's Lighthouse reef, which are protected, showed good results.

Other species that are protected in MPAs include molluscs, holothurians, seahorses, reef fishes and elasmobranchs (rays and sharks) (MITUR, 2004). Holothurians are now fully protected in BANP and it is generally accepted that the *Holothuria scabra* fishery has declined or collapsed in most areas outside MPAs (Marshall et al, 2001; Masquine and Torres, 2006).

Some data are available regarding fisheries resources inside BANP; it would seem that fishing effort has been increasing in recent years, as demonstrated by the increasing number of fishing vessels inside the park (Cumming et al, 1995; Mackie, 1999, 2001; Reina, pers. comm.). Mozambique's Institute of Fisheries Research initiated a monitoring programme on the mainland coast opposite BANP (which covers catches inside the park), which suggests that in Inhassoro District the catch increased between 2002 and 2005, even with a decrease in beach seining – which is thought to be due to the closed season initiated by local fishermen. In contrast, line fishing has been increasing, as has the total catch, but catch per unit effort in the last year of monitoring decreased. On the Vilankulos coast, the beach seining catch also decreased between 2002 and 2005, while both catch and effort have increased in the line fishery (Masquine and Torres, 2006).

Illegal fishing vessels from outside the country are a constant threat to the marine environment, including protected areas, and BANP has been particularly badly affected. An agreement between the tourism ministry and the navy for joint patrolling of protected areas has improved the situation (park warden, pers. comm.).

QNP was created in 2002 and was zoned to include some multiple use areas and some fully protected areas. A monitoring programme started in the fully protected areas in 2004 to look at the impacts of these areas in terms of fish biomass, diversity and the spillover effect. Preliminary results show increased fish biomass and diversity inside the fully protected area and adjacent fishing areas, while the spillover effect needs more data to be conclusive (Costa, pers. comm.). The results of the most recent coral reef monitoring also show that corals are recovering at a very fast pace in the area (Pereira et al, 2003). QNP is also implementing the MOMS, as well as a fisheries monitoring system.

Economic impacts

The main commercial activity occurring in the MPAs is tourism, with well-established tourism developments in BANP, Inhaca Island and on the islands within QNP. These islands are international tourist destinations and the attractions available include game fishing, snorkelling and scuba diving (Rodrigues et al, 2000).

In BANP, an economic impact analysis conducted in 2001 showed that the tourism industry was generating substantial financial revenues. At that time, three lodges and one campsite were offering a total of 150 beds, and approximately 6400 tourists were visiting the archipelago each year, with at least as many tourists staying in the Vilankulos area and making day trips to the archipelago (Engdahl et al, 2001). Those tourists who stayed in the park generated about US$6.1 million in direct expenditure per year, of which US$4.4 million was prepaid and US$1.9 was estimated to be spent locally (on the islands) (Engdahl et al, 2001). The park is part of a larger tourist area that includes the mainland coasts of Inhassoro and Vilankulos districts with a total of 1800 beds, and it has been estimated that the total turnover from tourism in this larger area is around US$17 million per annum (SAL, 2006).

Prior to 2001, an informal arrangement had been made where some hotel operators collected fees from tourists and transferred 60 per cent of these to BANP and 40 per cent to local communities. Between 1997 and 2001, approximately US$21,180 was transferred to local communities and used for various purposes (Engdahl et al, 2001). The system suffered from several disruptions until new regulations concerning the collection of tourist entry fees were passed by the Government of Mozambique in 2003. According to these regulations, the Ministry of Tourism is now responsible for the collection of tourist entry fees, 20 per cent of which are allocated to communities and 80 per cent to the state, via a tourism development fund. Although a smaller percentage is now being given to the communities than was the case prior to the regulation, the system is now formal, which means that all hotels must comply and so more money has been collected. In 2005 at least 12,000 people paid the entry fee to visit the park and around US$25,000 was transferred to communities on the islands (park warden, pers. comm.).

In BANP, the tourism industry is comparatively well developed, and benefits should flow to park residents from these enterprises. In 2001, job opportunities for the islanders at the local lodges were very limited, with only 10 per cent of the 283 full-time jobs created by these lodges filled by local employees, though approximately 25 per cent of the population benefited directly from tourism in other ways (Engdahl et al, 2001). By 2003, it was estimated that the percentage of local islanders working at the lodges in the park was closer to 7 per cent, and that female island employees probably constituted about 1 per cent (Taylor, 2003). In 2005, there were around 500 people working in hotels in the park (park warden, pers. comm.) but the proportion of local employees had not changed much. This may be the result of the low education level of islanders compared to the job seekers coming from the mainland, mostly due to the fact that the schools in the park do not provide education beyond the fifth grade.

In QNP, more than 300 jobs have been created by the tourism industry and the park, while the entry fee collection system is still in the process of implementation (park warden, pers. comm.). It is not known how many of these jobs have been filled by local residents.

Further analysis is needed to assess which benefits to the local, regional and national economies are actually derived from MPAs. There is generally a lack of input by economists in determining the costs and benefits of biodiversity conservation, particularly relating to non-market costs and benefits.

Meeting social, educational and cultural objectives

As noted above, MPAs were created with people living inside them, and their goals state the need for communities to benefit from improved natural resource management and the development of livelihood alternatives, such as involvement in tourism operations.

One of the main subsistence activities of the islanders in Mozambique's MPAs is artisanal fishing, the success of which is often threatened by the activities of non-resident fishermen and industrial fisheries (though the latter are excluded from fishing legally in these areas). In QNP, local fishermen have a say about the number of migratory fishermen allowed to fish in Quirimbas National Park (MITUR, 2004), but in BANP, fishermen coming from the mainland have the same rights as local fishermen (and both groups are excluded from areas of total protection, such as coral reefs).

BANP residents have a generally good awareness about the park and its conservation policies, and understand the reasoning behind conserving resources in order to benefit financially from tourism (Taylor, 2003). Most people interviewed commented that the park worked together with the community and spoke of it in a positive light, though it was noted that the way people spoke about the park made it appear paternalistic, laying down firm rules, but also guiding and helping the communities as necessary (Taylor, 2003). With support from the lodges, two community associations were created, are registered, have their own bank accounts and are functional institutional structures – the Bazaruto Island's Tomba Yidhu Association and Benguerua Island's Khani Kwedho Association. The main objective in the creation of these associations was to guarantee the sustainable use of their share of tourism revenues, with their funds being used for development activities intended to benefit the communities (van Vugt-Chilaule, 2007). However, relations between communities and lodges in the park have been sometimes very tense as conflicts have arisen from the use of land and the disruption or lengthy delays in distributing tourist entry fees by the lodges (Ricardo, 2004; Cunliffe et al, 2005).

With around 500 jobs available to a population of 3500 people, BANP could have a much better local employment rate if a higher proportion of residents were employed. Additionally, the percentage of the entry fees that are allocated to the communities seems very small, not only because the 20 per cent that is collected (as established by law) is very small, but also because this percentage is only applied to park entry fees, not to all revenue generated in the park. On the positive side, BANP has been pioneering a partnership between communities, the private sector and the state, and the legal framework has been established to improve community benefits. Similar studies have not been carried out in Quirimbas, so it is not yet possible to determine the impact of the park on local livelihoods, or whether the park is meeting its social and economic objectives.

Meeting political objectives

The most challenging issue that the Mozambican government needs to address is poverty. In doing so, it has already indicated, through enacted policies and legislation, that it intends to follow the path of the sustainable use of natural resources.

In the province of Cabo Delgado, where the QNP is located, life expectancy is just 37.8 years, while the under-five mortality rate is 295 per 1000 (Gabinete do Governador da Província de Cabo Delgado, 2001; MITUR, 2004). These shocking indicators of poverty are compounded by the impacts of human–wildlife conflict and the erosion of traditional livelihoods, diseases affecting cashew and coconut cash crops, and from the exhaustion of fragile tropical soils due to overuse. A particular challenge to the creation of QNP was the fact that inshore fisheries accessed by artisanal fishermen were on the verge of collapse due to overfishing, which was mainly due to the exhaustion of fish stocks in neighbouring areas in Nampula and Tanzania and the resultant and recent influx of migratory fishermen into the area (MITUR, 2004). In the face of such difficulties, protected areas must, and can, demonstrate that improved ecological services and sustainable use of natural resources through tourism, forestry, fisheries, agriculture and others can contribute to human welfare.

Financial sustainability of MPAs

Mozambican MPAs remain very dependent on external funding for their operation. At present, BANP is receiving external funding of approximately US$157 per km^2 per year, including support to community projects (WWF, 2006). Support from the state budget has increased in recent months, to the level of US$48,000 per year (park warden, pers. comm.). Unfortunately, the decree (27/2003) establishing entry and user fees for protected areas has not been fully implemented so far, and only the entry fee is being collected by the park, 80 per cent of which still goes back to central government coffers.

Current donations to the QNP are estimated at around US$173 per km^2, including support to community projects (WWF, 2006). Considering that the park is a new one and that infrastructure is still being developed, this is not a lot of money. Central government is not contributing substantially to this park (park warden, pers. comm.), and the collection of entry and user fees does not yet occur across the whole of the park.

Mozambique is one of the poorest countries in the world, and one cannot expected the government to set aside the considerable resources necessary to maintain conservation areas in the face of urgent human needs. Tourism is an industry that can support conservation and that is the reason that the management of conservation areas is now the responsibility of the tourism ministry. However, at present, only a small proportion of tourism revenues generated inside protected areas are being retained for conservation, and tourism itself should not be regarded as the only resource supporting conservation as there will be high conservation value areas that are poor tourism destinations. Alternative mechanisms for the sustainable financing of protected areas need to be considered.

Conclusions

With the transfer of protected areas to the Ministry of Tourism, there seems to be an overall improvement in the management of marine protected areas, the creation of much larger areas has occurred in recent years, management plans have been approved and park wardens nominated. This has demonstrated the political will of the government. The marine parks were created with people living inside them, and mechanisms are in place to improve and implement co-management practices for the benefit of people and the environment, in a context of restrictive legislation. However, the larger MPAs in Mozambique receive the majority of their funds from external donors, and government-led mechanisms of financing and enhancing capacity inside protected areas have been implemented very slowly.

There are already signs of recovery and maintenance of some habitats and species within some MPAs, though there is a perception that some resources inside protected areas remain in decline. As monitoring programmes have only recently been improved, there is still a long way to go before conclusive data are available to confirm or deny these perceptions.

Looking at the challenges ahead for marine conservation, there are several issues that need to be addressed. Because protected areas are now the responsibility of DNAC, there is the risk that economic activities and stakeholders with important roles (other than those involved in tourism) may be excluded from the development and management of MPAs. Further, while it makes sense to think that tourism contributes to the sustainability of protected areas, this may not be applicable everywhere – not all biologically important marine areas have the same tourism potential, and the industry must be well managed to ensure that its activities do not degrade protected areas.

The existing network of protected areas is in a fragile position politically and economically. While the central government will not be capable of fully funding protected areas, it should recognize that protected areas contribute to the national economy. Protected areas need to demonstrate their direct value and that ecosystem services and species survival contribute to human welfare. In order to achieve this, economic valuations should be undertaken (recognizing their limitations), and there is an urgent need for the implementation of comprehensive socio-economic monitoring programmes.

The participation of communities in MPA management, and ensuring that benefits flow to these communities, is still new and very much at an experimental stage. Successes and failures are still being experienced, and the full benefits of participation are yet to be fulfilled. More tangible benefits of protected areas to communities living inside and around them are needed. Perhaps most important of all, legislation needs to reflect the social context of conservation areas in Mozambique. That the country is going through a process of redefining its conservation policy through open discussion is very positive.

Acknowledgements

I gratefully acknowledge the assistance and insights from discussions with Russell Taylor, Marcos Pereira and Alda Salomão, as well as information provided by Rafael Funzana, César dos Santos and António Reina.

Notes

1   A marine protected area is defined as any area of inter- or sub-tidal terrain, with its over-lying water and associated flora, fauna, historical and cultural features, that is protected by legal or other effective means (Kelleher and Kenchington, 1992).

References

CNP (1990) 'Plano de Desenvolvimento Integrado da Ilha de Inhaca', report prepared for Comissão Nacional do Plano/UNDP, Maputo

Cumming, D. H. M., Mackie, C., Dutton, P. and Magane, S. (1995) 'Aerial census of dugongs, dolphins and turtles in the proposed greater Bazaruto National Park, Mozambique: April 1995', report prepared for WWF-SARPO, Harare

Cunliffe, R., Taylor, R., Motta, H., Borner, M. and Martinussen, A. (2005) 'Mid-term review of Bazaruto Multiple Resources Uses Project, 2001–2005 and Bazaruto CBNRM Project, 2003–2005', internal report for WWF Mozambique, Maputo

Dutton, P. and Zolho, R. (1990) 'A conservation master plan for the sustainable development of the Bazaruto Archipelago', report prepared for WWF/EWT/ORI, Maputo

Dutton, T. P. (2004) 'Dugong (*Dugong dugon*) population trends in the Bazaruto Archipelago National Park, Mozambique, 1990–2003', *Sirenews*, vol 41, pp12–14

East African Marine Ecoregion (2004) *Towards the Establishment of an Ecologically Representative Network of Marine Protected Areas in Kenya, Mozambique and Tanzania*, WWF, Dar-es-Salaam

Engdahl, S., Bjerner, M. and Enosse, C. (2001) 'Review of local community participation and the contribution of the tourism industry: The case of Bazaruto Archipelago, Mozambique', report for UNESCO/WWF, Maputo

Findlay, K., Guissamulo, A. and Bickerton, I. (2006) 'Specialist study: Marine ecology', in Environmental Resources Management Southern Africa Pty (Ltd) *Environment Impact Assessment of Offshore Exploration in Block 16 and 19, Inhambane and Sofala Provinces, Mozambique*', report for SASOL/ENH, Maputo

Gabinete do Governador da Província de Cabo Delgado (2001) 'Plano Estratégico do Desenvolvimento 2001–2005', paper prepared for the Second Conference on the Development of Cabo Delgado Province, Pemba, Mozambique

Guissamulo, A. T. (2004) 'Mozambique', in East African Marine Ecoregion *The Eastern African Marine Ecoregion: Towards a Western Indian Ocean Dugong Conservation Strategy. The Status of Dugongs in the Western Indian Ocean and Priority Conservation Actions*, WWF/UNEP, Nairobi

IUCN (1990) *Guidelines for Protected Area Management Categories*, IUCN, Gland

Kalk, M. (1995) *A Natural History of Inhaca Island, Mozambique*, 3rd edn,

Witwatersrand University Press, Johannesburg
Kelleher, G. and Kenchington, R. (1992) *Guidelines for Establishing Marine Protected Areas*, IUCN, Gland
Louro, C. M. M., Pereira, M. A. M. and Costa, A. C. D. (2006) 'Relatório sobre o estado de conservação das tartarugas marinhas em Moçambique', Grupo de Trabalho Tartarugas Marinhas de Moçambique, Maputo
Mackie, C. (1999) 'Aerial census of dugongs, dolphins and turtles in the proposed greater Bazaruto National Park, Mozambique: April, 1999', report prepared for WWF-SARPO, Harare
Mackie, C. (2001) 'Aerial census of dugongs, dolphins and turtles in the Bazaruto National Park, Mozambique: May 2001', report prepared for WWF-SARPO, Harare
Marshall, N., Milledge, S. A. H. and Afonso, P. S. (2001) *Stormy Seas for Marine Invertebrates: Trade in Sea Cucumbers, Seashells and Lobsters in Kenya, Tanzania and Mozambique*, TRAFFIC East/Southern Africa, Nairobi
Masquine, Z. A. and Torres, R. A. (2006) 'Specialist study: Fisheries', in Environmental Resources Management Southern Africa Pty (Ltd) *Environment Impact Assessment of Offshore Exploration in Block 16 and 19, Inhambane and Sofala Provinces, Mozambique*', report for SASOL/ENH, Maputo
MITUR (2002) 'Plano de Maneio do Parque Nacional do Arquipélago do Bazaruto 2002–2006', Ministério do Turismo, Maputo
MITUR (2004) 'Plano de Maneio do Parque Nacional das Quirimbas 2004–2008', Ministério do Turismo, Maputo
Motta, H., Pereira, M. A. M., Gonçalves, M., Ridgway, T. and Schleyer, M. (2000) 'Coral reef monitoring in Mozambique. II: 2000 Report', report prepared for MICOA/CORDIO/ORI/WWF, Maputo
Muacanhia, T. (2000) 'Marine Protected Areas of Inhaca and Portuguese Islands', paper prepared for the International Coral Reef Action Network, Maputo
Pereira, M. A. M., Motta, H., Videira, E. and Schleyer, M. (2003) 'Coral reef monitoring in Mozambique. III: 2002 report', report prepared for MICOA/CORDIO/ORI/WWF, Maputo
Ricardo, G. (2004) 'Sustainable tourism development: A case study of Bazaruto Island in Inhambane, Mozambique', MA thesis, University of Western Cape, Cape Town
Rodrigues, M. J., Motta, H., Pereira, A. M., Gonçalves, M., Carvalho, M. and Schleyer, M. (1999) 'Reef monitoring in Mozambique. I: The monitoring programme and 1999 report', unpublished report
Rodrigues, M. J., Motta, H., Whittington, M. W. and Schleyer, M. (2000) 'Coral reefs of Mozambique', in McClanahan, T., Sheppard, C. R. C. and Obura, D. O. (eds) *Coral Reefs of the Indian Ocean: Their Ecology and Conservation*, Oxford University Press, Oxford
SAL/Consultoria em Desenvolvimento Social (2006) 'Specialist study: Socio-economy', in Environmental Resources Management Southern Africa Pty (Ltd) *Environment Impact Assessment of Offshore Exploration in Block 16 and 19, Inhambane and Sofala Provinces. Mozambique*', report for SASOL/ENH, Maputo
Salomão, A. (2003) 'Reflexões sobre os direitos das comunidades nas áreas de conservação', unpublished report for WWF Mozambique, Maputo
Taylor, J. (2003) 'A brief social overview of the island communities in Bazaruto Archipelago National Park', report prepared for WWF Mozambique, Maputo
Videira, E. (2004) 'Parque Nacional do Arquipélago do Bazaruto: Análise das monitorias dos ninhos das tartarugas marinhas, 1999–2003', report prepared for WWF/PNAB/FNP, Maputo

Videira, E. J. S. and Louro, C. M. M. (2003) 'Análise dos estudos feitos no Parque Nacional do Arquipélago do Bazaruto', unpublished report for BICO/FNP/WWF, Maputo

van Vugt-Chilaule, A. (2007) 'Análise da capacidade organizacional e do funcionamento das Associações *Khani Kwedho* e *Thomba Yedyo*', draft report prepared for WWF/FNP, Maputo

WWF (2006) 'WWF Mozambique Country Coordination Office: Annual report to the Government of Mozambique', unpublished report, Maputo

WWF (2008) 'Mozambique's protected areas', WWF Mozambique, Maputo

22

Towards Transformation: Contractual National Parks in South Africa

David Grossman and Phillipa Holden

South Africa's transition from a minority government to a democratic order explicitly required the transformation of all sectors of society; parks and other conserved areas were not excluded from this process. In the past, the majority of South Africans were effectively prevented from enjoying the benefits of formal conservation areas, often bearing the costs associated with their removal and exclusion from parks. However, the 'fortress' or 'fence and fines' approach became politically and socially unacceptable, and there were calls for more innovative means of ensuring that the dual goals of biodiversity conservation and social justice were met.

Various models aimed at achieving these goals have emerged, ranging from the passive where a proportion of park entrance fees are paid to neighbouring communities (Hughes, 2002), through more active involvement of neighbouring communities, for example on local management boards, to emerging models where communities own and manage conservation areas.

Co-management of parks in situations where local people have strong historical linkages to the land is seen as a means of achieving biodiversity conservation, ecological sustainability, social equity and economic benefit. Fortuitously, South Africa's national parks legislation provides for the establishment of contractual national parks, whereby landowners enter into a contract with the relevant government minister and the area is formally proclaimed as a contract park, with the owners retaining title as well as negotiated rights. But how easy is it to achieve an equitable, mutually satisfying relationship?

In what follows, the South African experience with three contractual parks involving local indigenous people is described. Tentative 'lessons' are briefly discussed, bearing in mind that these are emerging and dynamic models with relatively short histories, and any conclusions drawn today will quite possibly be found wanting as events unfold.

The emergence of contract parks

During the build-up to South Africa's transition from a racial minority government to a democratic order, the conservation establishment was rattled by a provocative paper in which Derek Hanekom and Louis Liebenberg of the African National Congress (Hanekom and Liebenberg, 1993) argued that there was merit in considering the use of national parks for livestock grazing. This, they proposed, would make parks more relevant to people whose experience of them was largely one of conflict. In many cases people had been removed from parks, losing ancestral lands and livelihoods in the process.

The paper evoked an immediate response from the conservation establishment, with opinions justifying parks ranging from their being 'in the national interest', to cigarette-box type calculations intended to demonstrate the economic benefits of park activities versus those of livestock grazing. The debate, mirroring that in other parts of the world, focused attention on park–neighbour issues and it was generally accepted that, in line with the transformation of society and its structures, the conservation establishment would need to change in order to reflect the new dispensation and to become relevant to the majority.

Several steps were taken in this regard. At a national level, the National Parks Board (NPB) was restructured to reflect the demographics of society; new senior personnel were appointed from historically disadvantaged sectors of society; and the name was changed to South African National Parks (SANParks), with the motto '*Xamina Xawena*' ('for me and for you' in the XiTsonga language). In order to explicitly address park–people issues, a Social Ecology Unit was established (now referred to as 'People and Conservation').

The land restitution initiative, aimed at redressing the dispossession of land during the apartheid era added impetus to the transformation process, and SANParks' official approach to land claims and land restitution is that they do not necessarily jeopardize the conservation of biodiversity (Social Ecology and SANParks, 2000). Land claims made include those in the Kruger National Park by the Makuleke clan and in the Kalahari Gemsbok National Park (now Kgalagadi Transfrontier Park) by the ‡Khomani San (Bushmen as they call themselves) and Mier communities.

These claims were recognized as valid and were not contested by the state. The Makuleke people received full ownership and title deeds to some 25,000ha of their former land in the far north of Kruger National Park, and the ‡Khomani San and Mier communities each received 25,000ha in the extreme south of the Kgalagadi Transfrontier Park. In all cases, the claimants decided to maintain their land under conservation and entered into contractual agreements with the Minister of Environmental Affairs and Tourism to establish a formal contractual national park, legally registered as such in terms of the National Parks Act of 1976, as amended.

The biodiversity-rich mountain desert Richtersveld National Park, owned by the Nama-speaking people who entered into a contractual agreement with the state in 1991, provides a further example of a contractual park, in this case established shortly before the democratic order.

The contract parks

Richtersveld National Park

The Richtersveld National Park (RNP) covers an area of some 162,000ha south of the Orange River along the border with Namibia, in the arid north-west corner of South Africa. Essentially a mountain desert region, rainfall varies from about 15mm to around 300mm per annum. Extremes of temperature, ranging from below 0°C to more than 40°C, coupled with aridity have given rise to high biodiversity, particularly floral diversity, with almost 50 per cent of plant species endemic to this region (van der Walt, 1991).

Approximately 7000 people live in four villages in the Richtersveld area; mainly descendants of Nama-speaking Khoe pastoralists. The people are largely dependent on local jobs in the diamond mining industry or on small stock farming; highly mobile flocks of livestock have been grazed within the current park boundaries for centuries. Mining and livestock grazing (as stock numbers increased) posed real and perceived threats, and formal proclamation was mooted during the 1970s, as the natural richness of the area became better known to conservationists from outside the area.

Formal negotiations between the NPB (as it was then) and the local authority (the Northern Richtersveld Management Board – a structure of the so-called 'coloured parliament' under the then prevailing apartheid tri-cameral parliamentary system) gained momentum during the late 1980s. However, the people of the Richtersveld opposed the negotiations and the impending procla-mation as a national park on the grounds that they had been excluded from the process, and they were granted an interdict preventing proclamation by the High Court in 1989 (Surplus Peoples Project, 1995). Negotiations recom-menced after the court ruling, a contract was eventually signed and the area proclaimed as a national park in 1991.

In terms of the contractual agreement, the NPB agreed to pay an escalat-ing lease fee of R0.50 (US$0.18) per ha per annum to the Richtersveld Trust (to be distributed annually to fund local development projects). Local people were to be treated preferentially when employment opportunities became available, though the NPB appointed a park manager and other personnel. The NPB also undertook to establish an indigenous plant nursery with profits accruing to the Trust. An upper limit was set for the number of livestock permitted to graze within the boundaries of the park. A Management Planning Committee or *Bestuursplankomitee* (in Afrikaans, BPK), comprising one representative from each of the four villages, one park farmer representative and four members of the NPB was to be constituted. This committee was charged with drawing up a management plan and with the initial park management decisions.

The first draft of the management plan was prepared by NPB personnel and tabled in 1995, but failed to gain acceptance and was never ratified by the BPK. A stalemate ensued, with a succession of park managers failing to achieve consensus with community representatives on the BPK. Furthermore, community support for the park dwindled as very few jobs and secondary economic activities emerged, and the nursery was not successful. Failure by members of the BPK to report back to their constituencies and an abundance of rumours further weakened efforts at joint management, with the NPB forced to go it alone. In 2001, SANParks agreed to the appointment of independent experts to participatively prepare a second version of the management plan, supported by donor funding. This version was eventually ratified by the BPK in 2002.

The second edition of the management plan clearly describes the powers and functions of a new structure to replace the BPK – a Joint Management Committee (JMC) – designed to give effect to real co-management as well as to ensure the accountability of the community representatives. It also clearly allocates tourism concession rights to the community and day-to-day conservation management to SANParks, under the control of the JMC.

Effective co-management remains difficult and despite attempts to facilitate effective implementation of the management plan, in 2006 the Richtersveld members tabled notice of intent to renegotiate the 1991 contractual agreement. It remains to be seen whether the situation will improve and whether effective co-management will in fact be achieved, along with a more significant flow of benefits to the greater community.

A number of factors impacting on the success or otherwise of this particular case have emerged to date. The rapidly changing macro-political environment between signing of the contract in 1991 and the emergence of a democratic order in 1994 has influenced people's perceptions of their rights – people have become less inclined to be merely co-opted and have begun demanding full rights, challenging the current agreement. SANParks, with its skill levels and resources, has remained the dominant partner to date and has made only limited attempts to build the capacity of the BPK/JMC and the broader community, partly because of the historically 'rocky' relations between the parties.

The interests of the livestock owners have in the past been well represented on the BPK/JMC. There are still major disputes relating to the number of livestock in the park, with indications of localized heavy overgrazing and accelerated soil erosion, particularly in the vicinity of the (perennial) Orange

River, with obvious impacts on biodiversity in the area. It would also seem that this stock is owned by a small number of people and, as such, cannot be seen to be of benefit to the greater community.

In addition to localized degradation from overgrazing by stock, mining interests further complicate matters, with significant areas negatively impacted by unrehabilitated mine scars and access to operational mines still required.

Importantly, internal community dynamics and power relations play a major role, with SANParks increasingly frustrated at lack of community cohesion and at 'go-slow' tactics employed by members of the JMC when power relations are threatened. In most, if not all rural communities, existing power relations and economic activities (some nefarious) are threatened by external interventions aimed at equitable and democratic sharing of benefits. In extreme cases (not necessarily the current case), local power-brokers resist and actively undermine such interventions (a common theme in community–park relations).

A community conservancy has been established on land adjacent to the park, partly in frustration at the lack of benefits flowing to the community from the park, and has gained significant financial support from international donors, though there have been a series of disputes about how some of the funds should be allocated, hampering its implementation.

The recent emergence of youth with a voice seems likely to influence further developments along the road to co-management. Seemingly no longer satisfied with the way the older generation and the BPK went about matters, some youth are becoming increasingly vocal, and articulately so, with clear views on how the conservation estate should be developed and managed.

Although SANParks' top management has again expressed support for the effective co-management of the RNP, this will have to be backed up with action at ground level. Park management positions and appointments need to be reviewed, day-to-day management activities need to be assessed (it is understood that a great portion of resources are presently devoted to maintaining boreholes at stock posts), tourism development in the park needs to be prioritized (especially in relation to the opportunities offered by the formation of the proposed transfrontier conservation area with neighbouring Namibia) and the terms and provisions of the contract between SANParks and the community need to be reviewed and modernized.

The Makuleke Region of Kruger National Park

Historically, members of the Makuleke clan lived between the Luvuvhu and Limpopo Rivers along the borders with Mozambique and Zimbabwe. Livestock and rainfed agriculture along the floodplains, supplemented by wildlife products, formed the basis of the people's livelihoods. In 1969, some

3000 members of the clan were removed from their ancestral land in line with 'grand' apartheid policy and settled in the homeland of Gazankulu. The boundaries of the Kruger National Park were then extended to the Limpopo River, to include the Makuleke land. Justification for the inclusion of this land was based on the area's high biodiversity and conservation value. Located in the sub-tropical north-easternmost region of South Africa, the area contains species of flora and fauna not found elsewhere in South Africa. Makuleke people still wryly refer to the somewhat ironic situation whereby land they had lived on and used for over a hundred years, came to be regarded as 'pristine wilderness' by conservationists.

Shortly after the election of the democratic government in 1994, the Makuleke people initiated a land claim and entered into protracted negotiations with SANParks for the return of their land. An agreement was reached between the parties and ratified by the Land Claims Court in 1998. According to the settlement, ownership and title were returned to the Makuleke Communal Property Association (CPA) established in terms of the relevant land act, on behalf of the Makuleke people, who agreed to use the land in a manner compatible with conservation objectives and not to occupy it, nor use it for agriculture or mining. The Makuleke retained full rights to commercial (tourism) development, while SANParks retained the rights to future gate fees.

In terms of the contractual agreement between the Makuleke CPA and the Minister of Environmental Affairs and Tourism, the land was deproclaimed as part of a Schedule 1 national park and reproclaimed a contractual national park in terms of Section 2(b) of the National Parks Act. As owners of the land, the CPA has exclusive commercial and use rights of the Makuleke Region, while conservation management falls under the control of a Joint Management Board (JMB). The JMB consists of six members – three from each of the parties – and the chairmanship rotates annually. Decision making is by consensus and the agreement describes a series of deadlock-breaking mechanisms, including referral to the parties' principals, mediation and final arbitration. The parties have the right to invite technical advisers to the JMB meetings.

Throughout the land claim process, during the drafting of the master plan, and continuing to the present, the Makuleke have been supported by a range of NGOs and technical advisers, the Friends of Makuleke. In order to further build local capacity, several students were selected from within the community to undergo tertiary training in conservation, tourism and business management, supported by the one of the NGOs.

The JMB is responsible for the management of the Makuleke region in accordance with the 'Master Plan for the Conservation and Sustainable Development of the Makuleke Region of the KNP', as accepted by both parties. The first edition of the plan was produced under the guidance of the JMB and ratified by the parties in 2000. Conservation functions are performed by SANParks under the direction and control of the JMB. Commercial activities fall under the jurisdiction of the Makuleke CPA, all activities of an income-generating nature are to the benefit of the CPA (though gate entrance fees accrue to SANParks), and all commercial activities must be in line with the provisions of the master plan. A decision of the CPA is deemed a decision of the JMB when it is tabled at the JMB meeting, immediately following the CPA decision (Reid et al, 2004).

The Makuleke agreement has been widely regarded as an innovative and fair model for co-management. The rights and responsibilities of the parties are clear and unambiguously spelled out in the agreement as well as the master plan. Conservation principles are entrenched and the rights to use the resources sustainably and to develop tourism facilities are also clearly stated. The JMB met on a regular basis soon after constitution, and the early agendas reflect the Makuleke intent to develop the area for tourism as well as to derive income from trophy hunting. Early meetings also demonstrate the Makuleke delegates' full use of technical advisers from the Friends of Makuleke. Over time these advisers featured less and less as Makuleke confidence and capacity grew (Reid, 2001a).

Although JMB meetings were generally constructive, tensions soon emerged (Steenkamp and Grossman, 2001). The first major dispute arose over the Makuleke intention to market the hunting of two buffalo and two elephant. Although tabled and agreed to by the JMB, and in line with the provisions of the master plan, the media soon heard of the intention and a veritable furore erupted, with animal rights lobbyists and the concerned public expressing outrage. SANParks personnel, including the official spokesperson, were reported in the media as stating that the hunt was illegal, as hunting was not allowed in a national park. The Makuleke stood by their rights, supported by the Chief Executive of SANParks as well as conservation NGOs such as the Endangered Wildlife Trust. The hunt proceeded and realized more than R500,000 (US$47,500) for the Makuleke CPA, with each household in the community receiving a share of this; the rest going to much-needed community development projects. Over the next two years, SANParks staff again objected to proposed hunts although the numbers to be hunted were well within sustainable limits. The Chief Executive of SANParks again intervened and hunts went ahead, generating close to R1 million (US$142,500).

The CPA also called for proposals to develop and manage a 24-bed lodge in the Makuleke Region. SANParks participated in the bid evaluation process, the tender was awarded and the first lodge started trading, with income and employment accruing to Makuleke based on an agreed formula with the developer and operator. Again the Makuleke experienced opposition from some quarters within SANParks, with allegations that the lodge impinged on the wilderness quality of the adjoining area. However, the legally required environmental impact assessment process was duly followed, the development approved and SANParks was appointed to monitor compliance with the terms of development. A second lodge has since opened for trading.

Although SANParks are responsible for maintenance of conservation infrastructure, management tracks in the area initially received little or no maintenance; the boundary fence with Zimbabwe was wrecked but was not removed or replaced, providing snare material for poachers. SANParks repeatedly blamed the situation on a lack of funds, and the Makuleke placed SANParks on notice regarding track maintenance. Furthermore, and contrary to the spirit and letter of the agreement, there was no systematic transfer of skills or training provided by the park authority. Ad hoc training was provided, mainly in security matters, and Makuleke students (sponsored by NGOs) are in various stages of preparation for positions in conservation and tourism management.

Makuleke members also expressed their dissatisfaction at being excluded from early discussions around the creation of the Great Limpopo Transfrontier Park straddling South Africa, Mozambique and Zimbabwe, despite being a key landowner at the very point where the three countries meet. SANParks members of the JMB acknowledged and undertook to rectify these omissions.

The question arises as to why some members of SANParks have consistently tried to block the Makuleke from exercising their rights, especially to hunt on a sustainable basis. It would seem this was a result of unfamiliarity with the terms of the settlement agreement on the part of some, but there are indications that there has been a reluctance to relinquish the power enjoyed under the previous Schedule 1 national park status (Reid, 2001a, 2001b; Steenkamp and Grossman 2001).

In an attempt to facilitate smoother relations, a workshop was held to familiarize JMB members – including SANParks personnel – with the terms of the contractual agreement and the contents of the management plan. An outcome of this workshop was the appointment of a full-time implementation officer to carry out the day-to-day obligations of the JMB, as the task proved too demanding for SANParks personnel and the volunteer Makuleke members of the JMB.

To date, conservation objectives in the Makuleke Region are being met, sustainable use in the form of hunting has contributed over R2 million (US$299,550) and two lodges developed in conjunction with the private sector are operating. Maintenance of the conservation estate – the responsibility of SANParks – lags, and despite full endorsement of the co-management model by senior SANParks personnel, many SANParks staff members have yet to accept that the power they once enjoyed no longer applies.

The Makuleke people have displayed a high level of cohesiveness, often absent from other communities (see below), and through their self-initiated tertiary training of students and use of external experts where needed, have a high level of capacity compared to many rural communities in the region. Importantly, there appears to be a strong and mutually supportive relationship between the CPA executive committee and the traditional authority. This is in stark contrast to the tensions often experienced between traditional and civic structures elsewhere.

!Ae!Hae Kalahari contract park in the Kgalagadi Transfrontier Park

The South African portion of the Kgalagadi Transfrontier Park (KTFP) is located in the northernmost part of the Northern Cape Province, bordering Botswana (with whom it shares the transfrontier park) in the east and Namibia in the west. It lies in one of the hottest regions of South Africa, with annual rainfall of 150mm per annum in the south-western part where the contract park is located, increasing to 350–400mm per annum in the north-eastern part of the KTFP. Few vegetation types or ecological processes in South Africa are as well preserved as those of the park. Ownership of land in the contract park is shared between the ‡Khomani San and the Mier communities in the south-ernmost extremity of the South African side of the KTFP.

The original claimant group, the ‡Khomani San, were gradually forced out of the Kalahari Gemsbok National Park after its formation in 1931, some of them living on the side of the road to the park and in the township of Welkom, 10km south of the park, some of them living and working at cultural tourism facilities in the Northern Cape and others spreading out into the diaspora with other 'Southern Kalahari San' in South Africa, Botswana and Namibia. As with other displaced indigenous peoples, the San had largely been assimilated into or dominated by local pastoralist groups, their ancient cultural practices were sporadically maintained in isolated groups and their hunter-gatherer lifestyle was severely compromised or effectively destroyed.

Their land claim was settled by negotiation with the South African government (rather than being adjudicated by court), and the historic settlement agreement was signed on 21 March 1999, Human Rights Day. It is one of only two successful aboriginal land claims in southern Africa; the other is in the Richtersveld. The agreement constituted 25,000ha in the southern part of the KTFP (which was deproclaimed as a national park and reproclaimed as a contract national park), six private farms totalling roughly 37,000ha located 50km south of the park, and 'traditional and symbolic' rights and access to a further large portion of the park. The last of these rights were not spelt out in detail, but were in lieu of receiving full restitution in respect of their authenticated land claim. It was thought that, in light of national biodiversity interests (among other considerations), the majority of the park should remain under the custodianship of the state. Government also felt that the claim should be expanded from the original group to include other persons who could prove San heritage – an as yet undefined number.

As the negotiations had been rushed and further detailing of the agreement was required, title to the land in the park was not handed over to the community until August of 2002. In the interim, registered members of what was now named the ‡Khomani San Community, commenced the novel process of managing communally owned farms according to the Communal Property Association Act. This proved to be challenging, given that the 'community' was now artificially constructed from both original claimants and San drawn from the diaspora, many without any formal education or previous experience of owning and managing land. In the absence of a functioning 'community council' or other authoritative body, legislation required the San to operate in accordance with perceived western notions of 'representative democracy' (Chennels, 2002). However, this complex process was seriously undermined by a lack of adequate post-restitution support from government and NGOs. To date, it has not been achieved, with mismanagement and corruption prevailing as the 'empowered' few have enriched themselves at the expense of the still impoverished majority.

While the process of community rebuilding was theoretically under way, negotiations around the park recommenced, with the interests of the San community being represented by their lawyer and an elected park subcommittee. The agreement that was finally reached can be summarized as follows:

- Ownership of 25,000 hectares on the southern boundary of the park, within which area the San will be relatively free (within the limits of a negotiated contract park) to carry out cultural practices, to hunt, collect bush foods and conduct ecotourism ventures (including walking and overnight trails and 4x4 routes). It is accepted by the San that no permanent residence will be allowed in the park.
- Priority commercial use of the area between the owned area and the Auob River. In this zone, the ‡Khomani will be entitled to formulate and conduct ecotourism projects in partnership with SANParks or otherwise, in addition to all cultural practices.
- Symbolic and cultural use of the rest of the park. This right means in effect that the San are able to utilize a large portion of their ancestral land in the park for activities that are not commercial. What is envisaged here is the taking of groups of elders and youth deep into the park, where they can experience the Kalahari as it was, and live off the land as they once did.
- SANParks have recognized that San heritage is, and should be, inextricably linked with the identity of this section of the Kalahari and intend to find ways to give substance to that notion. A jointly owned (San, Mier and

SANParks) commercial lodge was agreed to and built. The San will be employed as trackers and in other capacities, with training and a view to managing the lodge in the long term, should they wish.

World Heritage Site listing will be applied for in due course, to register the interaction between the ancient culture of the ‡Khomani San and the conservation of the unique Kalahari ecosystem.

The Mier people, too, have a history of dispossession, as they also inhabited at least ten areas of approximately 90,000ha within what now constitutes the park. Following the proclamation of the Kalahari Gemsbok National Park, some of their farms formed part of the park, while others reverted to the state. The Mier, like the San, were originally allowed to continue living within certain areas of the park, but they lost their rights to hunt and manage stock on the land. They also lost 200,000ha of land for hunting and traditional use, and were eventually forcibly removed to state land, a 'coloured' area reserved for them under the settlement schemes of the 1930s (DANCED, 2000). The Mier community now live in six poorly serviced settlements south of the park. Education and job opportunities are limited, and literacy is estimated to be 10 per cent. Most of those without land rely on farming for their income, but only have access to about 150 farms, which is not enough to meet their economic needs (Chennels, 1999).

In December 1998, the Mier community, represented by the Mier Transitional Local Council, lodged its own land claim for land both inside and outside the Kalahari Gemsbok National Park (Chennels, 1999; Reid, 2001b). Their claim was based on the fact that the whole Mier area had been used for habitation, religious and cultural practices, grazing, cultivation, hunting, water trekking, harvesting and the use of a wide range of natural resources that were originally recognized and managed under a system of exclusive indigenous law (DANCED, 2000). The claim was particularly complicated, as the area being claimed by the Mier overlapped substantially with land claimed by the San, and many San would also be partial beneficiaries of the Mier claim having since become Mier residents (Chennels, 1999; Reid, 2001b). Numerous urgent discussions followed, aimed at trying to resolve all the issues and negotiate a creative solution that would enhance benefits for the San and Mier communities and SANParks (Chennels, 1999).

Both land claims were resolved in March 1999, the day before the scheduled handover ceremony. A settlement agreement was drawn up with the following conditions – the official granting of the land had to be negotiated by the Department of Land Affairs and the San and Mier communities, and parliamentary approval was necessary for deproclamation of the area as a Schedule 1 national park and reproclamation as a contractual national park. The settlement stated that the land would be managed in line with a contract to be drawn up between the three parties, and that it may not be used for residential or farming purposes. It was agreed that half of the 50,000 hectares released would be given to the San (the eastern side) and half to the Mier community (the western side).

The !Ae!Hae Kalahari contract park is to be co-managed by a JMB comprising three SANParks officials (including the park warden as an ex officio member), and three to five representatives of both the San and the Mier communities. In general terms, the JMB is responsible for the formulation, implementation and monitoring of an effective framework for the management and development of the park, and decisions are to be made by consensus. The principle of adaptive management has been adopted and the park will be managed in line with the officially accepted 'Master Plan for the Management and Development of the Kgalagadi Contract National Park' and its stated objectives. In terms of the agreement, SANParks will provide the necessary staff for the day-to-day conservation management function of the park.

A substantial period of time elapsed between the deproclamation of the park as a Schedule 1 national park and its reproclamation in terms of Schedule 2(b) of the National Parks Act as a contractual national park. The signing ceremony took place in August 2002, while South Africa was hosting the World Summit on Sustainable Development. This was viewed by some as being rather opportunistic, especially given that while the event was touted as indicating the success of land reform and conservation initiatives in the country, the 'on the ground' reality of the matter was, and continues to be, somewhat different. Furthermore, until very recently, little more transpired with respect to implementing the terms of the agreement and initiating co-management of the park.

The situation in the Kalahari is characterized by complex dynamics both within and between the communities and progress with park matters cannot be seen in isolation from the greater CPA issues. While the Mier community tends to be more cohesive and institutionally functional, the San community is constituted of a widely dispersed range of people with very different aspirations, and is not yet even fully defined. Great tensions exist within the San community, and the lack of progress on many issues and the degeneration of the situation since the transfer of land to the San in 1999 have exacerbated these problems. Despite numerous contractual commitments, support from the Department of Land Affairs has been very limited. A lack of resources and capacity, coupled with corruption within the CPA management committee(s) has added to the frustration. Against this backdrop, progress in implementing co-management in the park has been, and will continue to be, slow unless proactively addressed. While it might have been expected that SANParks would take the lead in this respect, there seems to have been reticence at a local park management level to do so. Insufficient personnel and resources were allocated to the task, and some personnel seemed reluctant to accept the 'new order', in a similar manner as experienced by the Makuleke.

A further factor undermining progress has been the lack of substantial and suitable NGO support to the community. Of the NGOs involved, the Southern African San Institute has traditionally focused on cultural and linguistic issues, while Farm Africa has played a smaller role, focusing on land-care and livestock projects. While other NGOs were involved in funding investigations into suitable settlement and land use options in the early stages of the negotiations, NGO interest and funding for projects has not been forth-coming since the agreement was reached.

The more 'traditional' members of the CPA, largely members of clans that were the last residents of the park, are strongly attached to the park and place a high value on rekindling their material, cultural and spiritual connections to the land. For them, perhaps more than anyone else, it is important that the terms of the agreement are properly met and that justice is done by the effec-tive and successful implementation of their vision in a mutually constructive and cooperative manner.

In 2006, some positive steps were taken to address this situation, including the appointment of a dedicated 'People and Conservation' officer based at the park. One of the objectives of creating such a position was to improve the flow of information between the parties and to facilitate the implementation of the agreement. With the financial support of SANParks, a group of San elders and youths, who had not yet set foot on their land since the settlement of the claim, were transported to, and spent the night on, their land some seven years after receiving their land back. A workshop aimed at unbundling the formal settle-ment agreement and clearly articulating the rights and responsibilities of each of the parties has taken place and is a further step in the right direction. As in the case of the Makuleke, an implementation officer is needed to follow up on the work of the JMB on a day-to-day basis, and funds are being secured for this position.

The situation with respect to the Mier community is simpler in that, as previously mentioned, the community is more cohesive, more institutionally functional (a local municipal council is in place), has greater capacity and expe-rience, with fewer expectations and less reliance on the outcome of what happens in the park. They have, for a number of years, successfully managed a relatively lucrative hunting and tourism (4x4 trails) operation on their land bordering the park, and there are a number of successful small stock farmers and entrepreneurs in the area.

It remains to be seen how this theoretically fair co-management model will develop over time, but it is fairly certain that without concerted efforts to build the capacity and empower all of the parties, as well as the will to work constructively and cooperatively in the spirit of the agreement, progress towards the implementation of the agreement and the achievement of the objectives laid out in the master plan will be protracted and limited.

Conclusions

Is genuine co-management effectively being achieved in these examples of South Africa's contractual national parks? The answer to date has to be 'not yet'. However, promising advances have and are being made in this regard as lessons are learned and system 'drivers and parameters' change. As Magome and Murombedzi (2003) contend, the contractual national park model devised under the apartheid regime in an attempt to expand national parks by entering into agreements with politically powerful private landowners, was not meant for the disadvantaged majority of black people. The result of unequal treatment of private and communal landowners in their contracts with the state represents a new form of 'ecological apartheid' in the democratic South Africa, perpetuating a dual tenure system and preventing communities from reaching the full potential of possible resource utilization.

While the conservation objectives of contract parks are rarely directly compromised, at least not in the short term (the Richtersveld perhaps being an exception, though large parts of the park remain inaccessible to stock farmers and the diamond deposits tend to be alluvial and concentrated along the river), in the medium to long term they may be compromised if anticipated social and economic benefits do not accrue to the communities involved, and if areas of conflict are not adequately addressed. How fast true power sharing and the realization of rights can be achieved depends on the local circumstances as well as the prevailing political climate.

While the process of achieving power sharing and genuine co-management tends to be long, it is affected by a number of different factors. A clearly defined contract that accurately accounts for and equitably attributes short- and long-term benefits and costs, as well as the rights and responsibilities of each party is vital. It must also be accompanied by a clear understanding of the letter and spirit of the contract by each of the parties. Strong, committed characters and champions are needed from all parties, as the 'right' model with the 'wrong' people will invariably struggle or fail.

A willingness to devolve authority and embrace the principle and ethic of co-management needs to exist within the parks authority at both the park and head office levels. Strong institutional arrangements must exist in both the park authority and communities, and provision must be made for the fact that building and equipping such institutions usually involves much time and high transaction costs – necessitating the availability of resources to build capacity and transfer skills, and the willingness to do so.

Cohesive communities with few internal issues tend to be more successful at the outset, and the differences between communities mean that situation-specific, tailor-made approaches are likely to be more successful. This should include a willingness to renegotiate contractual details should initial contracts prove difficult to implement. Rigid 'blueprint' approaches should be avoided.

Strong NGO support (both technical and with respect to capacity-building) providing good groundwork and follow-through with community members, is essential.

The empowerment of communities is necessary to ensure their equal weight in the attainment of rights and acceptance of responsibilities and co-management. However, co-management is based on the assumption that communities do in fact want or need to be actively involved in management. Whether this is so is a moot point, and in some cases, such as the Mier case presented above, the community may well be satisfied with a less active landlord role.

Sensitivity to local politics and internal community dynamics needs to be displayed by governments and park authorities, and external factors (for example regional development issues and different levels of political will at different levels of government) also need to be accounted for. Finally, rushed processes and negotiations resulting from political pressure seem to have been recurring problems and should be guarded against, as the time saved at the outset often results in conflict and delays further down the line.

In time, issues of financial sustainability will in all probability emerge, given the growing pressure on parks to pay their way. Parks such as the Richtersveld and Kgalagadi have limited income-generating potential, given their locality, extreme climates and limited carrying capacities. Of course, dependency on income generated could change if government accepts that commitment to international conventions and to the conservation of at least 10 per cent of the country is likely to require some form of subsidization by society.

References

Chennels, R. (1999) 'What have we achieved?', unpublished report for the South African San Institute, Cape Town

Chennels, R. (2002) *South African San Institute Annual Report*, South African San Institute, Cape Town

DANCED (2000) *Status Report. Pilot Parks: Kalahari Gemsbok National Park. DANCED: Capacity building in SANParks*, SANParks, Pretoria

Hanekom, D. and Liebenberg, L. (1993) 'Livestock grazing in national parks', ANC occasional paper, African National Congress, Johannesburg

Hughes, G. R. (2002) 'Democratization: Biodiversity conservation for all people – A case study from KwaZulu-Natal', in Pierce, S. M., Cowling, R. M, Sandwith, T, and MacKinnon, K. (eds) *Mainstreaming Biodiversity in Conservation*, World Bank, Washington DC

Magome, H. and Murombedzi, J. (2003) 'Sharing South African national parks: Community land and conservation in a democratic South Africa', in Adams, W. M. and Mulligan, M. (eds) *Decolonizing Nature: Strategies for Conservation in a Post-Colonial Era*, Earthscan, London

Reid, H. (2001a) 'Contractual national parks and the Makuleke community', *Human Ecology*, vol 29, no 2, pp135–155

Reid, H. (2001b) Unpublished research paper, University of Kent, Canterbury

Reid, H., Fig, D., Magome, H. and Leader-Williams, N. (2004) 'Co-management of contractual national parks in South Africa: Lessons from Australia', *Conservation and Society*, vol 2, no 2, pp377–409

Social Ecology and SANParks (2000) *Visions of Change*, Development Communications Corporation/South African National Parks, Pretoria

Steenkamp, C. I. and Grossman, D. (2001) 'People and parks: Cracks in the paradigm', IUCN South Africa Policy Think Tank Series No. 10, IUCN South Africa, Pretoria

Surplus Peoples Project (1995) *Land Claims in Namaqualand*, Surplus Peoples Project, Cape Town

van der Walt, P. T. (1991) 'Waarom dié bergwoestyn bewaar?', *Custos*, vol 20, no 6, pp18–22

23

Transfrontier Conservation Initiatives in Southern Africa: Observations from the Great Limpopo Transfrontier Conservation Area

Webster Whande and Helen Suich

The positioning of transboundary conservation approaches since the mid-1990s in the midst of southern African governments' conservation, regional economic integration and social development objectives heralded a remarkable convergence of interests in international relations. Transfrontier conservation initiatives have also been strongly supported by Southern African Development Community (SADC) wildlife-related policy and protocol, which seeks to promote these initiatives as a means for interstate cooperation in managing and sustainably using ecosystems that transcend political boundaries, and to develop a common framework for natural resource conservation (see SADC, 1999). The political interest aroused is demonstrated by the nine (current and former) presidents of the region, who are patrons of the Peace Parks Foundation, a South Africa-based NGO dedicated to raising funds for, and facilitating the implementation of, transboundary initiatives (Hanks, 1997). Further support to cross-border conservation occurred when, in the late 1990s, following years spent funding community-based natural resource management (CBNRM) programmes, foreign donors financing environmental initiatives shifted en masse to funding transboundary conservation activities (Hutton et al, 2005; Frei, 2007).

The central premise for transboundary approaches is that co-management of natural resources occurring along geopolitical boundaries (through national-level international agreements) can contribute to the peaceful resolution of interstate conflicts, promote regional economic development and integration, contribute to efforts to conserve globally significant biodiversity and to address a number of social issues. Yet, as Jacobsohn (undated) warns, their size presents a challenge in developing effective management tools and their success depends on being able to manage smaller units within the whole in a coherent manner.

There are several different types of transboundary approaches. A starting point is transboundary natural resource management (TBNRM), for which two different uses can be discerned in the literature: first, it is used to denote a 'more holistic approach' in processes 'across boundaries that facilitates or improves the management of natural resources (to the benefit of all parties in the area concerned)' (Griffin et al, 1999, pp2–3); and second is in relation to an up-scaling of CBNRM approaches across geopolitical boundaries (Jones and Chonguiça, 2001). TBNRM can therefore imply a range of different activities and processes, for example, from managing or facilitating local collaborative management of wildlife and other natural resources through the relaxation of geopolitical boundary restrictions in certain areas, to the harmonization of national natural resource management policies and legislation (Griffin et al, 1999).

A transfrontier park (TFP, also known as a transboundary protected area or a 'peace park') involves a network of formally proclaimed state or provincial/regional protected areas straddling international boundaries and is subject to a shared management agreement among the countries involved. A transfrontier conservation area (TFCA), by contrast, can incorporate multiple use zones on state, communal and/or privately owned land as well as strictly protected areas. The implementation of all transfrontier conservation initiatives has, to date, focused on conserving biodiversity through the designation or extension of conservation areas across geopolitical boundaries – with implicit emphasis in southern Africa on re-establishing migratory routes for wildlife. Other objectives relate to regional economic development and integration (through conservation-driven tourism development)[1] and the promotion of peace and cooperation between neighbouring countries (see Griffin et al, 1999). Reuniting local communities estranged by colonial boundaries also forms an objective of many transboundary initiatives.

A TFP and a TFCA can be implemented simultaneously, as examples of the Great Limpopo Transfrontier Park (GLTP) and transfrontier conservation area (GLTFCA), and the |Ai-|Ais–Richtersveld TFP and the proposed Greater !Gariep TFCA demonstrate. However, different categories of transboundary initiatives have different emphases on the various 'generic' objectives. In spite of the interest they have generated, the implications of transboundary approaches remain poorly understood, partly because they are still in their formative years (Dzingirai, 2004), and partly because they are pursued along narrow interests (such as the breakdown of fences along the Mozambique–South Africa border to allow animal movement, which later impacted on livestock–wildlife diseases), even as their implementation continues at an unprecedented pace (Katerere et al, 2001). This chapter uses the case of the GLTP and GLTFCA to demonstrate some of the implications of the implementation of transfrontier conservation, and that despite the convergences implied, there are significant difference between TFPs and TFCAs in practice.

The Great Limpopo Transfrontier Park and Conservation Area

South Africa's General Jan Smuts first proposed the idea of 'a great fauna and tourist road through Africa' that would link Kruger National Park and the then Southern Rhodesia (National Archives of Zimbabwe quoted in Wolmer, 2003). In 1938, Gomes de Sousa, a Portuguese ecologist requested that the Portuguese colonial government in Mozambique enter into negotiations with South Africa about linking part of the adjoining areas of the two countries (Munthali and Soto, 2002). Efforts were made to rekindle these ideas in the 1930s with the establishment of the Gonarezhou Game Reserve in Southern Rhodesia, and in the 1970s when a Mozambique-based conservation biologist proposed a Mozambique–South Africa conservation area (Wolmer, 2003), both of which failed.

In the early 1990s, after the end of the civil war in Mozambique and the imminent end of apartheid in South Africa, the idea was raised once again. Anton Rupert, then President of the Southern African Nature Foundation (now WWF South Africa), is widely credited as having advanced the idea to the Mozambican president (PPF, 2006), leading to the commissioning of feasibility studies of the proposal (Jones and Chonguiça, 2001). The real impetus for the initiative came with the active involvement of the World Bank.

From around 1993, the World Bank actively supported efforts to establish the GLTFCA, funding millions of dollars worth of feasibility studies and consultancies, and by heavily influencing the Mozambican government (Anstey, pers. comm.). The World Bank's Transfrontier Conservation Areas Pilot and Institutional Strengthening Project for Mozambique made an important conceptual shift from strict protected areas to including multiple resource uses, in particular by local communities (World Bank, 1996). This reflected the broader World Bank policy which prevented it from funding activities that involved forced resettlement. At about the same time, Rupert requested another meeting with the President of Mozambique to emphasize the benefits of nature-based tourism growth, if the transfrontier initiative was implemented (Hanks, 1997), all of which subsequently led to an agreement on the need for cooperation among Mozambique, South Africa and Zimbabwe in order to realize TFCA-related economic benefits.

In a parallel process in the mid-1990s, the premier of Limpopo Province in

South Africa met with the governor of Gaza Province in Mozambique to explore means of developing and improving economic linkages between the two provinces. While their motivations might have been different to those for establishing a transfrontier conservation initiative, their interest in building inter-provincial economic linkages gave credibility to the conservation initiative, which was framed in terms of developing the tourism industry as a means to stimulate local economic activity (Braack, pers. comm.). The interest in cooperation at the provincial level highlighted the growing political interest and support for transboundary initiatives, even as the understanding of what form these initiatives were to take diverged. For WWF South Africa, this was predominantly for marketing southern Africa as an integral tourist destination (Hanks, 1997), while the World Bank wished to shift from command-and-control to incentive-based conservation practices, encouraging the participation of local communities (World Bank, 1996). The more 'preservationist' aims of the Peace Parks Foundation (PPF) have greatly impacted on the design of the Great Limpopo transfrontier initiative with their more dominant role since the late 1990s (Anstey, pers. comm.).

The signing of a trilateral agreement to the establishment of the GLTP occurred in November 2000. This agreement facilitated the formation of a technical committee and working groups (guided by a ministerial committee) to prepare the terms and conditions for an international treaty establishing a transfrontier park. In 2002, the presidents of Mozambique, South Africa and Zimbabwe signed an international treaty establishing the GLTP, a culmination of these historical events.

The GLTP, measures an estimated 35,000km^2 (see Figure 23.1) and incorporates the Kruger National Park and the Makuleke Contractual National Park in South Africa, the Limpopo National Park in Mozambique and Gonarezhou National Park, Manjinji Pan Sanctuary and Malipati Safari Area in Zimbabwe. The communal lands of the Sengwe corridor in Zimbabwe are also included, and are regarded as an important link between Kruger and Gonarezhou national parks.

According to the GLTP treaty, the transfrontier park aims to foster transnational collaboration and cooperation among the parties to facilitate effective ecosystem management in the area comprising the park. Additional objectives are to encourage social, economic and other partnerships among the private sector, local communities and NGOs to manage biodiversity, to harmonize environmental management across borders and remove artificial barriers to the movement of wildlife. The GLTP also aims to facilitate the establishment and maintenance of a sustainable sub-regional economic base through appropriate development frameworks, strategies and work plans, with cross-border tourism anticipated to foster regional socioeconomic development (Governments of Mozambique, South Africa and Zimbabwe, 2002).

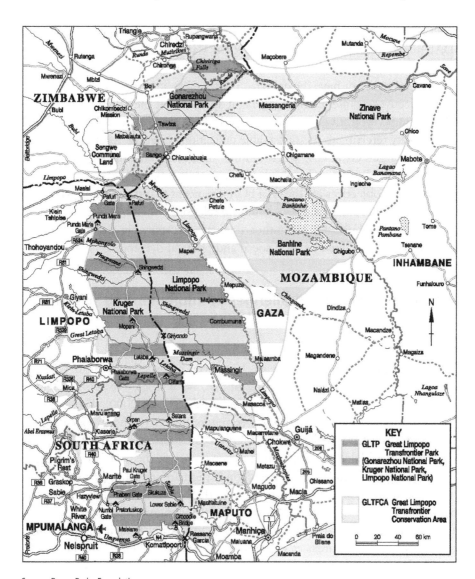

Source: Peace Parks Foundation

Figure 23.1 *The Great Limpopo Transfrontier Park and Conservation Area*

The GLTFCA is not as yet subject to an official agreement, but is described in the GLTP treaty as including compatible conservation areas adjacent to the GLTP (and indeed has been described in many maps, including that of Figure 23.1). It is thought to incorporate approximately 100,000km^2 (the core of which is the GLTP), and include privately owned land (mostly game reserves and conservancies) in South Africa and Zimbabwe as well as communal lands in

Mozambique and Zimbabwe, and other protected areas in the three countries (DAI Impacto, undated; DEAT and PPF, undated). Given the lack of formal development of the GLTFCA, the objectives of the initiative have not been publicly defined, nor have the envisaged differences with the GLTP in management objectives or interventions. So far, it appears that the GLTP agreement is taken to imply objectives for the entire GLTFCA and that the sections of the GLTP treaty referring to the sustainable utilization of natural resources apply to the much larger GLTFCA, where human habitation and sustainable natural resource use is allowed and multiple land uses are recognized. The lack of clear and separable objectives for the GLTFCA has meant that the multiple use activities in communal areas are often defined in relation to their impacts on the core protected area of the GLTP, rather than as a sustainable natural resource management determined by local social and political dynamics.

Achievements and challenges in implementing the GLTP

Biodiversity conservation

A variety of plant and animal species have been recorded within the GLTP – around 2000 plant species, 49 species of fish, 34 of frogs, 116 reptile species, approximately 505 of birds and 147 of mammals (DEAT, 2000). These species are supported by various vegetation communities, including mopane woodland and shrubland widely distributed in the northern half of the TFP, specifically the Makuleke contractual national park in the north of the Kruger National Park, as well as the Gonarezhou National Park. Mixed bushveld and riverine woodland are widely distributed in the Kruger and Limpopo national parks, sandveld to the east in Mozambique occurring both in protected areas and communal lands. These vegetation zones are variously located in the three constituent protected areas as well as the surrounding TFCA.

Hanks (2003) notes that transboundary approaches can facilitate the conservation (or re-establishment) of migratory and/or wide-ranging species. The GLTP – specifically the Limpopo National Park – is viewed as a way of dealing with the high elephant population numbers within the Kruger National Park by providing an expanded area into which they can easily move. Yet, the actual biodiversity impacts of linking and extending protected areas within the GLTP are not yet clear – the lack of baseline studies further presents difficulties for ascertaining the 'true biodiversity benefits to the transfrontier park' (Schoon, 2007, p7) and the JMB notes that as yet there are no monitoring studies to assess the impacts of the GLTP on biodiversity (JMB members, pers. comms).

From a protectionist perspective, the most important impact on biodiversity of the GLTP to date has been the proclamation of the Limpopo National Park in 2001 (formerly *Coutada* 16), and the improved management of the area resulting from its proclamation and subsequent funding being poured in to its rehabilitation and development. Other activities have involved the transloca-tion of more than 4000 animals from Kruger National Park since 2001, while others have moved out of their own accord (see www.peaceparks.org).

One of the most important biodiversity-related activities is the dropping of fences between the three countries. To date, however, only a small length of fence has been dropped between the Kruger and Limpopo national parks. Part of the reason relates to concerns regarding disease transmission between wildlife and domestic livestock, and it has been suggested that further fence removal will have to be associated with the fencing of the (currently unfenced) Limpopo National Park in order to prevent disease transmission, to control poaching and restrict human movement (Spenceley, 2005). Concerns about disease transmission from wildlife to domestic livestock have also prevented the establishment of a link between Kruger and Gonarezhou national parks through the Sengwe corridor (Daconto, 2003).

Since the GLTP agreement was signed, parts of the Gonarezhou National Park in Zimbabwe have been resettled (Ferreira, 2004; Spenceley, 2005) and there are unconfirmed allegations that hunting concession holders from the Sengwe corridor in Zimbabwe use helicopters to drive wildlife from South Africa into Zimbabwe, and Zimbabweans are said to be hunting and snaring wildlife within the Makuleke section of Kruger (Makuleke Contractual Park rangers, pers. comms). Thus, any assessment of the impact of the GLTP on biodiversity necessarily needs to factor in social and political issues as indica-tors of long-term sustainability, rather than focusing on indicators such as the length of fences dropped, the area dedicated for protected areas or increases in ranges for single species such as elephants.

Security, peace and cooperation

In terms of the regional- and national-level institutional structures set up for the implementation of the GLTP, the Ministerial Committee provides policy and political guidance to the implementation of the GLTP (Mombeshora, 2005), which is interpreted by the JMB into implementation guidelines and action plans. The JMB consists of government officials from various ministries (for example those responsible for wildlife, security and agriculture) in each of the countries and also provides technical information to the Ministerial Committee. Other platforms for cooperation and communication are the sub-committees, structured according to specific issues within the GLTP – including conservation, veterinary, security, community relations, finance, tourism and human resources subcommittees.

While many of the institutions set up for the GLTP encourage communication among government officials, similar levels of communication are not visible between government and local resource-dependent residents. For instance, local people were once represented in meetings and workshops for the GLTP but this decision was reversed on the pretext that governments represented their citizens (JMB member, pers. comm.; Whande, 2007). The result of this has been that local residents in already marginal areas (Katerere et al, 2001) are experiencing new forms of exclusion from the policy- and decision-making processes that impact on their access to and use of land and natural resources.

In an overlap with the objective of removing fences to facilitate wildlife movement, much of the public discourse surrounding TFCAs relates to their supposed effect on breaking down geopolitical boundaries as countries co-operate to manage common natural resource areas (Godwin, 2001). Ramutsindela (2004) notes the portrayal of TFCAs as leading to the breakdown of boundaries paints a picture of a decolonizing concept and process, thereby appealing to the post-Independence political establishment. In reality, however, there is little break with these established boundaries, as independent African countries perceive this as potentially leading to territorial conflicts (Mbembe, 2000).

The GLTP treaty indicates that transboundary cooperation and collaboration will occur in the pursuit of effective ecosystem management. However, in the process of planning, development and implementation of the GLTP, issues of national sovereignty and border security have come to prominence. The top-down, politically driven process of implementation (Gwature, 2003) has, on occasions, created conditions for conflict among different government agencies and between the government and local people (Whande, 2008).

Chidziya (2003) notes that the GLTP has resulted in conflicts among sectoral agencies, as security agencies felt their involvement was to legitimize the environmental agenda, while environment agencies were not regarded as having the mandate to negotiate agreements with potential impacts on national security and sovereignty. The result has been that geographical areas that have previously been politically and economically marginalized – yet regarded as important for biodiversity conservation – now assume importance for national security (Duffy, 1997; van Ameron, 2002). Thus, instead of leading to a breakdown of boundaries, the increased national security interests in these regions has had the opposite effect – territorial integrity is now of primary concern, effectively increasing control over the movement and activities of local people. For instance, the cooperation between police units from Masvingo and Limpopo provinces in Zimbabwe and South Africa respectively (Malelo, pers. comm.) is not meant to facilitate local people's movement but to intercept them.

The GLTP treaty recognizes the centrality of national sovereignty and territorial integrity, which has had significant effects on the GLTP. Article 6(2c) of the treaty emphasizes the need for parties to harmonize policies and legislation, while Article 5 emphasizes issues of national sovereignty in managing constituent protected areas of the GLTP according to existing management plans. In practice, this has meant that differences in national wildlife policies and legislation continue (Munthali and Soto, 2002), and the implementation of the GLTP has not resulted in any significant moves towards harmonization of policies and legislation, even for shared natural resources. Instead, issues of national security and sovereignty have continued to dictate the implementation of the GLTP, delaying the implementation of work that raises national security concerns (Braack, pers. comm.).

Concerns about the possibility of an escalation in poaching and smuggling of cars, drugs and even weapons resulting from relaxed border controls (Duffy, 1997; Mombeshora, 2005) have meant that security agencies are now involved in the planning of activities within the GLTP. It is not only government officials that are concerned about the potential for increased illegal activities, a senior ranger at the Makuleke/Kruger Contractual National Park noted that local transport operators should not be allowed to conduct their business across boundaries as they are likely to engage in illegal activities such as smuggling of drugs (Greefe, pers. comm.).

In terms of human security – which is understood as providing voice for the politically marginalized (Brauch, 2005) – the GLTP has arguably had a negative impact on the residents of the Limpopo National Park, who have been inadequately consulted during the implementation of GLTP activities and whose fate – in terms of resettlement out of the park – has still not been resolved (Huggins et al, 2003; Spenceley, 2005; Spierenburg et al, 2006). The international coordinators of the GLTP have recognized the weaknesses in dealing with improving local livelihoods, and have indicated that the next phase of implementation will focus on understanding how local people can be more meaningfully engaged in the process of decision making and the management of the initiative.

Economic development and regional integration

Regional economic integration has received widespread support, specifically for opportunities for investment in nature-based tourism development (Wolmer, 2003). Considerable investment has been made in tourism and infrastructure development in the three parks by governments and foreign donors, a considerable proportion of which would probably not have occurred without the GLTP initiative. At least US$30 million was spent on tourism infrastructure in the South African portion of the GLTP between 2000 and 2004, and almost US$9 million in the Limpopo National Park alone. Since the Limpopo National Park was proclaimed, a number of tourism trails and facilities have been developed, and the tourism 'access facility' at Giriyondo, between Kruger and Limpopo national parks has been opened. In contrast, little tourism infra- structure development has taken place in Zimbabwe in recent years due to the unsettled political situation (Spenceley, 2005; Spenceley et al, 2008).

The private sector has taken the opportunities presented by the new approaches and often portrays itself as a primary engine of the success of such initiatives. Indeed in the South African and Zimbabwean parts of the GLTFCA, tourism is estimated to have generated approximately US$144 million in 2007, and provided approximately 8900 jobs (Spenceley et al, 2008). Approximately 70 per cent of the US$30 million spent on tourism infrastructure development in South Africa in the early 2000s was made by the private sector in developing tourism concessions within the Kruger National Park (Spenceley, 2005). It is likely that additional tourism investments have been made in South Africa in areas adjacent to the Kruger National Park – particularly on private game reserves – however, the extent of these investments is not known.

National security concerns have influenced investment in the GLTP; the current political situation in Zimbabwe discourages private investment in tourism facilities (ex-manager Pafuri River Camp, pers. comm.; Ferreira, 2004), foreign donors have pulled out of the country, and local NGOs there- fore face funding deficits and are unable to facilitate or implement GLTP activities. The reduction in resources for the implementation of the Zimbabwe component has resulted in official perceptions that Mozambique and South Africa are proceeding without consideration for the Zimbabwean component (Pienaar, pers. comm.).

The importance of tourism investment figures is in assessing the pace and direction of regional economic integration in terms of a preferred land use. However, rather than paint a picture of regional economic integration, these investment figures highlight the dominance of South Africa over its weaker neighbours, specifically in terms of tourism benefits. The current imbalance is illustrated by the US$137 million of tourism revenue earned in the South African section in 2007 and the US$7.1 million earned in the Zimbabwean section over the same period (Spenceley et al, 2008). A criticism of the tourism industry is that it can perpetuate and reinforce regional and international inequality (Ferreira, 2004), which bodes ill for the GLTP unless proactive steps are taken to address the issue.

The economic integration objectives of the GLTP should have received a significant boost from the geographical overlap with the Limpopo spatial development initiative (SDI), though evidence on the ground indicates that the SDI appears to have been abandoned. Officials in Mozambique noted that the SDI had quietly been grounded to focus on the development of the Limpopo

National Park, with the focus on tourism and protected areas at the expense of more localized integration, which would have involved the upgrading of infrastructure such as the railway line linking Mozambique and Zimbabwe (Mozambique Department of Forestry and Wildlife official, pers. comm.).

Local level impacts of the GLTP and implications for the implementation of the GLTFCA

Much of the research into the local-level impacts of the Great Limpopo has highlighted negative aspects, noting that local resource-dependent communities have been excluded from policy- and decision-making processes and that, to date, transfrontier initiatives have tended to replicate historically dispossessory approaches to conservation (Dzingirai, 2004; Spierenburg et al, 2006; Whande, 2007; Büscher and Dressler, 2007).

As noted above, it is not yet clear if the GLTP treaty will also apply to the GLTFCA or a separate agreement will be negotiated, but the negative perceptions associated with the marginalization of local people from the GLTP will need to be addressed – the future participation of the (freehold and communal) residents in the development and decision making regarding the GLTFCA is likely to have a considerable impact on the success (or otherwise) of the initiative.

Those involved in the development of the GLTFCA should learn lessons from the implementation of the GLTP, as the implementation of the larger conservation area will become more complicated with respect to social and economic objectives, given that an estimated 500,000 people live on the communal lands of Mozambique and Zimbabwe that fall within the proposed GLTFCA boundaries (Cumming et al, 2007).

Biodiversity conservation

The inclusion of non-state land and multiple use zones within the proposed GLTFCA suggests that implementers accept that some communities will not wish to take part in the initiative, impacting on the success of biodiversity conservation efforts (especially if the land is considered to be an important wildlife corridor). Conservation initiatives in southern Africa have often been contested at the local level, though not always successfully, and the establishment of many protected areas has resulted in the dispossession of local people – including Gonarezhou and Kruger national parks (Ferreira, 2004). Thus it should be recognized that while some communities may choose to be part of the GLTFCA initiative, others may not want to cede their land for conservation activities. The inclusion or omission of areas of land within the TFCA also has implications for cultural restoration objectives (see Box 23.1).

Box 23.1 *Madimbo corridor*

The Madimbo corridor is a piece of land along the Limpopo River (in South Africa), immediately to the east of the Makuleke Contractual National Park. The residents of the corridor were forcibly removed in the 1960s to make way for the South Africa National Defence Force (SANDF) and the establishment of the Matshakatini Nature Reserve. SANDF is still stationed along the Madimbo corridor, ostensibly for military training purposes, though they are more frequently engaged in intercepting Zimbabweans crossing illegally into South Africa.

Following the end of apartheid in South Africa and the passing of land claim legislation, the people who were removed from the Madimbo corridor have made a claim to have the land returned to them. The future use of the land is a source of local contestations between those in support of grazing and crop farming, conservation-driven tourism and human settlements.

Decisions over land use are complicated by the support for nature-based tourism as a land use by many stakeholders in the area – such as the local municipality, which borders the Kruger National Park, Makuya Park and the Matshakatini Nature Reserve, and the PPF that identified the area as a strategic link to the Zimbabwean portion of the GLTFCA, as well as to the Limpopo–Shashe transfrontier conservation initiative to the west of the GLTP (PPF, 2006). A local leader for the land claim for Madimbo corridor notes this is part of 'conservation's strategy as a land grabbing approach' (Vhembe Communal Property Association leader, pers. comm.).

Because of its strategic location, its inclusion in, or omission from the GLTFCA will have significant impacts on the ability of the TFCA to meet biodiversity objectives (for example, reestablishment of migratory routes for wildlife), social objectives (for example, reuniting local communities estranged by colonial boundaries) as well as economic objectives (for example, regional economic integration).

In terms of regional economic integration, the Zimbabwean members of the security subcommittee of the GLTP noted in 2002 that the fence between South Africa and Zimbabwe should be maintained, except that part where the Kruger National Parks adjoins the proposed Sengwe wildlife corridor, in order to control illegal movement across the border. It is clear, therefore, that while the facilitation of regional economic integration through nature-based tourism development is supported, this does not apply to the localized economies that depend on informal flows of goods and services across borders.

In terms of the restoration of cultural integrity, the commitment of the Madimbo corridor to conservation would preclude a physical link between the Venda families in South Africa and Zimbabwe, perpetuating the current situation of separation by the Matshakatini Nature Reserve and the presence of the SANDF.

The lesson that can be learned from this situation is that land use options proposed by centralized planning initiatives are often at variance with local realities. Without the provision of mechanisms and incentives for communities to participate in planning and decision making, they are unlikely to support such initiatives, which will undermine their ability to meet set objectives. Indeed, given the experience of land dispossession resulting from the declaration of protected areas and ongoing conflicts over land, it is unlikely that the residents of the Madimbo Corridor will opt to use the land for conservation should their land claim be successful.

Source: Based on Whande (2007)

Buzzard (2001) argues that the different policy frameworks on, and approaches to, the use of natural resources act as a hindrance to joint management of natural resources. However, the GLTFCA could facilitate the harmonization of many natural resource management activities, particularly those undertaken by residents on communal lands, where cultural similarities frequently mean similar natural resource management practices are undertaken. Indeed it is likely that the most significant differences in resource management activities occur at the provincial and national levels. It is not yet clear how any national-level policy and legislative harmonization would impact on local people and their natural resource management practices, particularly as recent experience in southern Africa demonstrates the insignificant role that local residents tend to play in resource management initiative design or implementation, processes led by national governments, NGOs and foreign donors (Simon, 2003).

There have been recent moves to start addressing the impacts of the implementation of transfrontier initiatives on the social–ecological systems they incorporate (Cumming et al, 2007), measuring biodiversity impact by considering the sustainability of the whole system, specifically the 'importance of wildlife/livestock/human/ecosystem health (the concept of one health) in sustaining large landscapes such as the GLTFCA' (Cumming et al, 2007, p2). Focusing on human and non-human factors allows an assessment of the problematic areas between local livelihoods and efforts to achieve biodiversity conservation through expansion of protected areas. The incorporation of other aspects of biodiversity such as ecosystem health are not yet reflected in practical terms.

Security, peace and cooperation, social and cultural reunification

A lack of explicitly stated objectives for the GLTFCA means it is difficult to differentiate between the GLTP and GLTFCA when considering security issues. While TFCAs are understood as multiple use zones and hence can accommodate local people's livelihood needs, in reality it is not clear how this will be balanced with the TFCA's role as extending conservation activities beyond protected areas. As noted above, the implementation of the GLTFCA provides opportunities for cooperation for harmonizing (and improving) the natural resource management practices of residents of the TFCA, improving relationships among residents, and also between residents and the managers of private and state-protected areas.

The prominence of sovereignty and border security issues is likely to have negative impacts at the local level. The GLTFCA is likely to result in new forms of control over human movement, specifically through the increased interest of state agencies involved in security issues (Dzingirai, 2004). The signing of the

GLTP treaty has meant these agencies have become increasingly interested in the region, in terms of issues related to maintaining territorial integrity or reinforcing geopolitical boundaries, which can negatively impact on informal cross-border livelihood activities (for example, small-scale trading). The recent xenophobic attacks on foreigners in South Africa have put the issue of border control into the spotlight, with parliamentary portfolio committees on home affairs and security receiving submissions on the need to increase police patrols and to bring in the army. For an area such as the Madimbo corridor, currently occupied by SANDF, this could reduce the chance that local land rights will be restored, which would reduce the control of the state over sensitive border regions.

The reluctance to facilitate local-level social and cultural reunification across political borders is exacerbated by the economic disparities among the parties to the GLTFCA and the high levels of illegal migration from Mozambique and Zimbabwe to South Africa. The GLTP joint management plan makes specific recommendations regarding the definition of border access, securing the entire periphery of the GLTP and limiting or preventing commercial traffic (excluding tourist traffic) (Governments of Mozambique, South Africa and Zimbabwe, 2002). Thus, in practice, claims that transboundary approaches will lead to a breakdown of fences are only partial, as controls over the movement of people within the wider GLTFCA are likely to be tightened. In addition to the human security issues already present in the GLTP, the implementation of the GLTFCA will raise additional issues – particularly in Mozambique and Zimbabwe where residents within the GLTFCA reside almost entirely on state land, often with poorly defined tenure over land and natural resources. These issues can only be resolved satisfactorily in consultation with these communities, and with their full participation in the decision making over, and implementation of, TFCA activities. The experience of the exclusion of communities from representation on the JMB of the GLTP is not encouraging.

Economic development/regional integration

Very little economic impact has been felt at the local level from developments associated with the GLTP. Though a recent study of some of the tourism enterprises in the GLTFCA estimates at least 8900 people are employed within the Zimbabwean and South African sections of the proposed TFCA, with local employees sharing approximately US$25 million in wages between them (Spenceley et al, 2008), the bulk of these tourism enterprises were established prior to the implementation of the transfrontier initiatives. There is potential for involvement in tourism developments in the future, but it is likely that these communities will need significant technical advice and capacity building in order to maximize the benefits arising from these opportunities. The principles of sustainable tourism (emphasizing the need to achieve a balance between the industry's environmental, social and economic impacts to achieve long-term sustainability) will also need to be adopted by those involved in the initiative to ensure that local benefits are maximized.

Experiences from villages along the Madimbo corridor (see Box 23.1) indicate that local people are often suspicious of tourism initiatives as a result of their experiences with conservation in the past and because a focus on tourism tends to simplify their often diverse and complex livelihood strategies (Whande, 2007). As mentioned, prospects for tourism development along the Madimbo corridor are a source of conflicts among locals who stand to benefit from different land uses such as conservation and tourism, crop agriculture and livestock production. Where local communities have already started tourism initiatives, they have not received much support from organizations such as PPF, as is the case with the Bennde Mutale youth initiative and the Pafuri Lodge, both along the Madimbo corridor (pers. obs.).

The dominance of conservation activities within the Limpopo region, after initial attempts at implementing an SDI, has also limited other possibilities for economic integration based on infrastructure development. While the proposed construction of a bridge across the Limpopo River linking South Africa and Zimbabwe would facilitate the movement of people and provide opportunities for localized cross-border trade, the proposal has already become a centre of conflict locally, due to uncertainty regarding whether it is predominantly for the movement of tourists or local people as well.

Conclusions

The evolution of the GLTP has important lessons for the future of both the GLTP and the GLTFCA as well as for other TFCAs being developed across southern Africa. Widespread and high-level political support can justifiably drive initiatives involving state-protected areas; however, when private and communal lands are involved, then mechanisms and incentives for landholders and residents to participate in the decision making and implementation of the TFCA need to be in place to ensure they act to support rather than undermine the initiative.

While those involved in the GLTP claim that they are only concerned with issues regarding the core protected area, as there is no agreement or set objectives for the GLTFCA, the divisions between the two initiatives are not so clear cut. Those driving the GLTFCA initiative need to recognize that processes for the establishment and management of core protected areas impact on local communities in several ways, most prominently through land alienation. The observed resistance to conservation-driven tourism along the Madimbo corridor and alleged acts of poaching highlights the continued conflicts between protected areas and local resource-dependent communities. Rather than plan further expansion of protected areas through TFPs, transboundary initiatives need to revisit the potential for up-scaling local-level initiatives, putting emphasis on getting the smaller units within the whole managed in a coherent manner. A starting point is to pay more attention to objectives of cultural integrity and to explore local transboundary collaborative processes and how they relate to formal government agreements.

Nature-based tourism development is viewed as necessary for the success of transfrontier conservation initiatives. However, the prescription of conservation-driven tourism as a means of economic development disregards the inequalities between investors and local communities, between the countries involved, as well as the impacts of these inequalities on the distribution of benefits. Experiences from the GLTP indicate that the shift in focus from sustainable resource use in the 1990s to a focus on the benefits of tourism and increasing the role of the private sector, in particular in relation to the Mozambique component, has contributed to a growing sidelining of local concerns. Clear strategies to facilitate the participation of communities in tourism development and an equitable formula of sharing benefits accruing from conservation-driven tourism have to be found if the GLTFCA is to make a meaningful contribution to local people's lives.

The paradox of the increased focus on previously politically and economically marginal areas that accompanies transfrontier conservation initiatives is that they are viewed by some residents in negative terms – interpreted as the beginning of new constraints on local people's access to and use of natural resources as a result of increased state presence (Hughes, 2002; Dzingirai, 2004). In order to counter these perceptions, the emphasis on national security within the GLTFCA zone needs to be revised, as does the manner in which communities are involved in planning- and decision-making processes related to the GLTFCA. In particular, consideration needs to be taken of the security concerns of local people, which include articulating their views on the ongoing planning and implementation of the GLTFCA, not as mere observers of interventions developed at national and regional levels. To date, community issues have been excluded from the joint GLTP agenda, and the drivers of the GLTFCA are the same actors that drive the GLTP, which is not encouraging. The importance of the inclusion of residents is unquestionable given that at least 500,000 people live within the GLTFCA (Cumming et al, 2007).

Notes

1   Several transfrontier conservation initiatives in southern Africa have also been linked to spacial development initiatives (SDIs), particularly where (nature-based)

tourism development opportunities exist. SDIs were first conceptualized to address inequalities within South Africa's industrial and infrastructural development through targeted corridor developments. They have since evolved into regional initiatives to address transboundary economic cooperation and integration (Simon, 2003). Examples of SDIs that coincide with transboundary natural resources management include the Limpopo SDI (GLTFCA), the Orange River SDI (|Ai-|Ais–Richtersveld TFP and proposed Greater !Gariep TFCA) and Lubombo SDI (Lubombo TFCA).

References

Brauch, H. G. (2005) *Threats, Challenges, Vulnerabilities and Risks in Environmental and Human Security*, United Nations University, Bonn

Büscher, B. and Dressler, W. (2007) 'Linking neoprotectionism and environmental governance: On the rapidly increasing tensions between actors in the environment–development nexus', *Conservation and Society*, vol 5, no 4, pp586–611

Buzzard, C. H. (2001) 'Policy environment governing the Great Limpopo Transfrontier Park and Conservation Area: A review of relevant international agreements, SADC protocols, and national policies', Gaza-Kruger-Gonarezhou Transboundary Natural Resources Management Initiative, Development Alternatives Inc (DAI)

Chidziya, E. (2003) 'Country experiences and perspectives on the GLTFCA', in Gwature, C. (ed) *Supporting Community Development through Transboundary Natural Resources Management in Southern Africa*', Proceedings of the exchange visit seminar for directors, Letaba Rest Camp, Kruger National Park, 20–24 January

Cumming, D., Biggs, H., Kock, M., Shongwe, N. and Osofsky, S. (2007) 'The animal health for environment and development (AHEAD)–Great Limpopo Transfrontier Conservation Area (GLTFCA) programme: Key questions and conceptual framework revisited', www.wcs-ahead.org/documents/gltfca_revisited.pdf (accessed 8 October 2008)

Daconto, G. (2003) 'Implications of the GLTFP joint management plan for the Sengwe corridor', in G. Daconto (ed) *Proceedings of the Seminar on the Development of Collaborative Management for the Sengwe Corridor*, Southern Lowveld Project, CESVI, Harare

DAI Impacto (undated) *Overview of the Great Limpopo Transfrontier Park*, GLTP Publication Series Booklet 2, Impacto, Maputo

DEAT (2000) 'Conceptual plan for the establishment of the proposed GKG Transfrontier Park', www.environment.gov.za/Documents/Documents/GreatLimpopoTP/ConceptualPlan.htm (accessed 28 March 2008)

DEAT and PPF (undated) 'Great Limpopo Transfrontier Park', PPF, Stellenbosch

Duffy, R. (1997) 'The environmental challenge to the nation-state: Superparks and national parks policy in Zimbabwe', *Journal of Southern Africa Studies*, vol 23, no 3, pp441–451

Dzingirai, V. (2004) Disenfranchisement at large: Transfrontier zones, conservation and local livelihoods', IUCN-ROSA, Harare

Ferreira, S. (2004) 'Problems associated with tourism development in southern Africa: The case of transfrontier conservation areas', *GeoJournal*, vol 60, pp301–310

Frei, J. (2007) 'Is transboundary natural resource management working against community based conservation?', MiniReview Tropical Ecology and Management

Term Paper NATF350 2007:4, www.umb.no/ina/studier/soppgaver/2007-Frei.pdf (accessed 28 March 2008)

Godwin, P. (2001) 'Without borders: Uniting Africa's wildlife reserves', http://www7.nationalgeographic.com/ngm/data/2001/09/01/html/ft_20010901.1 .fulltext.html (accessed 28 March 2008)

Governments of the Republic of Mozambique, Republic of South Africa and Republic of Zimbabwe (2002) 'International treaty on the establishment of the Great Limpopo Transfrontier Park', signed 9 December 2002, Xai-Xai, Mozambique

Griffin, J., Cumming, D., Metcalfe, S., t'Sas-Rolfes, M., Singh, J., Chonguiça, E., Rowen, M. and Oglethorpe, J. (1999) *Study on the Development of Transboundary Natural Resource Management Areas in Southern Africa*, Biodiversity Support Programme, Washington DC

Gwature, C. (ed) (2003) *Supporting Community Development through Transboundary Natural Resources Management in Southern Africa*, Proceedings of exchange visit seminar for directors, Letaba Rest Camp, Kruger National Park, 20–24 January

Hanks, J. (1997) 'Protected areas during and after conflict: The objectives and activities of the Peace Parks Foundation', *Parks for Peace Conference Proceedings*, Peace Parks Foundation, Somerset West

Hanks, J. (2003) 'Transfrontier Conservation Areas (TFCAs) in southern Africa: Their role in conserving biodiversity, socioeconomic development and promoting a culture of peace', *Journal of Sustainable Forestry*, vol 17, pp121–142

Huggins, G., Barendse, E., Fischer, A. and Sitoi, J. (2003) 'Limpopo National Park: Resettlement policy framework', unpublished report

Hughes, D. M. (2002) 'Going transboundary: Scale-making and exclusion in southern African conservation', paper presented at the Environment and Development Advanced Research Circle, University of Wisconsin, Madison, 19 April

Hutton, J., Adams, B. and Murombedzi, J. C. (2005) 'Back to barriers? Changing narratives in biodiversity conservation', *Forum for Development Studies*, no 2, pp341–369

Jacobsohn, M. (undated) 'Walking with nature: Protected areas, people and prosperity. CBNRM vs TBNRM – allies or enemies?' www.irdnc.org.na/download/ margis%20WPC%20article.pdf (accessed 28 March 2008)

Jones, B. and Chonguiça, E. (2001) 'Review and analysis of specific transboundary natural resource management (TBNRM) initiatives in the southern Africa region', IUCN ROSA Series on Transboundary Natural Resource Management Paper No. 2, IUCN-ROSA, Harare

Katerere, Y., Hill, R. and Moyo, S. (2001) 'A critique of transboundary natural resource management in southern Africa', IUCN-ROSA Series on Transboundary Natural Resources Management – Paper No. 1, IUCN-ROSA, Harare

Mbembe, A. (2000) 'At the edge of the world: Boundaries, territoriality, and sovereignty in Africa', *Public Culture*, vol 12, pp259–284

Mombeshora, S. (2005) 'Collaborative partnerships in transboundary wildlife management: A review of southern African experiences', IUCN-ROSA, Harare

Munthali, S. M. and Soto, B. (2002) 'Overt and latent conflicts associated with the establishment, development and management of the Great Limpopo Transfrontier Park', in IUCN, *Conflict Management in Transboundary Natural Resources Management in Southern Africa*, IUCN-ROSA, Harare

PPF (2006) News page of Peace Parks Foundation website, www.peaceparks.org/ news.php?pid=161andmid=598 (accessed 27 March 2008)

Ramutsindela, M. (2004) 'Glocalisation and nature conservation strategies in 21st century southern Africa', *Tijdschrift voor Economische en Sociale Geografie*, vol 95, no 1, pp61–72

SADC (1999) 'SADC protocol on wildlife conservation and law enforcement', signed 18 August 1999, Maputo

Schoon, M. L. (2007) 'Building robustness to disturbance: Governance in southern African peace parks', paper presented at the Parks, Peace, and Partnerships Conference, 9–12 September, Waterton, Canada

Simon, D. (2003) 'Regional development: Environment discourses, policies and practices in post-Apartheid southern Africa', in Grant, J. A. and Söderbaum, F. (eds) *New Regionalisms in Africa,* Ashgate, Aldershot

Spenceley, A. (2005) 'Tourism in the Great Limpopo Transfrontier Conservation Area: Relating strategic visions to local activities that promote sustainable tourism development', workshop proceedings, Wits Rural Facility, Kruger National Park, 14–16 April

Spenceley, A., Dzingirai, P. and Tangawamira, Z. (2008) 'Economic impacts of transfrontier conservation areas: Tourism in the Greater Limpopo Transfrontier Conservation Area', unpublished report for the Southern African Sustainable Use Specialist Group/University of Witwatersrand

Spierenburg, M., Wels, H. and Steenkamp, C. (2006) 'Resistance of local communities against marginalization in the Great Limpopo Transfrontier Park', *Focaal*, vol 47, pp18–31

Van Ameron, M. (2002) 'National sovereignty and transboundary protected areas in southern Africa', *GeoJournal*, vol 58, pp265–273

Whande, W. (2007) 'Transfrontier conservation areas: Historical and livelihoods considerations within the Great Limpopo Transfrontier Conservation Area', Programme for Land and Agrarian Studies Research Report No. 25, Programme for Land and Agrarian Studies, Cape Town

Whande, W. (2008) 'Breaking down or consolidating boundaries? Re-imaging geopolitical boundaries in a time of transfrontier conservation in Southern Africa', paper presented at the Nurturing Culture, Conserving Nature: Heritage Governance and Local Participation in Kenya, South Africa and Sweden workshop, Centre for Public Sector Research, Goteborg, 7–8 February

Wolmer, W. (2003) 'Transboundary conservation: The politics of ecological integrity in the Great Limpopo Transfrontier Park', Sustainable Livelihoods in Southern Africa Research Paper 4, Institute of Development Studies, Brighton

World Bank (1996) 'Mozambique: Transfrontier Conservation Areas Pilot and Institutional Strengthening Project. Report No. 15534-MOZ', Agriculture and Environment Division, World Bank, Washington DC

24

Making 'Conventional' Parks Relevant to All of Society: The Case of SANParks

Guy Castley, Christopher Patton and Hector Magome

This chapter outlines the recent developments and new directions that SANParks is taking to ensure that the management of their parks is in keeping with their mission and vision – to be the pride and joy of all South Africans (SANParks, 2002). SANParks has adopted an adaptive management approach that has seen the organization set a number of strategic objectives that reflect the evolving conservation mission within the organization. Not only have a number of conservation achievements been made in recent years, a number of social milestones have also been achieved – national parks have been proclaimed on the basis of their cultural value (for example, Mapungubwe) and co-management agreements have been established with communities in areas where portions of national parks have been reclaimed by these communities to ensure the continued conservation of these landscapes (for example, the Makuleke region in the Kruger National Park). The organization has also been operating on a positive financial footing in recent years (SANParks, 2002, 2003, 2004, 2005, 2006a) and has expanded economic opportunities for entrepreneurs and local communities through the commercialization of non-core activities, and by implementing conservation activities through poverty relief and similar programmes.

SANParks' contribution to national biodiversity conservation

Historically, national parks were established in a largely ad hoc fashion based on limited scientific information, contemporary opinion of the value of the areas as national parks and, in some cases, political pressure. However, with the new fields of conservation planning and gap analysis available to park planners today, there is a much greater focus on designing a park network to meet strategic conservation goals and objectives (Lombard et al, 1999; Nel et al, 2002; Younge and Fowkes, 2003). Though the conservation targets established in these planning processes are often focused on habitats, the area requirements of viable populations of large herbivores highlight the need to establish large conservation areas to protect biodiversity components.

Only about 6 per cent of South Africa's natural heritage is currently protected under existing legislation, in areas managed by national and provincial agencies (Rouget et al, 2004; DEAT, 2005a), and it is recognized that some of the areas listed within South Africa's protected area database may only represent 'paper parks', i.e. those that are not adequately protected. SANParks, through the present network of 21 national parks, contributes almost 62 per cent to the country total for formal protected areas or 3 per cent of South Africa's terrestrial landscapes (based on the percentage contributions to biome conservation by Type I protected areas, as depicted by Rouget et al, 2004). The contributions made by private reserves and natural heritage or conservancy areas remains to be determined, though indications are that their contribution could be significant (see Chapter 10).

National parks currently protect 126 of the 440 vegetation types (29 per cent) identified for the country (Mucina and Rutherford, 2006), the proportions protected ranging from less than 1 per cent to almost 100 per cent. Many of the remaining vegetation types are captured within provincial and informal protected areas, though a number have been listed as critically endangered habitats (Rouget et al, 2004; DEAT, 2005a). It is evident that there are still gaps in reserve design given that there are biomes with less than 1 per cent of their area protected within national parks – although for some of these biomes (for example the Nama karoo) this represents the only habitat protected nationally in formal reserves (see Table 24.1). SANParks contributes significantly (more than 50 per cent in parks) to achieving conservation objectives at a national level of the Nama Karoo, Succulent Karoo, savanna, desert and wetland biomes; however, the fynbos, Succulent Karoo and grassland biomes have been identified for urgent conservation action.

One cannot assess the conservation performance of a single organization, or the country, without considering the threats that face much of the natural environment. Humans have impacted heavily on a number of habitats resulting in ecosystem degradation and species decline (see Table 24.2) (WWF-SA, 2002). The extinction risk of vertebrates is said to be underestimated for both the Cape Floristic region and Succulent Karoo hotspots (Brooks et al, 2002) and there are clearly a number of priority areas that can be identified for conservation action, given that some areas are under considerable threat from ongoing land transformation (agriculture, afforestation, urbanization and so on). These priority areas include the grassland, Nama Karoo, Succulent Karoo

Table 24.1 *Contribution of national parks to the formal conservation of South Africa's vegetation biomes*

Biome	South Africa area (km²)	South Africa % protected	SANParks area (km²)	SANParks % protected	Proportion protected in SANParks (%)
Forest	4730	39.6	213	4.5	11.4
Fynbos‡	84,580	11	1231	1.5	13.2
Nama Karoo	250,069	0.7	1717	0.7	114.4*
Succulent Karoo	85,207	3.1	1927	2.3	73.0
Grassland	373,984	1.9	610	0.2	8.5
Albany thicket	30,256	6.3	872	2.9	40.8
Savanna	412,753	8.9	28,999	7.0	78.9
Desert	8548	12.5	1079	12.6	101.0*
Wetlands	16,790	4.6	557	3.4	74.3
Total	1,266,917	6	37,205	2.86	

Note: ‡ The fynbos lowland areas were identified as being under-protected; * Figures that exceed 100 per cent are due to increases in the national park estate after 2004 that would not be reflected in the total percentage of biome area protected in South Africa.

Source: Adapted from Low and Rebelo (1996); Rouget et al (2004); Driver et al (2005); Mucina and Rutherford (2006); SANParks (unpublished data)

and thicket biomes, all of which currently have substantially less than the IUCN's recommended 10 per cent of area protected.

SANParks is able to highlight a number of conservation success stories, most notably for certain keystone or threatened species. Examples include the Cape mountain zebra, whose numbers have increased significantly (the bulk of this population being found within national parks), and the recovery of elephant, buffalo and black rhino in the Addo Elephant National Park. The need to undertake frequent live capture operations from many of the smaller parks is also evidence of increasing population growth trends.

Table 24.2 *Proportion of land degradation and transformation within the South African vegetation biomes*

Biome	Natural (%)	Degraded (%)	Transformed (%)
Forest	68.57	2.87	28.53
Fynbos	83.26	0.93	15.74
Nama Karoo	93.83	4.84	1.29
Succulent Karoo	93.18	2.05	4.57
Grassland	65.34	11.09	23.56
Thicket	73.36	12.08	14.48
Savanna	66.79	11.81	21.17

Source: WWF-SA (2002)

Although it is often the larger charismatic species that are the focus of attention, SANParks meets its mandate to conserve South Africa's biodiversity with the many smaller fauna and flora that are conserved in the expanding park network. Species inventories illustrate that national parks support populations of 70 per cent of the threatened mammal species listed in the South African Red Data Book (Friedmann and Daly, 2004), increasing to 75 per cent if oceanic species are not considered. Those species listed but not yet protected within the national parks system include primarily bats, golden moles and other endemic species with restricted distributions, highlighting the importance for additional research efforts on these species, but also emphasizing the importance of smaller reserves (under various management options) dedicated to the management of such species.

Maintenance of ecosystem services and ecological functioning is also of critical importance. In meeting these objectives, SANParks has initiated a number of government-supported programmes within national parks to address aspects such as rehabilitation, alien vegetation removal and infrastructure development. Funded to the tune of R292 million (US$43.7 million), these programmes have been highly successful. To date, activities in 14 parks have contributed to the clearing of over 45,000ha of land, with follow-up clearing on almost another 86,000ha. Infrastructure improvements include the construction of 19 new rest camps, removal of redundant structures such as derelict farmhouses and dams, and the upgrading or construction of roads and perimeter fences (SANParks, 2005).

Of concern is the paucity of marine protected areas along the South African coastline and SANParks currently has plans to establish a number in both the south-west and south-east to protect coastal and marine biodiversity. The national government has acknowledged marine protected areas as a key area for implementation, with plans to increase marine protected areas from 5 per cent to 20 per cent. The first steps to achieving this positive change occurred when the Algoa Bay islands were gazetted as part of the Addo Elephant National Park in November 2004 (see Chapter 19). SANParks will be instrumental in assisting the further realization of this vision, as three parks – Addo Elephant, Table Mountain and Namaqua national parks – have been earmarked for the establishment or expansion of marine protected areas within their boundaries (DEAT, 2003) (see Chapters 19 and 20). It is expected that significant progress in meeting these objectives will be made within the next five years.

Increasing the size of the protected area network

One of the most significant contributions made by national parks, supported by DEAT, is the increase in land purchases to expand and consolidate existing national parks and establish new ones. There are few protected area management organizations able to exhibit similar growth.

Although a large proportion of this new land has been acquired using government funds, many areas have been funded using non-governmental and/or private donor funds. Namaqua National Park is an example of park development funded by non-governmental sources – the initial core of the park (the Skilpad Nature Reserve) was purchased and initially managed by WWF South Africa, and the Leslie Hill Succulent Trust has funded subsequent land purchases. The expansion would not have been possible without the continued support of De Beers Consolidated Mines, which owned significant portions of the landscape. Another example is the expansion of the Marakele National Park where the involvement of private individuals such as Paul van Vlissingen (through Marakele Parks (Pty) Ltd, see Chapter 25) culminated in the finalization of successful public–private contractual partnerships. SANParks is also entering into new and innovative PPPs to increase the land under conservation in the vicinity of the Addo Elephant National Park, where a number of contractual agreements have been signed and wildlife has been relocated into the new areas.

It should be noted that the focus of these park expansions has not been placed on one individual park, but rather significant progress has been made in consolidating areas around a number of parks, including Addo Elephant and Mountain Zebra in the Eastern Cape, Namaqua and Augrabies Falls in the Northern Cape, Mapungubwe and Marakele in the Limpopo Province, and Agulhas and Table Mountain national parks in the Western Cape.

Another means of expanding the protected area network has been land transfers from other agencies (and there are a number of areas that are still in the process of being transferred to SANParks). Recent examples include the newly proclaimed Camdeboo National Park (previously known as the Karoo Nature Reserve), the transfer of the Qwa-Qwa National Park from the Free State Department of Environment Affairs and Tourism for consolidation within the Golden Gate Highlands National Park, and the transfer of Department of Water Affairs and Forestry landholdings into the Tsitsikama Coastal National Park (DEAT, 2005b). Although these transfers do not necessarily reflect an increase in the actual land held under conservation, in many cases the management effectiveness of the land in question has been improved as a result of these agreements.

There are five basic principles that drive SANParks' land acquisition strategy (referring to land that is purchased outright or incorporated into national parks through contractual or other partnership arrangements). They are:

1   ecological integrity and the need to conserve entire ecosystems;
2   representativeness and the need to conserve biodiversity hotspots;

3   enhancing biodiversity;
4   enhancing economic viability (as national parks must be sustainable); and
5   social responsibility.

Best practice management

Central to conservation best practice within SANParks are a number of key components, including a fundamental understanding of the ecological patterns and processes that inform the goals of the organization, and adequate and concise management plans that detail the specific conservation requirements for individual parks, based on clearly defined objectives that address critical threats. Monitoring systems to assess ecosystem health and management impacts are vital, as is the necessity of integrating science and management to inform of the need for and the consequences of interventions. Management also needs to be driven by conservation and must utilize the most appropriate tools to achieve its objectives.

As noted above, SANParks has five principles driving their land acquisition strategy, with non-conservation factors affecting land incorporation and acquisition strategies, including market supply and demand for appropriate conservation land, economic incentives, social responsibility, political pressure and operational requirements. SANParks has been instrumental in developing procedures – following the Global Environment Facility best practice guidelines – to facilitate the purchase of land while ensuring that the needs of any affected farm workers are considered. Property purchases are guided by resettlement action plans that seek to provide employment or a range of other capacity development programmes to resettled workers. One of the key tenets of this policy is that no farm workers are disadvantaged as a result of SANParks' land purchases.

SANParks is currently in the process of implementing Integrated Environmental Management Systems (IEMS) within national parks to address the need to be accountable for conservation and management actions taken. The organization has also recently completed – for all parks – draft park management plans that have been subject to public comment, as required by the Protected Areas Act and its amendments. The IEMS facilitate the ongoing improvement of management efficiency within parks and strengthen the adaptive management approach to meeting corporate and conservation values. In addition to the strategic park management plans, a series of Conservation Development Frameworks that incorporate sensitivity mapping and zonation schemes, are currently being drafted for a number of parks and will form an integral part of park development processes. Underpinning these plans and frameworks is the use of biodiversity- and conservation-related information, as well as cultural and heritage data, and additional value systems such as aesthetics.

In an effort to fully integrate the services provided, the Conservation Services Division has been reorganized. A number of significant departures from the previous organizational structure are evident, including the focus on integrating active adaptive management principles (Walker, 1998; Enck et al, 2006) across the various programmes with specific focuses on biodiversity hotspots and climate change, transfrontier conservation areas, conservation planning, policy development and ecological functioning. Current scientific support services are being assigned to various programmes under these focal areas to streamline the provision of scientific support in an integrated manner.

SANParks commitment to social and economic objectives

Accompanying the expansion of the national parks system in the democratic environment of the post-1994 South Africa was the need to address the social injustices of the past through land restitution. Historically, local communities were not consulted when areas were set aside for conservation, and in some cases they were dispossessed of their land. SANParks' participation in land restitution programmes is viewed as complementary to meeting conservation objectives, as every effort is made to ensure the continued conservation of the land affected. The land claims in the Kruger National Park and the Kgalagadi Transfrontier Park have seen land titles returned to local communities, while there has been little change to the conservation status of these areas (see Chapter 22). However, the land restitution and subsequent co-management of these areas are not without challenges, and it may yet be too early to determine the success of these ventures. Undoubtedly there are lessons to be learned from these processes, which will be integral to improving the levels of community involvement in achieving conservation and social goals in the future. A proactive and adaptive approach should be adopted when settling such claims, as prescriptive agreements may not be relevant in all possible co-management scenarios. An opportunity exists for improvement in these areas, which should be addressed by SANParks, government departments, NGOs and constituent communities in a transparent consultative fashion.

In other areas, where the anticipated future land use options are not compatible with conservation efforts (for example mining), land claims have resulted in SANParks establishing new parks in the region (for example the impending deproclamation of Vaalbos National Park and the acquisition of a new area to replace these landscapes).

As mentioned above, as part of SANParks' efforts to maintain ecosystem and ecological services, a number of rehabilitation, alien vegetation removal and infrastructure development programmes have been implemented, largely through Working for Water, Poverty Relief, CoastCare and Working for

Wetlands programmes. These activities were conducted across a number of parks and were designed to address not only biodiversity and conservation requirements, but also to meet social responsibilities in terms of employment generation, capacity-building and poverty alleviation. By March 2005, poverty relief projects had created 42 small, medium and micro enterprises (SMMEs) and used a further 194 for supplying services (SANParks, 2005). The Expanded Public Works Programme employed almost 8000 people, while training was also an important focus of these initiatives. The long-term viability of these SMMEs is yet to be determined, but with a further R328 million (US$45 million) for the next round of public works programmes in national parks, the continued development of these enterprises should be secure (SANParks, 2005).

Many of these projects are also the focus of SANParks' People and Conservation constituency building programme. One of the current approaches is the establishment of park committees for each national park. The purpose of these committees is to consider the values of surrounding constituents in setting management objectives for the national parks. The committees are elected by the communities and liaise with park officials on matters of interest to the communities. While some committees have already been established (for example at Mapungubwe and Table Mountain), many have yet to be constituted, though the continued inclusion of communities and other stakeholders is embedded within SANParks' management strategy.

While SANParks is primarily responsible for conserving South Africa's natural heritage for future generations, it also by necessity performs a tourism and hospitality function to ensure its financial sustainability. Although the organization has been through troubled financial periods in the past, the most recent financial review is indicative of the strength of the organization. Gross revenue increased to R486.1 million (US$66.3 million) in 2006 from R211.5 million (US$24.9 million) in 2001 (see Figure 24.1), an increase of approximately 18 per cent per year.

There was a decrease in the operating loss from R98.4 million (US$11.6 million) in 2001 to R75.6 million (US$10.1 million) in 2003 as a result of increases in gross revenues and decreases in human resource expenditures due to the organizational restructuring of 'Operation Prevail'. However, these gains have more recently been offset by increases in general operating expenses, human resource costs, grants to special projects and maintenance costs, which all contributed to an increase in the operating loss to R134.7 million (US$17 million) in 2006. Adjustments to take account of operational grants from government and live game sales results in a net positive income to expenditure ratio.

There has been an increase in the income received from operations from R5.52 million (US$650,000) in 2001 to R39.25 million (US$6.3 million) in 2006, though the net surplus of R41.35 million (US$5.6 million) reported for 2006 has shown a steady decline since the peak surplus in 2002 (R52.93 million, US$5 million), but was still well above the figure for 2001 (see Figure 24.1) (SANParks, 2006a). There are a number of areas where SANParks can improve its financial standing, and one means of achieving this would be to leverage funds from donor agencies locally and internationally, particularly in the US.

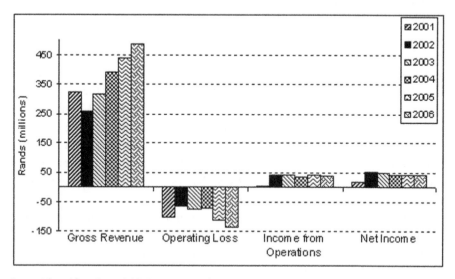

Source: Adapted from SANParks' 2001–2006 annual reports

Figure 24.1 *Financial position of SANParks, 2001–2006*

Since 1999, when SANParks adopted commercialization as a conservation strategy, there has been a drive to leverage expertise and capital from the private sector, while enabling SANParks to focus on the core conservation function of the organization (Fearnhead and Mabunda, 2003). To this end SANParks established a number of privately run concession areas within the national parks system. The process, although not well accepted in some circles, was expected to contribute significant financial benefits to SANParks over the long term without the financial risk associated with operating non-core tourism operations such as upmarket lodges (Fearnhead and Mabunda, 2003). That these expectations were met has been demonstrated by the 12 lodge concessions currently operating within national parks countrywide. In the 2005/2006 financial year some R10.2 million (US$1.4 million) was generated for SANParks through these concessions, though not all were fully operational (SANParks, 2006a).

Critical to the success of the concessions is close monitoring by SANParks to ensure that acceptable service levels are maintained at all times, thus their management is carefully controlled by contractual agreements aimed at meeting sustainability objectives, and contravention of these agreements is closely monitored by SANParks to ensure that potential conflict situations on both operational and ecological levels are dealt with quickly. As part of the agreements governing the operation of the concessions, and in addition to the income generated, these ventures also contribute to job creation and regional economic development.

The concession strategy also saw the commercialization of all shops, restaurants and cafeterias within national parks in 2001. Once again, the interest shown by the private sector in taking over these operations has allowed SANParks to focus on its core conservation function, and the net income generated from these operations exceeded R16 million (US$2.2 million) in 2006. There is also scope for future outsourcing of a number of operational activities including information technology, housekeeping, gardening, hygiene and security services. If this outsourcing is to take place, these agreements will also need to be carefully monitored by SANParks to ensure the maintenance and/or improvement of the service levels delivered.

The commercialization strategy is not without problems. In recent years, SANParks has reviewed various contracts and offered relief mechanisms to concessionaires (for example eliminating the need to pay fixed concession fees), made necessary by fluctuating tourism market conditions. Although this may result in some decline in concession revenues in the short term, the long-term benefits are still expected to outweigh such shortfalls (SANParks, 2005). SANParks has also recently reviewed their approach to commercialization and the establishment of PPPs, and the Strategic Plan for Commercialization focuses on the need for achieving responsible tourism and commercialization, rather than depending on commercialization solely as an alternative source of funding (SANParks, 2006a).

SANParks has also recognized the need to review the scope of their activities and to broaden their target markets. It has identified the need to develop and foster a stronger constituency among the South African public, particularly among black citizens. There have been significant advances on this front, with the number of black South Africans visiting national parks increasing from only 4 per cent in 2003 to almost 20 per cent of the domestic tourism market in 2006, which represents 75 per cent of all visits to national parks (SANParks, 2006a, 2006b).

Tourism development in SANParks

Effective conservation and other park responsibilities such as community empowerment are highly costly. In South Africa, where sectors such as health, education and housing demand attention from the electorate, the state cannot cover all the costs associated with conservation. Nevertheless the South African government is the biggest financial supporter of SANParks, providing an annual conservation grant that has increased from about R50 million (US$6.7 million) in 2003 to R73.6 million (US$11.6 million) in 2005, with a further R12 million (US$1.9 million) for road improvements (SANParks, 2005). In addition over the four years of 2002–2005, an additional R89 million (US$12.1 million) was allocated for land acquisition (SANParks, unpublished data). Yet despite this invaluable support, the onus is on SANParks to do its utmost to be as financially self-sustaining as possible, a challenge that is addressed primarily through the generation of income from tourism activities.

Tourism to national parks in South Africa is over three-quarters of a century old and has long been regarded as a mechanism to ensure the continued existence of these conservation areas. In 1925, the warden of the Sabi Game Reserve (the southern portion of the area proclaimed as Kruger National Park in 1926) James Stevenson-Hamilton noted, 'without public support and revenue from visitors, the Park's future was severely limited, if not doomed' (Paynter, 1986, p57).

Since the first democratic elections in 1994, SANParks has been faced with considerably increased domestic and foreign tourist markets. The massive increase in tourist numbers to national parks in South Africa is shown in Figure 24.2. In 1927 when Kruger National Park opened to the public, 27 guests in three vehicles entered the gates. In the 11 years prior to the 1994 elections, growth in visitor numbers was 3.3 per cent per annum, while in the following 11 years, tourism growth in the Kruger National Park was 4.3 per cent per annum. In 2002–2003, over one million visitors entered the park in a single year for the first time, and visitor numbers have continued to increase. Recent tourism statistics reveal that the total number of visitors to South Africa's national parks has increased substantially. It is evident that much of this increase can be accounted for by an elevated number of day visitors rather than those staying overnight, as there has been a steady decline in the unit and bed occupancy rates (see Figure 24.3). The 1.5 million visitors to the Table Mountain National Park contribute the bulk of this increase in day visitor numbers.

The ongoing development of tourism opportunities within protected areas can produce concomitant deterioration of both natural habitats and the tourism experience (Belnap, 1998; Manning, 1999; Leung and Marion, 2000; Newsome et al, 2002; Choi and Sirakaya, 2006). Thus the onus is on SANParks to protect and manage the natural environment, and also to encourage a sense of ownership and pride in the parks through education, economic opportunity and leisure through tourism.

The eruption of the nature-based tourism industry in South Africa has

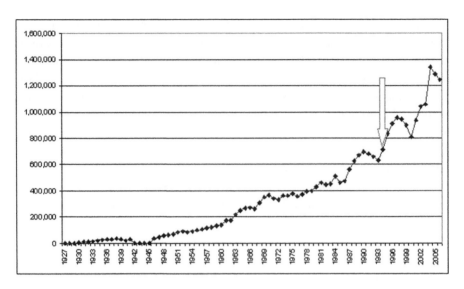

Note: The arrow depicts the democratic elections that took place in 1994

Source: Unpublished SANParks data

Figure 24.2 *Trend in visitor numbers, Kruger National Park, South Africa, 1927–2006*

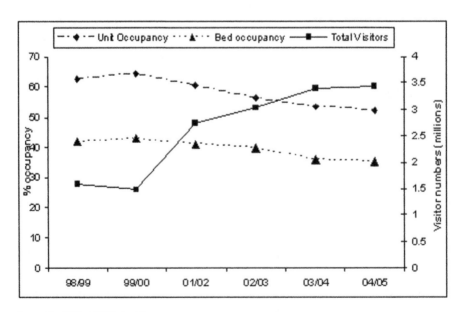

Source: Unpublished SANParks data

Figure 24.3 *Total visitor numbers to national parks, average unit and bed occupancy rates*

meant that national parks have to ensure they offer a competitive tourism product to guarantee their continued existence, while still meeting conservation and social objectives. Some of the approaches adopted to enhance SANParks' tourism competitiveness include product diversification, tourism plant upgrading and refurbishment, outsourcing of elements of non-core competency, channelling government funds into areas of community poverty relief, establishing loyalty programmes and re-evaluating park utilization fees.

Product diversification has seen the unbundling of rigid and restrictive rules that limited the ability of the visitor to interact with the park environment. In some parks SANParks has opened up activities such as guided drives at various times of night and day, guided walks into wilderness areas, mountain biking, horse-riding, abseiling, canoe trails and golf, though purists may argue that such activities constitute an environmental risk. Diversification has also manifested itself in the provision of accommodation of differing standards aimed at various socio-economic groups, from budget camping through to super-luxury establishments.

Charging a daily utilization fee has been another innovation. Historically, it had been accepted practice in South Africa to pay a nominal once-off fee to enter a national park, which did not take into account the amount of time a visitor spent within the park. It was acknowledged that such a utilization fee did not recognize that visitors have the opportunity to experience a wonderful diversity of fauna, flora, culture and landscapes and make use of infrastructure, all of which require considerable financial and human resources to conserve and manage. After many years of virtual 'free access' to SANParks, a new per diem utilization fee was introduced in 2003 and a loyalty programme aimed at facilitating park visitation at minimal cost was also launched. The loyalty programme, the WildCard, was pioneered at the Table Mountain National Park in 2000 and expanded to include all parks in 2003. It is hoped that the WildCard will increase domestic visitation to the parks, and it is believed that this innovation will create a heightened domestic awareness of national parks and of South Africa's protected biodiversity, particularly among the country's youth.

Conclusions

The types of initiatives described above all contribute to SANParks' strategy to build a constituency within the South African public. In order to achieve this, the benefits of national parks to local communities need to be strengthened and urgent attention needs to be given towards improving visitor demographics to be more representative of all South Africans, although recent figures suggest that these inequities are slowly being resolved. SANParks has also identified a need to drive environmental education programmes within the national parks and at the national level to promote and foster an appreciation for the natural environment, which will require the involvement of the South African public, all levels of government, the private sector and NGOs.

The success recorded by SANParks in recent years would not have been achieved without those who make it possible. SANParks has realigned itself within the new South Africa to meet employment equity targets while promoting skills development and retaining operational excellence, and in excess of 3000 people from all walks of life are employed as permanent staff to ensure that SANParks delivers on its objectives (SANParks, 2006a). From rangers and scientists to resource economists and accountants, the diversity of cultures and experience all contribute to SANParks' success. In addition to these staff are the dedicated efforts of a nationwide public volunteer force of more than 500 members of the Honorary Rangers Corps and over 1000 junior honorary rangers (between the ages of 12 and 18 years).

SANParks has met the challenges facing it head on and has emerged a stronger organization for it, making significant contributions to meeting its conservation mandate. Although there are improvements still to be made, and some of these have been highlighted, the successes just in the past few years have been considerable in a difficult conservation climate.

References

Belnap, J. (1998) 'Environmental auditing, choosing indicators of natural resource condition: A case study in Arches National Park, Utah, USA', *Environmental Management*, vol 22, pp635–642

Brooks, T. M., Mittermeier, R. A., Mittermeier, C. G., da Fonseca, G. A. B., Rylands, A. B., Konstant, W. R., Flick, P., Pilgrim, J., Oldfield, S., Magin, G. and Hilton-Taylor, C. (2002) 'Habitat loss and extinction in the hotspots of biodiversity', *Conservation Biology*, vol 16, pp909–923

Choi, H. C. and Sirakaya, E. (2006) 'Sustainability indicators for managing community tourism', *Tourism Management*, vol 27, pp1274–1289

DEAT (2003) *People, Parks and Transformation in South Africa: A Century of Conservation, a Decade of Democracy*, Department of Environmental Affairs and Tourism, Pretoria

DEAT (2005a) *South Africa's National Biodiversity Strategy and Action Plan*, Department of Environmental Affairs and Tourism, Pretoria

DEAT (2005b) *Annual Review 2004–2005*, Department of Environmental Affairs and Tourism, Pretoria

Driver, A., Maze, K., Rouget, M., Lombard, A. T., Nel, J., Turpie, J. K., Cowling, R. M., Desmet, P., Goodman, P., Harris, J., Jonas, Z., Reyers, B., Sink, K. and Strauss, T. (2005) 'National spatial biodiversity assessment 2004: Priorities for biodiversity conservation in South Africa', *Strelitzia*, vol 17, pp1–46

Enck, J. W., Decker, D. J., Riley, S. J., Organ, J. F., Carpenter, L. H. and Siemer, W. F. (2006) 'Integrating ecological and human dimensions in adaptive management of wildlife-related impacts', *Wildlife Society Bulletin*, vol 34, no 3, pp698–705

Fearnhead, P. and Mabunda, D. (2003) 'Towards sustainability', in Hall-Martin, A. and Carruthers, J. (eds) *South African National Parks: Celebration*, Horst Klemm, Johannesburg

Friedmann, Y. and Daly, B. (eds) (2004) *Red Data Book of the Mammals of South Africa: A Conservation Assessment*, Endangered Wildlife Trust, Johannesburg

Leung, Y-F. and Marion, J. L. (2000) 'Recreation impacts and management in wilderness: A state-of-knowledge review', in Cole, D. N., McCool, S. F., Borrie, W. T. and O'Loughlin, J. (eds) *Wilderness Science in a Time of Change. Volume 5: Wilderness Ecosystems, Threats, and Management*, US Department of Agriculture, Rocky Mountain, Colorado

Lombard, A. T., Hilton-Taylor, C., Rebelo, A. G., Pressey, R. L. and Cowling, R. M. (1999) 'Reserve selection in the Succulent Karoo, South Africa: Coping with high compositional turnover', *Plant Ecology*, vol 142, pp35–55

Low, A. B. and Rebelo, A. G. (1996) *Vegetation of South Africa, Lesotho and Swaziland*, Department of Environmental Affairs and Tourism, Pretoria

Manning, T. (1999) 'Indicators of tourism sustainability', *Tourism Management*, vol 20, pp179–181

Mucina, L. and Rutherford, M. C. (eds) (2006) 'Vegetation map of South Africa, Lesotho and Swaziland', *Strelitzia*, vol 19, pp1–816

Nel, J., Davies, S., Kotze, I., Gelderblom, C., van Wilgen, B., Schonegevel, L., Hughes, S., Kerley, G., Cowling, R. and Boshoff, A. (2002) 'Conservation planning framework for the greater Addo Elephant National Park', CSIR Report Number ENV-S-C 2002-010, CSIR, Pretoria

Newsome, D., Moore, S. A. and Dowling, R. K. (2002) 'Natural area tourism: Ecology, impacts and management', in Cooper, C. and Hall, M. (eds) *Aspects of Tourism 4*, Channel View Publications, Colchester

Paynter, D. (1986) *Kruger: Portrait of a National Park*, Macmillan Publishers, Johannesburg

Rouget, M., Reyers, B., Jonas, Z., Desmet, P., Driver, A., Maze, K., Egoh, B., Cowling, R. M., Mucina, L. and Rutherford, M. C. (2004) *South African National Spatial Biodiversity Assessment 2004: Technical Report. Volume 1: Terrestrial Component*, South African National Biodiversity Institute, Pretoria

SANParks (2002) *South African National Parks: Annual Report 2002*, South African National Parks, Pretoria

SANParks (2003) *South African National Parks: Annual Report 2003*, South African National Parks, Pretoria

SANParks (2004) *South African National Parks: Annual Report 2004*, South African National Parks, Pretoria

SANParks (2005) *South African National Parks: Annual Report 2004/2005*, South African National Parks, Pretoria

SANParks (2006a) *South African National Parks: Annual Report 2005/2006*, South African National Parks, Pretoria

SANParks (2006b) 'SANParks bursting through the demographics barrier', http://celtis.sanparks.co.za/about/news/default.php?id=150 (accessed 15 February 2006)

Walker, B. (1998) 'The art and science of wildlife management', *Wildlife Research*, vol 25, pp1–9

WWF-SA (2002) *The Biodiversity of South Africa 2002: Indicators, Trends and Human Impacts*, Struik Publishers, Cape Town

Younge, A. and Fowkes, S. (2003) 'The Cape Action Plan for the environment: Overview of an ecological planning process', *Biological Conservation*, vol 112, pp15–28

25

Privately Managed Protected Areas

Peter Fearnhead

Southern Africa has enjoyed much success in the biodiversity conservation sector with several important ecosystems secured and some endangered species having been saved from extinction. Financial success, however, has been largely limited to initiatives that have taken place on private land; innovation and success in the commercial and financial spheres has been rare in state-protected areas, despite the fact that many of these areas are the most important from a conservation perspective and offer some of the best commercial opportunities. Indicators of stagnation of the formal conservation sector include low employee morale and high staff turnover, low operational budgets and wages, few quality people being attracted into the industry (at least in comparison to other closely related industries), protracted and bureaucratic decision making, and sometimes, unfortunately, corruption. The consequences of stagnation have been a reduction in the size of the formal wildlife estate that is under effective management and the reversal of many early biodiversity conservation successes (as evidenced by declining wildlife populations in some protected areas). The resettlement of sections of parks and wildlife reserves by people and the consequent conversion of land use to subsistence agriculture is also an indicator of the failure of state-protected areas. Examples of such situations exist in Mweru Wantipa, Sumbu, Lusenga Plain, Isangano, Lavushi Manda, Blue Lagoon, Lochinvar, West Lunga and Sioma Ngwezi National Parks in Zambia; Zinave, Banhine and Limpopo National Parks in Mozambique; Kasungu National Park, Nkhotakota, Mwabvi and Majete Wildlife Reserves in Malawi and Gonarezhou National Park in Zimbabwe. Although depleted and run-down protected areas can be rehabilitated, the process is usually prohibitively expensive, takes long periods of time and is fraught with social and political challenges. Ironically, the decline in status of state-protected areas has been occurring when the size and quality of the private (and in a few exceptional cases communal) conservation estate has been increasing. When innovations have occurred in the state conservation sector, they tend to have been dependent on individuals, rather than as a result of an institutional culture that encourages and rewards innovation. This chapter explores some of the reasons for this, and provides some early examples of management of protected areas by the private sector.

'Legislated mediocrity'

The case has often been made that insufficient funding of protected area management agencies is most probably the single most important threat to biodiversity conservation in southern Africa (for example de la Harpe, 2003; Emerton et al, 2006). However, limited funding is also partially the result of the inability of protected area management agencies to innovate, and the underlying cause of the lack of innovation is the institutional nature of such agencies.

The wildlife legislation throughout southern Africa anticipates only government or state agencies undertaking protected area management. This implicitly prevents competition in the management of parks. With no competition in service delivery, there is little or no incentive for innovation. By contrast, in private organizations with the discipline of the bottom line and the pressures of competition, survival requires innovation in methods and mechanisms that are used to enhance revenue and control or reduce operating costs (de la Harpe, 2003). For example, in South Africa, despite the 11 different state authorities managing protected areas, the responsibilities of these agencies are geographically defined and there is no competition to deliver services among them. Further, there are often no legal mechanisms to terminate the services of a bankrupt or poorly performing management institution, resulting in 'legislated mediocrity'. With the survival of the institution delinked from its performance, and with the costs thereof borne by society, there is no economic reason to be anything other than mediocre. This is further complicated by the state being both manager and regulator, judging its own management performance. In a healthy and vibrant sector, these responsibilities are split.

Those who are uncomfortable with a competitive approach to the management of protected areas often focus on the fact that commercial companies seek to make a profit and that conservation is not about profit. This misses the point: profit is simply the measure by which performance in commercial companies is judged – in the case of protected area management, all that is required is a clear understanding of how to measure performance. A simple set of measures, including income generation, can be established to measure protected area management performance that can then be used to compare the management outcomes of one institution with another. Importantly, there should be a mechanism to transfer the management of specific protected areas from poorly performing institutions to better-performing ones.

Legislative changes that allowed the devolution of authority over wildlife to individuals and communities in countries such as Namibia and Zimbabwe has probably been the single most successful innovation in the conservation sector in the last 30 years. In effect, this introduced a level of competition between different landowners who chose wildlife as a land use option, where better management was rewarded with increasing income levels either from tourism concessions or from trophy hunting fees. Introducing competition into the management of formal protected areas is almost certain to yield similar results.

The establishment of parastatals

Some governments have established semi-autonomous agencies or parastatals to manage the conservation estate, which is done in the belief that they allow a more businesslike management approach and greater financial independence, as well as breaking with the bureaucratic requirements of governments' financial, accounting and employment requirements. Revenues generated from the wildlife estate are retained by the agency rather than going to state coffers, thereby introducing some incentive for conservation managers to generate income. As a result, some of the more commercially oriented agencies have diversified their funding sources and reduced their reliance on central transfers of tax revenues. However, the costs of establishing and maintaining a separate institution are often more expensive than anticipated, with little hope of being able to generate sufficient funds from operations. This was the case with both the Zambian Wildlife Authority and Zimbabwe's Parks and Wildlife Management Authority (de la Harpe, 2003). The need to generate income often results in the agencies attempting to provide commercial services (for example, accommodation, shops and restaurants), thereby competing with the private sector, or even crowding them out completely using artificial restrictions or subsidies. The provision of such services by a parastatal is economically inefficient and an impediment to the entire sector. Generally speaking, while parastatals tend to function better than government departments, taking on some of the characteristics of private corporations – such as flat management structures, flexibility in staff compensation and rewards for good performance – they still tend to operate outside of competition and are therefore institutionally inferior.

New models of protected area management

Over the last decade, there has been a proliferation of PPPs that endeavour to improve financial returns from state-managed protected areas, though they have generally related to opportunities for tourism enterprises. In South Africa this was led by the North West Parks and Tourism Board in 1994 (Davies, undated; see Chapter 18) and SANParks in 1999–2002 (Fearnhead, 2003; see also Chapter 24). In the case of SANParks, this involved the concessioning of lodges as well as the outsourcing of numerous restaurant and retail facilities. According to a formal audit following the outsourcing of the restaurant and retail facilities, net profit to SANParks in 2002 was R6 million (US$0.92 million) and the comparable net profit to SANParks in the first year post-outsourcing was R20 million (US$3.08 million) (Price Waterhouse Coopers, 2003, cited in Fearnhead, in prep.). The concessioning of lodges, mostly within Kruger National Park, also yielded exceptional results, with some lodges bidding as much as 22.5 per cent of their turnover as a concession fee that in 2006, for a single lodge of 42 beds, would yield an income to SANParks of US$1 million.

Traditionally PPPs involve the private sector generating an income from an opportunity owned or controlled by the state. However, the more interesting innovations in PPPs have involved the delegation of all or part of the non-income generating components to the private sector, such as general park and biodiversity management and community participation. Examples of such partnerships discussed in this chapter involve Marakele and Addo Elephant National Parks and the establishment of the African Parks Foundation, the first non-governmental multi-park management agency.

Marakele National Park

The system of protected areas inherited by post-apartheid South Africa was inadequate in terms of both size and representivity. As a result, SANParks had to identify target areas for expansion through the purchase of farmland. Since most land in South Africa is privately owned this had to be purchased on a willing buyer–willing seller basis. Lack of financial resources meant that park expansion was slow, with the process further frustrated by land speculators and developers that drove land prices upwards. Marakele National Park in the Limpopo Province of South Africa was no exception – it was fragmented and incomplete, with an additional 15,000ha identified as necessary to complete the core area of the park. Given the financial limitations of SANParks, only the occasional purchase of farmland was possible for incorporation into the park. In 2000, Paul Fentener van Vlissingen, a wealthy philanthropist offered to assist with the development of the park, and an agreement was concluded between a company formed for this purpose (Marakele (Pty) Ltd) and SANParks. In terms of the agreement, Marakele (Pty) Ltd was to purchase an earmarked 13,700ha, the bulk of the desired area, to be incorporated into

Marakele National Park as a contractual national park. Since SANParks wanted to own the core area of the park, it was granted an option to purchase the 13,700ha of land at the original purchase price, fixed in real terms. By doing so, Marakele (Pty) Ltd effectively financed the expansion and development of Marakele National Park for SANParks at a 0 per cent real interest rate and with no risk.

Marakele (Pty) Ltd is responsible for all infrastructure and development costs and SANParks is responsible for stocking the area with wildlife, introducing approximately R27 million (US$4 million) worth of game into the Marakele (Pty) Ltd section of the park. Game ownership remains with SANParks, which receives the proceeds from any sales of game. Marakele (Pty) Ltd has the tourism rights on its land (though the extent of development is limited to 100 beds) and a revenue-sharing agreement is in place – 12 per cent of turnover is paid by the tourism operator, one third of which goes to SANParks, one third to Marakele (Pty) Ltd and one third to the landowner (the latter being an additional incentive for SANParks to purchase the land). Revenue earned from entrance fees paid by tourists to the Marakele (Pty) Ltd land is also shared between SANParks and the company. The entire Marakele National Park – including the contractual section – is now governed by a single management plan, with SANParks and Marakele (Pty) Ltd each responsible for the management of their land. The company is also responsible for the management of a few of SANParks' properties located within the Marakele (Pty) Ltd section.

In retrospect it is difficult to believe that such an agreement – which has such a clear benefit to the state – could ever have been contested, but at the time it was being negotiated there was extreme scepticism within some quarters of SANParks' senior management and board. This was probably the result of a combination of the traditionally risk-averse nature of state conservation employees, the fear of being exploited by a private partner, a general reluctance to embrace partnerships, or possibly even simple 'turf' protection. In the execution of the agreement, culture differences still characterize the relationship between the two. SANParks views Marakele (Pty) Ltd as being commercially driven, with a focus on financial sustainability sometimes at the expensive of more traditional conservation principles (Knight, pers. comm.), while Marakele (Pty) Ltd feels that SANParks is slow to fulfil its contractual obligations and is slow and bureaucratic in its decision making (Schroder, pers. comm.).

Though comparisons between the management performances of each institution are largely anecdotal, one quantitative comparison is average monthly electric fence voltages for adjoining fences managed by the two institutions (see Figure 25.1). A consistently high voltage is an indicator of the quality of management (preventing wildlife breakouts, etc.). Over a 33 month period, Marakele (Pty) Ltd had a total of two months where a zero reading was recorded on at least one day, for a total duration of two days. In comparison,

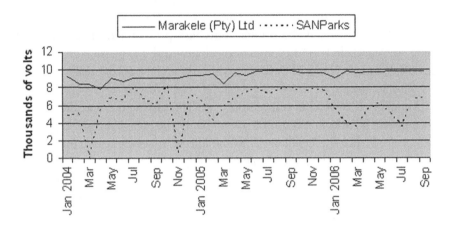

Source: Schroeder (pers. comm.)

Figure 25.1 *Comparison of average monthly voltages for an adjoining fence maintained by a private conservation entity and SANParks*

SANParks had a zero reading recorded in 18 of the 33 months with a total duration of 119 days.

Despite difficulties, the agreement accelerated the ecological and financial development of Marakele National Park. The land purchase, rehabilitation and restocking took approximately three years to complete, a process that would have taken SANParks considerably longer, if at all achievable. The first lodge opened in December 2005 and revenue was expected to start flowing to both Marakele (Pty) Ltd and SANParks in 2007, after the expiry of the agreed 18-month rent-free establishment period. The second lodge is scheduled to be open for trade in late 2008 and the third in late 2009. At 2007 land prices, the value of the option to SANParks is estimated to result in a windfall gain to SANParks in excess of R100 million (US$15.4 million).

Addo Elephant National Park

In a similar contractual arrangement to that of Marakele, SANParks made an agreement with a foreign consortium ('Kuzuko') regarding conservation land falling within the area identified for the expansion of the Addo Elephant National Park (AENP) in the Eastern Cape of South Africa (see also Chapter 19). The major difference in the agreement to that of Marakele was that the consortium did not want to grant an option to SANParks to purchase the land at a future point in time, so the contract incorporates the land into the park for 100 years. The Kuzuko land is privately managed at Kuzuko's expense, although in accordance with the overall management plan for AENP. Kuzuko has the rights to develop tourism lodges on its own land, although SANParks will receive a share of revenue earned from these enterprises, equivalent to 3.5 per cent of gross turnover from all lodges and 50 per cent of entrance fees. Kuzuko and SANParks share equally any income from hunting that arises out of the incremental increase in game numbers on their land. In the three years since the Kuzuko agreement was signed, the area has been fenced with 'Big 5' predator proof fencing and elephant, black rhino, buffalo and plains game have been reintroduced. Plans are under way for the development of a tourist lodge, which was expected to open in 2007.

In a separate case, a businessman purchased a farm that formed a critical link between two existing sections of the park and built an upmarket tourism lodge on the land. This apparent problem was solved by incorporating his land into the park, adding immediate value to the game-viewing experience offered by his lodge. In return SANParks receive park entrance fees from all guests to the lodge, 7.5 per cent of lodge turnover, and they have an option to purchase the land for R100 (US$15) after 50 years.

Another approach adopted by SANParks in the expansion of AENP was a R35 million (US$5.4 million) loan from the Industrial Development Corporation, granted on concessionary interest rates with a three-year window on principal repayments, which required the approval of the Ministers of Finance and of Environmental Affairs and Tourism. The notion of borrowing money to buy land for a national park caused a great deal of debate within the Ministry of Finance (MoF), which was understandably nervous of the financial consequences for SANParks arising from the loan repayments, although the intention had been to finance loan repayments from the proceeds of high-value game sales. The outcome of the debate was that the MoF acknowledged that it was governments' role to fund park development and hence fund the loan repayments, and the MoF introduced an annual grant to SANParks for this purpose. Initially this matched the loan repayments but the principle had been established and as a result the size of the annual grant began to grow to fund land purchase elsewhere. This grant was R36 million (US$5.5 million) in 2005 and R55 million (US$8.5 million) in 2006 (Knight, pers. comm.).

African Parks

In 1999, four individuals[1] conceived the idea for a private sector conservation organization and registered the African Parks Management and Finance

Company (Pty) Ltd in 2000. The group was familiar with the inefficiencies associated with state-driven conservation and anticipated the benefits of a more commercial approach, with parks being run on business principles. The founders believed that good natural resource management, a focus on income generation, and a healthy respect for costs would enable more parks to be financially self-sufficient than was the case at the time. They were also aware that governments did not have the financial resources to adequately fund their parks, let alone pay for the services of a professional park management company, so offering a total financial solution was deemed necessary if the company was to achieve its objectives. With a history of poor accountability for donor funds by some state institutions, it was expected that donors would welcome an alternative institution, which emphasized transparency and accountability. It was also assumed that donors, whether institutional or private, would be prepared to pay a management fee to ensure professional use of the funds.

In August 2000, during negotiations over the Marakele contractual park agreement, the concept of African Parks Management and Finance Company (Pty) Ltd (AP) was introduced to the Dutch philanthropist, van Vlissingen. Within four weeks, he asked to become an equal stakeholder in the company with a financial commitment of US$50,000 to cover the initial operating costs. This was unanimously accepted by the founders and he henceforth played a critical and influential role in the shaping of AP, eventually committing in excess of US$50 million by the time of his death in August 2006.

AP was originally established as a for-profit entity that anticipated earning revenue from park management fees. An endowment fund was also designed to be a mechanism for funding the operational costs of various parks, which was to be capitalized through the sale of a number of private leases to individuals who would be permitted to erect non-commercial, light-footprint camps for their private use in parks managed by AP. This endowment was to be professionally managed and the yield was to be used to fund park-operating costs.

The founders set out to persuade the formal conservation sector of the merits of two key philosophies: that a non-governmental institution can be an effective and responsible manager of a formal protected area, and that appropriate remuneration should be paid for professional park management services (recognizing that this would probably have to be paid by donors). It was thought that once this model was demonstrably successful, similar institutions would be attracted to compete in the market, which would lead to innovation and improved management, and conservation as a land use would become more competitive relative to alternative land uses. The ultimate success of AP was to be judged by the existence of competitors adopting a similar model – indicating that the mindset of park management would have shifted from one of 'legislated mediocrity' to a competitive one with institutional survival determined by performance.

The ideal situation for effective protected area management was thought to be one where the state was responsible for establishing policy and performing necessary regulation, and where AP was responsible for on-the-ground management. AP would finance the entire park from its own and donor resources, with all income generated in the park being retained at the park level. Should an operational profit be achieved, then a profit-sharing arrangement among the key stakeholders (including AP) would be developed. It was envisaged that a local (for-profit) company would be established as the legal entity to implement the project, contract with employees and service providers, open bank accounts and so on. This institutional design caused initial confusion in Zambia, with local stakeholders equating ownership of the management company with ownership of the land and wildlife – the Zambia Wildlife Authority (ZAWA) in particular felt that they should have a majority share in the local company as they were 'contributing the park and the animals'. However, the intention was to focus specifically on the management of national parks, and did not involve ownership of either the land or the wildlife, which was to remain state property.

A key aspect of achieving financial sustainability was to be through tourism. AP did not want to develop and operate tourism facilities, preferring this to be done on an arm's-length basis by professional operators. AP believed its responsibility was to be the rehabilitation and management of a park, making it an attractive destination for tourists and providing tourism operators with the stability and surety necessary to generate a commercial return on their investment, and who would pay royalties for the right to operate a lodge. There was also a strong emphasis on ensuring that local people were part of the project at all levels and should benefit from it. One of the criteria for selection of a park was the ability of the park to contribute meaningfully to the socioeconomic development of the region, and each park plan and budget included a number of practical mechanisms to achieve this from the outset of a project.

The early years – 2000 to 2003

AP selected its first target park as Liwonde National Park in Malawi, largely because of the long association of one of the founders with the park and the Malawian Department of National Parks and Wildlife (DNPW). The initial proposal to the Government of Malawi was well received by the Minister of Tourism and the Permanent Secretary of the DNPW, who both expressed their desire for such a project and eagerness to conclude an agreement. Unfortunately, the then President Muluzi dismissed his cabinet, appointed an entirely new one and discussions had to start again. As is often the case with innovations that challenge the status quo, there was a great deal of resistance from some stakeholders to the AP proposal, including from consultants advising the DNPW and Frankfurt Zoological Society who had been supporting law enforcement in Liwonde.[2] As a result, DNPW suggested that AP take on the management of Majete Wildlife Reserve that had been depleted of almost all wildlife.

By this time, engagements with other countries were beginning to reveal a trend in how the concept was being received – governments were assuming that the approach would fail and therefore were electing to minimize the costs of a failure, rather than assessing where the potential benefits of such a partnership could be maximized as quickly as possible. Some countries responded by saying that they wanted the money but without the participation in management; others that were willing to consider participation in management wanted to apply it to extremely marginal parks with very high rehabilitation costs and little chance of ever achieving financial sustainability. Hence, although Majete was far from an ideal opportunity, the board of AP decided that it was worth pursuing on the basis that if the model could be proven in Majete, then it could be successful almost anywhere. However, this was done with the full knowledge that the investment costs would be significant and that it would take a long time before cash inflows were equal to or greater than cash outflows.

The Malawian government decided that the award of Majete should be done through a public tender. When the bid box for tenders was opened and there was only one bid, the minister ordered a rerun on the assumption that other potential bidders had been unaware of the opportunity. Needless to say there was only one bid on the second occasion, but at least the authorities were convinced that there were no other bidders prepared to take on the significant rehabilitation and conservation costs of Majete, without any prospect of ever making a commercial return. AP signed an agreement in March 2003 and operations commenced within a month.

Simultaneously, discussions were being held with ZAWA during 2002 to take on one or more parks in Zambia. Once again, the parks that were initially suggested for partnership were ones that were severely depleted and further complicated with social issues, rather than ones where the benefits of good management coupled with the necessary financial investment could yield early benefits in terms of both biodiversity and socio-economic impacts. In a meeting in July 2002, ZAWA suggested AP consider Liuwa Plain and Sioma Ngwezi national parks in the west of Zambia.

In September 2002, an exploratory trip was planned with discussions at three levels – with the Barotse Royal Establishment (BRE), the traditional authority in the Western Province of Zambia, ZAWA and State House. It was APs' view that support from all three levels was critical to the success of the initiative. BRE remains a very strong traditional authority and operating in

Western Zambia without their consent and support is almost impossible. Liuwa Plain National Park was actually proclaimed as a protected area in the early 1880s by *Litunga* (King) Lubosi Lewanika, making it one of the oldest national parks in the world. Today, many of BRE's traditional ceremonies including the coronation of the *litunga* are still intricately linked to the park and make use of its resources. ZAWA by contrast was the official authority responsible for all parks in Zambia in accordance with the Zambia Wildlife Act (1998). President Mwanawasa was very welcoming and supportive of the concept of private management for protected areas. However, ZAWA did not appreciate the dialogue with and involvement of the BRE and State House – resenting what they perceived as an undermining of their responsibility. Unfortunately this led to the politicizing of the project at many levels, including between the ruling party and the official opposition; between ZAWA and BRE in the tussle between traditional and democratic governance; between factions within ZAWA that were fighting for control of the institution; and between rival factions within BRE. All of this made for difficult and protracted negotiations. However, in February 2003 a one-year MoU was signed between AP and ZAWA for the co-management of both Liuwa Plain and Sioma Ngwezi national parks. Although it was intended that the agreement would be for 20 years, the one-year MoU was a play-safe option for ZAWA who were nervous of both the potential consequences of signing a long-term agreement and of not signing an agreement. Regardless, the MoU allowed the project to commence.

Work on the ground only began in August 2003. The politicking had created a very unstable situation – certain opposition members of parliament, as well as disenfranchised members of BRE fuelled allegations that BRE had 'sold the parks' and that the people would be thrown off the land. As a result, BRE requested time for what they referred to as 'sensitization' that involved the deployment of senior *indunas* (headmen) to explain what the project was about and its benefits to the people of Western Province. AP's approach was simply to get on with the task at hand and ignore the banter. However, the climate of intrigue and suspicion made the negotiation of the long-term agreement very difficult. This was made worse by AP's decision in March 2004 to withdraw from Sioma Ngwezi National Park at the end of its one-year contractual commitment. Having been on the ground for eight months it had become clear that Sioma Ngwezi suffered from some major problems, including decimation of the wildlife populations beyond what was originally believed; the settlement of thousands of Angolan refugees along the Cuando River, adding to the 2000 or more Zambian citizens who were resident in the park when it was established in 1972, and the exclusion of the Zambezi River from the park, which was the only accessible source of permanent water for wildlife (Stichting African Parks Foundation, 2003). ZAWA was also reluctant to commit to assisting AP with addressing these major constraints and supporting the interventions necessary to make the park a success, including making game available to restock the park and the establishment of corridors to provide access to the Zambezi and Cuando rivers. As a result of the combination of the ongoing hostilities and ZAWA's reluctance to address the problems facing Sioma Ngwezi, AP decided to withdraw and reallocate its resources to another project where the objectives could be achieved, and which would yield higher ecological and socio-economic returns on investment. Perhaps understandably, ZAWA felt that this was sufficient grounds to terminate discussions on the contract for Liuwa Plain as well and it was only the direct political intervention of *Litunga* HRH Lubosi Imwiko, of Western Province, that ensured the conclusion of the long-term agreement for Liuwa Plain National Park.

AP's involvement in Ethiopia was the antithesis of its involvement in Zambia. Hearing an interview about AP on the radio in late 2003, the Ethiopian Ambassador to the UK requested a meeting with the chairman of AP to explore possibilities for its involvement in Ethiopia. As a result AP was invited to Ethiopia to assess potential parks for such a partnership. After a single reconnaissance trip and one round of negotiations, a contract for the management and financing of Nech Sar was signed in February 2004.

Progress with park management

Early progress in Majete and Liuwa was excellent. Management teams were put on the ground, radio-based communication systems were established, offices built or refurbished, equipment purchased and scouts properly equipped. Morale among law enforcement officers was very low in both parks before the implementation of the PPP agreement. In the eight months prior to AP's involvement, a total of three patrols were conducted in Liuwa Plain National Park; by comparison AP conducted a total of 3550 patrol days in the first eight months of management and 7200 patrol days during 2004 (Stichting African Parks Foundation, 2003, 2004). It was not necessary to train and employ more scouts for Majete National Park, but the 14 existing scouts responded extremely well to the new discipline and management, confiscating a total of 292 functional firearms, 70 gin traps and 250 wire and light cable snares within the first 20 months of operations (Stichting African Parks Foundation, 2004).

In Liuwa Plain National Park, the initial estimate of wildebeest numbers in 2002 was 15,000 and species such as roan antelope, eland, hartebeest, buffalo and wild dog were locally extinct (Stichting African Parks Foundation, 2004). The wildebeest responded well to improved law enforcement with numbers increasing to an estimated 23,000 in late 2004 (after calving) and 30,000 in April 2007. Subsequently roan antelope moved into the park and wild dog are now being seen on a regular basis (Turner, pers. comm.). In Majete Park, a 40km long and 1.8m high electrified game fence was erected within four months to create a 14,000ha sanctuary for reintroduced animals. A capture and translocation operation succeeded in introducing 599 animals into Majete Park, including two black rhino, 98 waterbuck, 99 sable antelope, 120 buffalo, 60 warthog, 211 impala and 9 nyala made available to AP by DNPW (Stichting African Parks Foundation, 2003). This was supplemented in 2004 with a further 100 buffalo and 15 nyala from DNPW, as well as 20 Livingstone's eland, 37 zebra and 4 Lichtenstein's hartebeest purchased by AP from private game ranches in Zambia (Stichting African Parks Foundation, 2004). Further reintroductions were made in 2005 and 2006 (Hall-Martin, pers. comm.).

At the socio-economic level, the number of full-time employees in Liuwa Plain National Park increased from 15 in 2002 to 94 (mostly scouts and community scouts) in 2004, with an additional 40 people employed part time. Three community-run campsites were built in the park and opened in 2004 with all proceeds being shared among the various communities. A monthly anti-poaching incentive of US$2000 per month was implemented, whereby a set amount is deducted per animal poached, giving an immediate sense of value to the wildlife. These funds were used by communities for a range of different community projects, including the purchase of agricultural implements, roofing for schools and clinics and the purchase of grain mills. In Majete National Park, nearly 100 local people are employed in full-time and part-time positions, and negotiations with adjacent communities were successfully concluded allowing the inclusion of the eastern bank of the Shire River into the park, in exchange for a fish farm and water troughs for livestock. This inclusion was by way of formal agreement with the community, although the community did not want the area to be formally gazetted.

From a funding perspective, Liuwa received financial support from the US State Department, the Peace Parks Foundation, the Dutch government, the Doen Foundation, the Walton Family Foundation and USAID. Entrance fees raised US$53,000 in 2005 (Stichting African Parks, 2005), up from less than US$100 in 2002. The increase in income was a result of active promotion of the park and although significant, there is still a long way to go before the park becomes financially sustainable. A daily vehicle limit of 25 vehicles has been implemented, which will ensure a limited environmental impact and the retention of the unique wilderness experience. In Majete, funding support was received from USAID, the Doen Foundation, the Walton Family Foundation and the Prince Bernhard Fund for Nature, and income from tourism remains at negligible levels.

Institutional changes

In mid-2003, the structure of AP and the equal shareholding began to be an issue for van Vlissingen, as he was willing to commit US$5 million to the company to fund the anticipated projects in Malawi and Zambia. As a result, the shareholding was changed to give him a 76 per cent controlling share of the organization. He also wanted the holding company to be moved from South Africa to the Netherlands for control purposes; hence the establishment of African Parks B.V.[3] Msimang, Hall-Martin, Eustace and Fearnhead became minority shareholders, but continued to be on the board of African Parks B.V. The commercial nature of AP and shareholding by individuals confused governments, who perceived that their parks were immediately commercially valuable, when in fact they required significant expenditure relative to the potential income streams. It was therefore decided to establish the African Parks Foundation in the Netherlands, which became the sole shareholder of African Parks B.V. The founders remained on the board of African Parks B.V. and remained responsible for the actual management of the parks. The board of the African Parks Foundation consisted of people known to and trusted by van Vlissingen, but they were not expected to make any technical contribution to the management of the parks. This structure continued until 2004 when it was decided that the two structures were duplicating each other, and African Parks B.V. was liquidated. The consequence of this change was that the entire organization was left without any African representation and without any conservation expertise or park management experience on the board.

During this period of institutional change, AP entered into three more contracts for four additional parks: Omo National Park in Ethiopia; Garamba National Park, an endangered World Heritage Site in the Democratic Republic of Congo; and two marine parks in Sudan (Dungonab Bay and Sanganeb Atoll). However, other than in the case of Omo National Park, the management agreements do not reflect the standard approach of the organization of having management control with full responsibility for operations for the long term.

Unfortunately van Vlissingen was diagnosed with cancer in December 2005 and passed away in August 2006. This served as a catalyst for a number of changes in the organization, both at the board level and at an operational level, allowing AP to survive the tragic loss and strategically reposition itself. Importantly, the organization is being relocated to Africa and a board of eminent Africans is currently being appointed. Most of the initial founders have also re-engaged in one way or another.

Conclusions

Successes in Majete are demonstrated by the confiscation of over 500 weapons, the reintroduction of over 1200 head of game and the fencing of a 14,000ha sanctuary. Over four years in Liuwa Plain, tourism income increased from less than US$100 to US$65,000, key species increased and patrol days have also increased. AP has thus demonstrated that national parks and other protected areas can be effectively managed by non-state institutions. It has shown that it is possible to bring the illegal utilization of wildlife and other resources under control, with the support of local people, within a very short timeframe. It has also shown that it is possible to rehabilitate parks, mobilize funds from a range of donors (institutional and private) as well as drive the level of income generated through tourism for the benefit of both conservation and local people. None of the parks under AP's management are as yet financially sustainable and many serious challenges lie ahead. In particular the wildlife product that is attractive to tourists and therefore commercial operators needs to be invested in and improved, especially in those parks where wildlife has been seriously depleted. Facilitating realistic CBNRM projects in areas adjacent to the parks will serve as a mechanism for extending both the positive ecological benefits and the social benefits associated with the effective management of a core protected area. The contentious issue of sustainable consumptive use of resources as a means of contributing to management costs of appropriate protected areas will also need to be addressed. Even so, the model has been sufficiently proven for it to be considered a serious option available for managing a park, and further adaptations of such a model are almost guaranteed to follow suit. The best sign of the success of the approach is its adoption (or similar) by other conservation institutions including by the African Wildlife Foundation and Peace Parks Foundation, which are becoming increasingly involved in the management of protected areas.

If government institutions realize that partnering with institutions like AP is not a weakness but a strength that involves leveraging available skills and resources, the impact of such partnerships could be considerable. The commercial world is comfortable with the concepts of core business, outsourcing, partnerships and contract management and these need to be understood, accepted and applied by conservation authorities. However, it is important that the roles and responsibilities of each party to the partnership are well defined and the measures of performance clear to all, and non-performance should result in termination of the management contract.

AP is still a young organization and, as is typical with innovations, it has adapted and changed as it deemed necessary. Some will criticize it for the early dominance of one personality, but only with time will the full impact of this innovation in the conservation industry be realized. If mistakes were made, they were the result of action.

Notes

1   Mavuso Msimang, Dr Anthony Hall-Martin and Peter Fearnhead were senior employees of SANParks while Michael Eustace was a merchant banker with an interest in conservation.
2   Frankfurt Zoological Society withdrew from Liwonde in 2005 without leaving a sustainable solution in place and as a result the park is in the process of collapsing again.
3   In the Netherlands, B.V. is the standard private limited company.

References

Davies, R. (undated) *Commercialization in North West Parks*, North West Parks and Tourism Board, Rustenberg
de la Harpe, D. (2003) 'Does "commercialization" of protected areas threaten their conservation goals?', paper presented to the World Parks Congress, September 2003, Durban
Emerton, L., Bishop, J. and Thomas, L. (2006) *Sustainable Financing of Protected Areas: A Global Review of Challenges and Options*, IUCN, Gland
Fearnhead, P. (2003) 'Tourism concessions: Public private partnerships for commercially sustainable conservation', paper presented to the World Parks Congress, September 2003, Durban
Fearnhead, P. (in prep.) *Concessions and Commercial Development: Experience in South African National Parks*
Stichting African Parks Foundation (2003) *Annual report 2003*, African Parks, Utrecht
Stichting African Parks Foundation (2004) *Annual report 2004*, African Parks, Utrecht
Stichting African Parks Foundation (2005) *Annual report 2005*, African Parks, Utrecht

Part VI

Conclusion

26

Innovations in State, Private and Communal Conservation

Brian Child

Environmental 'conservation' is often associated with keeping things the same, including preserving ecosystems from human depredation. And yet the environment is the contemporary product of evolutionary diversity and competition to yield the 'survival of the fittest'. The experiences in this book suggest that conservationists need to be innovative to cope with the massive demographic, climatic and economic changes we are facing; building adaptability is perhaps our most important goal. To do this, we need to invoke the qualities that Charles Darwin found essential to the adaptability of life on Earth – the variety and competition that drive evolution. Paradoxically, conservation is organized against Darwin's principles with centralization of decisions and power squeezing out variety and competition, the lifeblood of adaptability. Indeed, decentralized organizations are better at finding new solutions for complex problems (Peters and Waterman, 1980), while hegemonic nations often stagnate (Landes, 1998). Even economics is beginning to appreciate the importance of evolution. For instance, Beinhocker (2006, p187) suggests that 'evolution is a general-purpose … learning algorithm that adapts to changing environments and accumulates knowledge over time. It is the formula responsible for all the order, complexity, and diversity of the natural world'. However, conservation agencies are concentrating staff, money and power in head offices, when they should nurture the quality and independence of field managers, who are often the real drivers of innovation (see for example Peters and Waterman, 1980). It is no coincidence that most of the innovations that we describe in southern Africa are associated with highly motivated professionals working extensively in the field.

Aligning the goals of conservation with the lives and needs of ordinary people is also crucial. When people live with wildlife, our experience has shown that giving them formal rights to manage it in well-crafted ways usually has positive consequences; it is also socially just. Protected areas should ultimately serve the societies to which they belong; if these societies value them, the benefits to global society will automatically follow (the reverse of which is not true). Indeed, we need to ask if conservation can really be sustainable if it is does not also incorporate social justice. Our suggestion is, to be socially sustainable, conservation is ultimately a democratization process that safeguards (and even advances) the civil and political rights of individuals towards nature because people without legitimate aspirations or dignity are unlikely to be effective custodians of our planet (Anon, 2008). The alienation of people from parks and wildlife has already proven to be ineffective and problematic, and the policies described in this book are invariably an attempt to undo this alienation. Moreover, surveys conducted under the umbrella of the Southern African Sustainable Use Specialist Group (SASUSG) suggests that rural people value nature and wildlife far more than they are commonly given credit for. They like wildlife both for aesthetic reasons and because they see it as important to their economic future (views that are much stronger in young people), even in areas where elephants are numerous and problematic. However, people dislike the unjust wildlife management regimes imposed on them, ranging from the colonial past to elitist or predatory local management committees, and they undermine them using the 'weapons of the weak' (Scott, 1976), such as disinterest and poaching. However, we need to be careful not to confuse their resistance to institutions that alienate them from wildlife such as the top-down conservation strategies that emerged from the 1933 London Convention, from their positive, if complex, attitudes towards wildlife. There is more to be gained by including rural people in the process of conservation than by treating them as its natural enemies.

Even in its short history, modern conservation has seen remarkable changes. The history of state-protected areas (described in Part II) goes back as far as 100 years, but is often much shorter. Private conservation is newer still – the oldest examples are 50 years old, but the majority of experience has been gleaned in the past three decades. Community-based natural resource management (CBNRM) is even younger, and it is pleasing to be able to chart histories that are now two decades long.

Even in this short time, there have been three major transformations in the political economy of wildlife. Under a frontier mentality, wildlife was decimated by demographic change, new technology (for example, rifles and railroads) and markets acting on open-access property regimes at the frontier of white exploration. This heralded a period of state control that strictly controlled access to wildlife, banned the commercial use of its products and began to set aside protected areas. This was an effective response to the resource misuse associated with frontier economies, but its logic and efficacy fell rapidly away in the face of agricultural expansion and the enclosure of the commons. Under these economic conditions, the threat to wildlife and natural resources is mainly from the conversion of land to agriculture and domestic livestock production, requiring a new response and a new way of thinking about the relationship between protected areas, conservation and society. Certainly in southern Africa, conservationists have moved well beyond the idea that conservation is a struggle between the enlightened few and ordinary people. They have adopted a subtle (and much less dramatic) way of thinking – that tweaking the institutions of resource allocation can align the needs of people and wildlife. The examples in this book show how thinking in this way has brought some success. People in southern Africa increasingly want wildlife and view it as important to their future, and they are experimenting with myriad new ways to make this possible. The key, as mentioned, is to give wildlife as high a value as possible (in both monetary and non-monetary forms), to ensure that these values are captured at the level of the landholder (be this the state, an individual or community) through an appropriate combination of rights, and to empower people with discretionary choices over wildlife – accepting that people, given such responsibilities, are normally responsible.

Three conceptual models

Model 1: A model for wildlife's comparative advantage

At the point where rainfed crop production gives way to livelihoods that depend on using animals to harvest natural vegetation, there is a significant change in the ecology of human production systems. With agriculture, we are harvesting plants directly; livestock and wildlife systems, by contrast, represent a third trophic tier with energy being lost when plants are converted into animals. In southern Africa, the threshold between these systems occurs where rainfall is about 600–700mm per annum. Rangeland economies are prone to overgrazing, which presents serious limitations to the sustainability and viability of producing commodities such as meat. It is by breaking the direct link between economic productivity and the extraction of primary production that wildlife has allowed sustainable economic growth in these economies, largely through a variety of tourism and hunting enterprises (see Part III; Bond et al, 2004; Langholz and Kerley, 2006). Thus, we are selling experiences and services (including, in the future, ecosystem services) rather than agricultural commodities such as meat or grain. This situation is illustrated in Figure 26.1 that shows that wildlife is often more profitable than livestock in rangelands, but seldom competes with rainfed agriculture.

For historical reasons, wildlife has been undervalued by policy distortions. This includes banning commercial use of wildlife and restricting markets (for example, trade bans and veterinary restrictions), as well as bureaucratic regulations and fees (for example, hunting licences on private land) that seldom apply to agricultural commodities. More fundamentally, the 1933 London Convention introduced a policy regime that nationalized the wildlife resource, destroying most incentives for landholders to manage and conserve wildlife. Much of the progress in private and community wildlife conservation in southern Africa has been achieved by removing this economic undervaluation so that market prices approximate wildlife's economic comparative advantage (illustrated in Figure 26.1). However, in addition to wildlife being undervalued, agriculture and livestock production is often subsidized and protected, or has been in the past (see Part III), with wildlife becoming more competitive where these subsidies and regulations are reduced or removed.

This model also applies to communal and state land, where pricing signals are especially distorted by bureaucratic management and open-access property regimes and subsidies such as free fertilizer, where the fact that agricultural enterprises and livestock provide private benefits but their costs (for example, environmental degradation) are shared by society causing major economic distortions.[1]

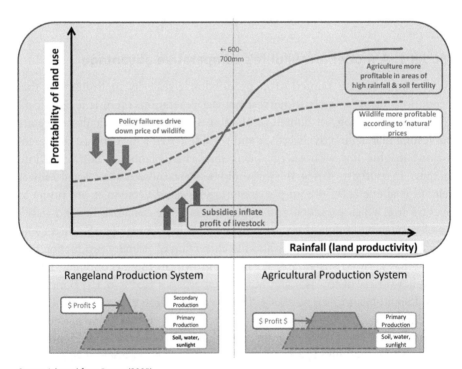

Source: Adapted from Brown (2005)

Figure 26.1 *The effect of the policy environment on returns to wildlife and agricultural land use systems*

This model may also apply to many arctic, mountainous and rangeland regions of the globe if similar institutional and entrepreneurial conditions are put in place (including maximizing landholder benefit). The increasing importance of global warming and fresh water will shift the curve for untransformed ecosystems (where activities such as wildlife production can occur) further upwards, provided we can learn how to capture these values and return them to the people who live with the resource.

Model 2: Value chains and the wildlife economy

The second descriptive model illustrates how the value from wildlife economies accumulates up the economic value chain (see Figure 26.2). This value chain is shaped like an inverted pyramid because, for example, when a tourist pays US$200 to stay in a lodge, this is usually associated with a US$20 park entry fee to the landholder, and considerable upstream and downstream economic activities including food, air travel, accommodation, shopping and the like.

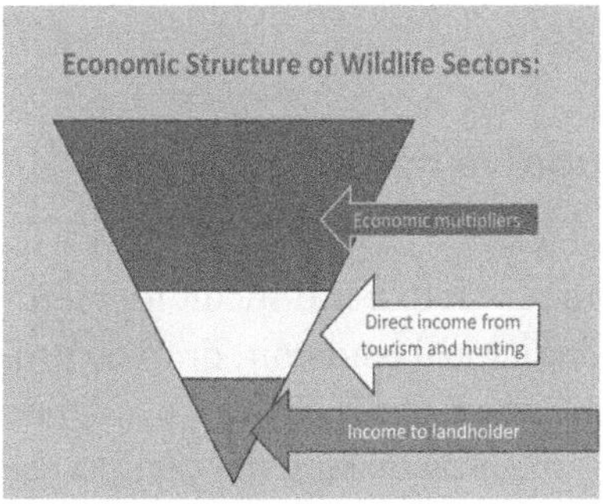

Who	Definition	Tourism Multiplier	Approx impact
	D+I+I+ Government Interacting	4.01	$30
	Direct + Indirect + Induced	2.57	$19
	Direct + Indirect	1.62	$12
Tour Operator	Direct (i.e. lodge income	(5-10 X park fees)	$5-10
Landholder/ Park			$1

Economic Structure of Wildlife Sectors:

Economic multipliers

Direct income from tourism and hunting

Income to landholder

<u>**Protect the base of the economic pyramid,**</u>
i.e. maximize returns to landholders (state parks; private property; community conservancies)

Note: The share of income captured by landholders consists of payments such as park fees (in parks) or royalties for trophies (for example, for game ranches or hunting concessions). Direct income to the economy includes, for instance, the money a tourist pays to a lodge or outfitter for guiding and accommodation services. In addition, there are all manner of economic multipliers associated with the linkages between tourism and other goods and services in the economy such as the purchase of food, air travel and curios (downstream), and the (upstream) expenditure by the lodge and its employees on housing, vehicles and other goods. The multipliers used in this example are indicative (we used those calculated for Kenya by TTC International, 1997). However, so important are the implications of this economic pyramid that it is an area that needs careful research.

Figure 26.2 *The value chain of wildlife economies*

The shape of the value chain emphasizes the importance of maximizing returns to landholders. If, for example, half the income from wildlife is captured by the district council or wildlife department through taxes and licences, we halve the profitability of wildlife. This slashes the incentives for landholders to conserve wildlife, and effectively cuts away that part of the economic pyramid that depends on these landholders managing wildlife. The failure to properly fund and manage protected areas can have enormous conservation and economic consequences. This is why the paradigm of sustainable use that is emerging in southern Africa emphasizes maximizing the returns from wildlife to landholders, be these state, private or communal actors, and studiously avoiding the differential taxation and bureaucratization of wildlife. Economic reasoning can be used to strengthen the argument for returning all benefits from wildlife to landholders in the language of economic growth, foreign exchange and employment growth that politicians are more likely to respond to. Even with weak data, this argument has been powerful in South Africa (see Part V), and in encouraging donor investment in Zambia's national parks (see for example Child, 2004). We are the first to admit that the data on which this model is based are thin. We need to strengthen our understanding of the economic and distributional structure of the wildlife sector if we are to argue convincingly that ensuring the viability of wildlife producers creates a much larger economy, creates jobs and expands the tax base.

Model 3: Institutions for managing fugitive natural resources such as wildlife

Wildlife is a good example of a fugitive resource where it is difficult to internalize the costs and benefits at the level of an individual land unit, or to control misuse (and free-riding) as wild animals move from one property to another. Namibia and South Africa dealt with this problem by making the right to use wildlife contingent on appropriate game fences. However, the intensive conservation area (ICA) movement in Zimbabwe and the proliferation of conservancies throughout southern Africa (see Part III) suggests that there is considerable opportunity to improve the efficiency of rural economies using collective action. While getting landholders to work together, particularly those with a highly individualist culture, involves significant transactions costs, collective action can add sufficient value by improving ecosystem connectivity and economies of scale to offset these costs.

Figure 26.3 illustrates how the trade-off can be managed between the relative effectiveness of smaller social jurisdictions and the need to scale up environmental management to take account of externalities. It reflects Murphree's (2000) argument that we need first to scale down by devolving use rights to the lowest appropriate land unit, and only then should we scale back up through a process of discretionary upward delegation. Figure 26.3 is a generic institutional proposal for natural resource management. Tier 1 represents the landholder or producer community. In this proposal, the rights to

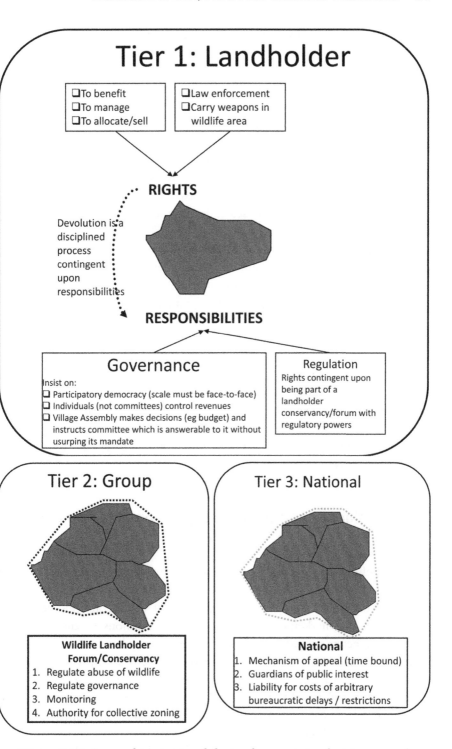

Figure 26.3 *A nested institutional design for managing fugitive natural resources such as wildlife*

benefit, manage and sell wildlife (and natural resources) are fully devolved to the landholder (or community), as are the rights of exclusion, including practical rights such as the authority to undertake law enforcement and carry weapons in wildlife areas.[2]

This model differs from conventional regulatory approaches by placing the power for regulatory control at the level of the landholder community (i.e. Tier 2) rather than in government agencies. Ideally, Tier 2 is comprised of elected landholder representatives, and its primary role is to manage the externalities associated with mobile wildlife (for example, overstocking and range degradation or the overhunting of a particular species of wildlife). Collective action can also incorporate efficiencies of scale, for instance for monitoring wildlife, setting quotas and or zoning land use[3] (see Chapter 11). In the case of CBNRM, Tier 2 is potentially valuable for monitoring and regulating the governance of its member communities using peer-review in the manner of the Grameen Bank (Yunnus, 1999). For example, the state might give the rights to use wildlife to, say, five neighbouring communities provided they ensure, through peer mechanisms, that they are all conforming to proper standards of governance and wildlife management.

The national level (Tier 3) still plays a critical role in providing an enabling environment for local forms of collective action to emerge. Key roles include arbitration (or facilitation) in the case of disputes, and acting as the regulator of last resort in the (rare) cases where local collective action fails. To avoid situations where bureaucrats impose costs on landholders thoughtlessly, such as whimsically banning hunting, such regulations should be subject to economic and legal recourse.

The resilience and adaptability of conservation policy and practice

A number of factors have been important for the resilience and adaptability of conservation practice in southern Africa. Innovations described in this book stem from the policy insight that the interests of wildlife are best served by placing it in the marketplace, and modifying institutions to ensure that its comparative economic advantage is reflected in the day-to-day decisions of landholders. Wildlife lives on people's land, and the best way to conserve it is to maximize the benefits that landholders are able to derive from it – obviously including financial benefits from hunting and tourism, aesthetic values and environmental services, and less obviously, but importantly, proprietary rights and discretionary choices over wildlife.

Progress is invariably associated with individuals. Southern Africa has been fortunate (or sensible) in creating a tradition of the scholar–practitioner. The individuals who have driven change, including many of the authors in this book, combine strong qualifications with a passion for wildlife conservation and rural livelihoods and a long track record of dedication to making their collective ideas work. Examples such as Pilanesberg, Addo, CBNRM and game ranching show that this has not been a quick process, but one requiring protracted tenacity.

Progress has also been accelerated where small groups of like-minded people have provided a critical mass for change, and by knowledge networks that have linked these groups across the region. The latter have been particularly influential when at least some of the participants are officials with power to implement the ideas arising from these networks. Prime examples are the emergence of devolved authority over wildlife on privately held land and the more recent emergence of CBNRM. However, the configuration of power and knowledge is changing. In the period between 1960 and 1990, many of the scholar–practitioners were themselves powerful bureaucrats. Nowadays there is a much more complicated brew of officials, NGOs, scholars, communities and the private sector, and new strategies are required to align innovative ideas and the bureaucratic power necessary to lead change.

The change processes described in this book have followed an adaptive approach. Much of the innovation has been built around hypotheses about the economics, political economy and organization of wildlife and park management. These concepts have been operationalized by park managers, farmers and communities and monitored by scholar–practitioners, often working within the processes. Introspection and analysis, often honed by oral peer-review, has then modified the process at the operational and conceptual level. It is interesting that much of the monitoring has comprised of experienced observation, sometimes supplemented by data, which has been peer-reviewed through personal relationships and through networks such as SARCCUS and SASUSG. Such qualitative and interdisciplinary expertise has been essential for monitoring complex conservation processes. However, adaptive management processes have invariably been strengthened by data, for instance the monitoring systems used for tracking recreational hunting (see for example Child, 1995) and CBNRM in Namibia and Zimbabwe (see Chapters 8 and 9). Nevertheless, high returns from investments are likely to be reaped in improving the monitoring of these programmes. In particular, more understanding is needed about the economics and governance of parks and wildlife programmes, as well as its impact on biodiversity and ecosystem services.

Conservation has benefited from a diversity of approaches and the competition of ideas unlocked by changes in parks and wildlife policy since the 1960s. We are learning how to build bigger and more valuable conservation landscapes using a matrix of state, private and community conservation areas depending on which model is best suited to the particular circumstances. New commercial models for 'useless' land (where many parks are situated) demonstrate that many parks should be able to pay for themselves if sound management systems are adopted. While any park that can pay for itself should do so, it is fully understood that many key sites of biodiversity may never be financially viable. These views are not in opposition. Resources to support conservation are scarce; they should not be squandered on parks that can pay for themselves but should be targeted to conserve important biodiversity areas that are not financially self-supporting.

Institutional ecology, resilience and adaptability

State-protected areas may be the cornerstone of conservation but they comprise only 10 per cent of what is often 'useless land' (Runte, 1979). Moreover, regulatory bureaucracies face little performance competition and historically have been slow to innovate. In the 1970s, the mission-driven field of conservation biology expanded beyond biology and began to include people, at least occasionally, in the conservation equation (Borgerhoff Mulder and Coppolillo, 2005). Indeed, Africans featured prominently in making this point at the 1982 World Congress on National Parks in Bali, Indonesia (Child, 1984; Lusigi, 1984). In the 1990s, concern about the inequities of conservation, and particularly the accumulation of benefits at the top and the imposition of costs at the bottom, led to the emergence of the field of 'political ecology'. Perhaps now is the time for 'institutional ecology' – to learn how to design human institutions and organizations to improve economic and biodiversity outcomes.

Outside of conservation, this is not a new idea. Recent scholarship, for example, suggests that economic growth is not constrained so much by physical and human capital as by institutional variables, especially those associated with economic and political governance (Chakravarti, 2007). The discipline of new institutional economics (see for example North, 1990, 2003) associates growth with the institutions of property rights, transactions costs, economic organization and stable, predictable laws, while Amartya Sen (Sen, 1999; World Bank, no date) 'thickens' the concept to include freeing people from the oppressive burden of the state to take advantage of their inherent capabilities (Anon, 2008). The case studies provided in this volume repeatedly demonstrate that effective conservation depends on improving the economic institutions of resource allocation; that it needs to unlock the energy and imagination of the thousands of individuals that live with wildlife and natural resources on their land; and that the sustainability of wildlife and natural habitats will ultimately be determined not only by ecology but by politics. It is questionable how long approaches to protected areas and wildlife conservation can survive if they are perceived as being socially unjust.

Given the importance we ascribe to the devolution and democratization of conservation, we speculate briefly on why devolution and market-led approaches are not used more broadly. To borrow from resilience theory (for example, Walker et al, 2004), we are concerned about how conservation policy can become locked into an unhealthy basin of attraction. Lock-ins are more likely when long feedback loops weaken accountability, as with the centralized control and regulation of natural resources. Moreover budget-based organizations (such as park or conservation agencies) tend to defend the status quo (for example non-use) and entrench their power (for reasons described by Drucker, 1973). By contrast innovation is commonly associated with decentralized structures and high levels of performance accountability. In most of the examples described, policy innovation has emerged where a small technical elite based in the field has been held accountable for performance, either by its peers or by its political constituency (for example, by white farmers in the case of game ranching). Commercial innovation has invariably been led by entrepreneurs acting at the level of a land unit, be this a park (for example Pilanesberg) or a private ranch. This leads us to float the hypothesis that the administrative structure of natural resource sectors – particularly whether they are hegemonic or decentralized – determines their capacity to adapt to rapidly changing local and global demands. It suggests that the current level of power held by global agencies and in capital cities is unbalanced and unhealthy; environmental conservation will suffer unless it is structured so that a far greater proportion of its talent is located and nurtured in the field.

Conclusions

The lessons that have emerged from 50 years of experimentation in southern Africa are relatively simple:

- Wildlife has a financial comparative advantage in many areas, especially on what was once considered 'useless' land where most parks have been designated.
- Because of the shape of the value chain, wildlife's economic comparative advantage is even greater than its financial advantages.
- Institutions and governance are critical if the comparative advantage of wildlife and protected areas is to be harnessed so that they can pay for themselves. Markets need to 'get prices right' in order to allocate resources to their highest-valued uses. This requires that costs and benefits are internalized as much as possible at the local level. In particular, rights need to be devolved to

landholders to ensure that they retain the benefits that are rightfully theirs, be they individual ranchers, producer communities or park managers.

- If people have reliable rights to manage and benefit from wildlife, they usually look after ecosystem processes, especially if these rights are accompanied by technical advice.
- Devolving discretionary rights to improve the economic allocation of resources to land managers also lays the foundation for the evolution of local democratic processes. Discretion (or the freedom to choose, see Sen, 1999) over the allocation of resources is a fundamental property of democracy.[4] Thus CBNRM is not only an institutional tool that uses collective action to manage and add value to fugitive natural resources such as wildlife, but the rights associated with CBNRM may be an important catalyst for rural democratization. Indeed, throughout this book we have emphasized the importance of choice and accountability, in other words the democratization of wildlife and protected area management.
- Innovation is associated with decentralized institutions and devolved agencies, especially when competent people are located in the field.

In presenting this new way of thinking about conservation, the real challenge is to design institutions to align these ideas in a unified approach. We need to maximize the accumulation of talent and value at the field level. At its peril, conservation ignores the importance of locating the best people in the field – the farmer must be on the farm.

Notes

1   It is sometimes difficult to see how people manage to survive in some remote rural areas (Kanapaux, unpublished data) without the provision of public services (schools, clinics, food aid and agricultural inputs, food relief), urban remittances and access to natural materials for their dwellings, food and medicine – agriculture alone is unable to feed them reliably.

2   Within the single land unit, these rights are contingent on mechanisms that hold managers fully accountable to communities and the landholder. This is seldom an issue on private land because the owner is often also the manager, but is critical in rural communities on communal land, which is why Part IV emphasizes the importance of processes of participatory democracy for ensuring that power (and accountability) originates in the people.

3   While there are many advantages that can be associated with the collective management of wildlife resources, commercial activities tend to be better managed by individual entrepreneurs.

4   Which implies that democracy may not be able to evolve if people do not have rights to the land and resources with which they live.

References

Anon (2008) 'Order in the jungle', *The Economist*, 13 March

Beinhocker, E. D. (2006) *The Origin of Wealth: Evolution, Complexity and the Radical Remaking of Economics*, Harvard Business School Press, Boston

Bond, I., Child, B., Harpe, D. D. L., Jones, B., Barnes, J. and Anderson, H. (2004) 'Private land contribution to conservation in Southern Africa', in Child, B. (ed) *Parks in Transition: Biodiversity, Rural Development and the Bottom Line*, Earthscan, London

Borgerhoff Mulder, M. and Coppolillo, P. (2005) *Conservation: Linking Ecology, Economics, and Culture*, Princeton University Press, Princeton

Chakravarti, A. (2007) *A Theory of Discontinuous Change*, working paper, USAID/BearingPoint South Sudan Program and Queen Elizabeth House, Oxford University, Oxford

Child, B. (2004) 'A financial and economic analysis of the costs and benefits of managing the protected area estate', UNDP/GEF Funded Project on Reclassification and Sustainable Management of Zambia's Protected Area Systems, Lusaka

Child, G. (1984) 'Managing wildlife for people in Zimbabwe', in McNeely, J. A. and Miller, K. R. (eds) *National Parks, Conservation and Development: The Role of Protected Areas in Sustaining Society*, Smithsonian Institution Press, Washington DC

Child, G. (1995) *Wildlife and People: The Zimbabwean Success. How the Conflict between Animals and People became Progress for Both*, Wisdom Foundation, Harare

Drucker, P. (1973) *Management: Tasks, Responsibilities, Practices*, HarperCollins, New York

Landes, D. (1998) *The Wealth and Poverty of Nations: Why Some Are So Rich and Some So Poor*, Abacus, London

Langholz, J. A. and Kerley, G. I. H. (2006) *Combining Conservation and Development on Private Lands: An Assessment of Ecotourism-Based Private Game Reserves in the Eastern Cape*, Center for African Conservation Ecology, Nelson Mandela Metropolitan University, Port Elizabeth, South Africa

Lusigi, W. (1984) 'Future directions for the Afrotropical realm', in McNeely, J. A. and Miller, K. R. (eds) *National Parks, Conservation and Development: The Role of Protected Areas in Sustaining Society*, Smithsonian Institution Press, Washington DC

Murphree, M. (2000) 'Constituting the commons: Crafting sustainable commons in the new millennium', paper presented at 'Multiple Boundaries, Borders and Scale', Eighth Biennial Conference of the International Association for the Study of Common Property (IASCP) Bloomington, IN, 31 May–4 June

North, D. C. (1990) *Institutions, Institutional Change and Economic Performance*, Cambridge University Press, Cambridge

North, D. C. (2003) *Understanding the Process of Economic Change*, Forum Series on the Role of Institutions in Promoting Economic Growth, Mercatus Center at George Mason University and The IRIS Center, Washington DC

Peters, T. and Waterman, R. H. (1980) *In Search of Excellence: Lessons from America's Best-Run Companies*, Harper Collins Business, London

Runte, A. (1979) *National Parks. The American Experience*, University of Nebraska Press, Lincoln and London

Scott, J. C. (1976) *The Moral Economy of the Peasant: Rebellion and Subsistence in Southeast Asia,* Yale University Press, New Haven

Sen, A. (1999) *Development as Freedom*, Anchor Books, New York

TTC International (1997) *The Impact of Tourism on the Economy of Kenya – 1996*, TTC International, Nairobi, Kenya

Walker, B., Holling, C. S., Carpenter, S. R. and Kinzig, A. (2004) 'Resilience, adaptability and transformability in social-ecological systems', *Ecology and Society*, vol 9, no 2

Yunnus, M. (1999) *Banker to the Poor: Micro-Lending and the Battle Against World Poverty*, Public Affairs, New York

World Bank (no date) *Community Driven Development in Africa: A Vision of Poverty Reduction Through Empowerment*, World Bank, Washington DC

Author Biographies

Simon Anstey is currently the Director of ResourceAfrica (UK) and previously worked for IUCN in Mozambique, Angola and West Asia–Central Asia between 1992 and 2007. He has (finally) submitted a doctoral thesis to the University of Zimbabwe on resilience and adaptive cycles in local governance and natural resource management, based on five years of applied practice and research in Sanga District, northern Mozambique.

Jonathan Barnes, PhD, is a specialist in environmental and resource economics based in Namibia. He has experience in economic and policy analysis related to tourism, wildlife, fisheries, forestry, rangelands and agriculture.

Willie Boonzaier, a well-known wildlife consultant in the southern African region, started in accounting and financial management – skills that he brought to bear in transforming the 'financial' model of wildlife management in Pilanesberg National Park in the late 1980s. He also played a leadership role in developing new economic approaches to protected area management in Africa, such as the Madikwe Game Reserve in the North West Province of South Africa. As a consultant specializing in protected area development and management, he has since been sharing his knowledge with conservation agencies, communities and private sector operators throughout southern Africa.

Professor **Jane Carruthers**, Department of History, University of South Africa, pioneered environmental history in South Africa. Her doctoral thesis was published as *The Kruger National Park: A Social and Political History*, which has become a standard reference work and she has authored numerous other books, book chapters and scholarly articles. She is a Fellow of the Royal Society of South Africa and Fellow of Clare Hall, Cambridge and has had visiting positions at the Australian National University, Canberra, and the University of Western Australia. Currently she is Vice-President of the South African Historical Society, an editor of the forthcoming Palgrave Macmillan *Dictionary of Transnational History* and serves on the editorial boards of a number of journals.

Guy Castley, previously employed by SANParks as a wildlife ecologist, currently lectures in ecology, wildlife management and conservation biology at

Griffith University in Australia. During his time at SANParks he was involved with a number of initiatives related to the management and expansion of national parks and how these processes could be conducted in a sustainable manner to achieve both conservation and social objectives.

Brian Child is Associate Professor in the Geography Department at the University of Florida, and editor of *Parks in Transition* (2004). He worked for wildlife departments in Zimbabwe for 12 years, and has also supported park management and CBNRM in Luanga valley, assisted with the restructuring of ZAWA, and worked within the wildlife sector in Uganda, Kenya, Namibia and South Africa.

Graham Child, PhD, is a wildlife ecologist who was an FAO Expert who advised the Botswana Government from 1965 to 1971, before rejoining the Department of National Parks and Wild Life Management in Zimbabwe as Director until 1986. He was then a freelance consultant advising on wildlife management and institutional development, from the Arctic to the tropics and on all continents except Antarctica.

Roger Collinson, MSc, is one of the pre-eminent 'visionaries' in southern African conservation, has been involved in wildlife conservation planning and management in southern Africa for the past 35 years. This involvement has entailed both 'hands-on' management of protected areas and 'arm's-length' technical advice to community-based, government, non-governmental and private sector organizations. During this period he has been involved in some of the most exciting protected area, ecotourism and CBNRM development projects in South Africa, Botswana, Zambia, Namibia, Madagascar and Lesotho. He was also involved in the planning and development of Pilanesberg National Park, Madikwe Game Reserve, the Greater St Lucia Wetland Park, the LIFE programme in Namibia and Lapalala Wilderness.

David Daitz describes himself as a 'rolling stone that gathers no moss'. He now works as a coach doing organizational development in a variety of contexts. His work in nature conservation between 1996 and 2005 included leading the team that negotiated the creation of the Table Mountain National Park in the Cape Peninsula and becoming the park's first manager, playing a leading role in the partnership that obtained the first Global Environmental Facility grant and leading the start-up of CapeNature, a new parastatal for conservation in the Western Cape established in 2000.

Richard Davies has worked in conservation since 1984 when he joined the Natal Parks Board based in Mkuzi Game Reserve, where he worked in the field on ecological issues and advising landowners on ecological management. He soon realized that many conservation problems revolved around business issues and completed a masters in business. While studying he moved to the North West Province where he worked with communities and also project managed the development of Madikwe. He is now a freelance consultant working at the conservation–business interface of park planning. He has worked extensively in southern Africa.

Jacobus du Plessis Bothma, PhD, became the incumbent of the postgraduate Eugéne Marais Chair of Wildlife Management at the University of Pretoria in 1970, and in 1988 he became the first Director of the postgraduate Centre for Wildlife Management at the same university, a position from which he retired in December 2005. He remains an Extraordinary Professor in the Department of Animal and Wildlife Sciences at that university and is currently the author of 107 refereed scientific publications, and the co-author, author or editor of 19 books on wildlife.

Raoul du Toit is a Zimbabwean ecologist whose professional career initially involved environmental planning of large hydro-power schemes but shifted into rhino conservation. His role in facilitating the development of Savé Valley Conservancy (and sister conservancies), with funding from international donors, has continued with an emphasis on hands-on rhino management and on land use policy issues.

Peter Fearnhead, MSc, was born and grew up in Zimbabwe. He was a management consultant with Deloitte before joining South African National Parks, where he held a number of positions including Resource Economist, Advisor to the Chief Executive Officer and Head of Commercial Development. He was also one of the founders of African Parks and worked for it from 2002 until 2005. He left to start an organization called Conservation Partnerships whose business interests were amalgamated with African Parks in February 2007, when he was appointed as the Chief Executive Officer.

Jill Gordon has a long history working with communities in South African National Parks. She made significant contributions to the People and Conservation Department before being appointed as the project coordinator for the greater Addo Elephant National Park project. Jill has recently taken up the position of park manager at Wilderness National Park where she remains involved with the Garden Route Initiative.

David Grossman is a South African-based savanna ecologist with experience in the planning of conservation areas and community-based natural resource management.

Phillipa Holden lives and works in southern Africa, focusing on protected area planning, development and management, ensuring that local and indigenous communities are integrally involved, and are beneficiaries of conservation activities and ecotourism development. Further work includes the promotion of community conservation areas, traditional and customary natural resource management, land use and environmental planning and management, ecotourism planning and development, education and awareness.

Steve Johnson is a well-known career conservationist in southern Africa and has worked extensively in protected area management and CBNRM in Zimbabwe and for the North West Parks Board in South Africa. Over the past 20 years his interest has tended more towards CBNRM having played an influential role as the SADC Regional Natural Resource Management Adviser while employed with IUCN–Regional Office for Southern Africa. A founder member of the SASUSG, he continues to influence the evolution of community-based conservation through his current work within the Department of Wildlife and National Parks in Botswana and as Vice Chairperson of the Botswana National CBNRM Forum, as well as through his membership of several of IUCN commissions.

Brian Jones, MPhil, is an environment and development consultant, working mainly in CBNRM Namibia and the SADC region. He worked for the Namibian Ministry of Environment and Tourism for ten years where he co-ordinated the Ministry's CBNRM programme. He is currently a policy advisor to the Ministry on CBNRM and collaborative management of protected areas.

Michael Knight, PhD, has 24 years of African conservation science experience. This includes large mammal ecology, with more recent attention focused on the identification, expansion and development of South Africa's national park system.

Peter Lindsey is a conservation biologist working in Zimbabwe on a range of key conservation issues, including the bush meat trade, human–wildlife conflict, land use planning for conservation and predator conservation.

Hector Magome, PhD, has 22 years of experience gained through practical involvement in environment and development issues including policy research, development and monitoring implementation of biodiversity conservation in South Africa. He has published papers and contributed to books on community participation in conservation and development.

Helena Motta is a Mozambican marine biologist who has worked in the fisheries and environment sector for more than 27 years. She is currently the country coordinator for WWF in Mozambique.

Brett Myrdal, 47, has managed the ten-year-old Table Mountain National Park for five years. He is a passionate conservationist and generalist with strong leadership qualities. He previously managed the Table Mountain Fund, a capital fund for the long-term biodiversity conservation of the Cape Floral Kingdom. Prior to that he received the national housing award for project managing the Lwandle Hostels to Homes project that converted the apartheid era, bucket toilet, single men's hostels into family units with solar-heated water. He served seven years in the ANC's military wing during the anti-apartheid struggle and lives to realize the dream of reconciliation ecology – people as part of nature in a city.

Isilda Nhantumbo, PhD, worked as a lecturer in the Department of Forestry, Universidade Eduardo Mondlane, before becoming Mozambique Country Programme Coordinator for the IUCN (2000–2005) and later Coordinator of the Regional Forestry Programme. She participated in the CBNRM pilot projects that were set up by DNFFB/FAO/Netherlands, providing technical assistance particularly in economic aspects, and she also engaged in CBNRM at policy and community level.

Christopher Patton has an MSc in Environmental and Geographical Science from the University of Cape Town. He currently works in the Business Development Unit of South African National Parks.

Alistair Pole undertook his doctoral studies on the ecology of African wild dogs in Savé Valley Conservancy, Zimbabwe. He spent four years running a game ranch in the Savé Valley Conservancy and then two years as a consultant. He is currently a partner in a safari company based in Zimbabwe.

Stephanie Romañach is an ecologist conducting wildlife research. She has worked on herbivore ecology in the US, southern and East Africa, and on carnivore conservation in southern and East Africa. Stephanie has worked with rural communities and conducted wildlife research through the Mpala Research Centre in Kenya and currently in the Savé Valley Conservancy in Zimbabwe.

Nico Rozemeijer worked for SNV/Netherlands Development Organisation as Senior Advisor to the CBNRM Support Programme in Botswana between 1999 and 2003. His previous work in Botswana was as a District Officer,

Lands in Ghanzi (1986–1989) and as a consultant in policy development and CBNRM-related fields for the NRMP and SNV. He currently works as Senior Advisor for Wageningen International, a consultancy and capacity-building unit of the Wageningen University and Research Centre in the Netherlands.

Bartolomeu Soto, MSc, was appointed the Head of the Mozambican Wildlife Department in 1994, in the National Directorate of Forestry and Wildlife. During this period he was involved in the Gorongosa Marromeu Integrated Management Plan, the Bazaruto Community Based Conservation project, administration of hunting safaris in *coutadas* and game farms, conceptualization and implementation of the 'Tchuma Tchato' programme, the first community-based natural resource management programme in Mozambique. He is currently working at the Ministry of Tourism as the National Director of Conservation Areas and Head of the Transfrontier Conservation Areas project.

Anna Spenceley is an independent consultant and researcher based in South Africa who focuses on responsible tourism and sustainable development issues, mainly in emerging economies. She is a former post-doctoral researcher at the Transboundary Protected Areas Research Initiative of the University of Witwatersrand, and is a Director of the International Centre for Responsible Tourism – South Africa.

Helen Suich is a development and resource economist. She works on issues relating to poverty alleviation, sustainable economic development, natural resource management and valuation, primarily in southern Africa.

Russell Taylor, formerly Conservation Director for WWF in southern Africa and now Chief Operations Officer with the Venture Philanthropy Company, has over 25 years of experience in applied conservation research and management. An early architect of CAMPFIRE, he places much importance on partnerships between communities and public and private sectors to achieve lasting conservation impact.

L. Chris Weaver is a rangeland ecologist who has been employed by the WWF as the Director of the Namibian Living In A Finite Environment (LIFE) project for the past 15 years. He has more than 30 years of practical experience in the promotion of improved management and conservation of communal natural resources (rangelands, wildlife, livestock, land and tourism resources) in both south-western US and southern Africa. He is based in Windhoek, Namibia.

Webster Whande has worked on regional southern Africa CBNRM projects. He is currently conducting doctoral research on human and environment security within transfrontier conservation areas, with case studies from the Great Limpopo Transfrontier Conservation Area (GLTFCA). He is also a member of a regional research team working with local communities on local level scenario planning within the GLTFCA.

Index

Printed and bound by CPI Group (UK) Ltd, Croydon, CR0 4YY

01/11/2024

01782629-0020